Optimization in Locational and Transport Analysis

Optimization in Locational and Transport Analysis

A. G. Wilson,
J. D. Coelho,*
S. M. Macgill, *and*
H. C. W. L. Williams

School of Geography
University of Leeds
Leeds LS2 9JT, UK

**Now at Department of Mathematics,*
University of Lisbon

JOHN WILEY & SONS
Chichester · New York · Brisbane · Toronto

British Library Cataloguing in Publication Data:

Optimization in locational and transport analysis.
1. Mathematical optimization
I. Wilson, A. G.
515 QA402.5 80-42068

ISBN 0 471 28005 4

Phototypeset by Macmillan India Ltd, Bangalore.
Printed in the United States of America.

Acknowledgements

Chapter 3 is partly based on H. C. W. L. Williams (1977) On the formation of travel demand models and economic evaluation measures of user benefit, *Environment and Planning A*, **9**, 285–344.
Reproduced by permission of Pion Ltd., London

Chapters 4 and 8 are partly based on H. C. W. L. Williams and M. L. Senior (1978) Accessibility, spatial interaction and the spatial benefit analysis of land-use transportation plans, in A. Karlqvist, L. Lundqvist, F. Snickars and J. W. Weibull (eds.) *Spatial Interaction and Planning Models*, North Holland, Amsterdam, 253–88.
Reproduced by permission of North Holland Publishing Co., Amsterdam.

Chapter 5 is partly based on S. M. Macgill (1978) Rectangular input–output tables, multiplier analysis and entropy maximising principles; a new methodology, *Journal of Regional Science and Urban Economics*, **8**, 355–70.
Reproduced by permission of North Holland Publishing Co., Amsterdam.

Chapter 6 is partly based on S. M. Macgill (1977) The Lowry model as an input–output model and its extension to incorporate full intersectoral relations, *Regional Studies*, **11**, 337–54.
Reproduced by permission of Pergamon Press Ltd., Oxford.

Chapter 7 is partly based on J. D. Coelho and A. G. Wilson (1976) The optimum location and size of shopping centres, *Regional Studies*, **10**, 413–21.
Reproduced by permission of Pergamon Press Ltd., Oxford.

Chapter 7 is also partly based on J. D. Coelho and A. G. Wilson (1977) An equivalence theorem to integrate maximising models within overall mathematical programming frameworks, *Geographical Analysis*, **9**, 160–72; and J. D. Coelho, H. C. W. L. Williams and A. G. Wilson (1978) Entropy maximising models within overall mathematical programming frameworks: a correction, *Geographical Analysis*, **10**, 195–201.
Both reproduced by permission of Ohio State University Press, Columbus. Copyright © 1977 [1978].

Chapter 8 is partly based on J. D. Coelho and H. C. W. L. Williams (1978) On the design of land use plans through locational surplus maximisation, *papers, Regional Science Association*, **40**, 71–85.
Reproduced by permission of the Regional Science Association, Philadelphia.

Chapter 10 is partly based on S. M. Macgill and A. G. Wilson (1979) Equivalences and similarities between some alternative urban and regional models, *Sistemi Urbani*, **1**, 9–40.
Reproduced by permission of Guida Editori.

Chapter 10 is also partly based on A. G. Wilson and S. M. Macgill (1979) A systems analytical framework for comprehensive urban and regional model building, *Geographia Polonica*, **42**, 9–25.
Reproduced by permission of Polish Scientific Publishers, Warsaw.

Figures 9.3, 9.4, 9.18, 9.19 and 9.21 are based on Figures 2, 3 and 7 from J. C. Amson (1975) Catastrophe theory: a contribution to the study of urban problems? *Environment and Planning B*, **2**, 177–221; Figure 1 from A. G. Wilson (1976) Catastrophe theory in urban modelling; an application to modal choice, *Environment and Planning A*, **8**, 351–6; and Figure 5 from B. Harris and A. G. Wilson (1977) Equilibrium values and dynamics: an attractiveness term in production-constrained spatial interaction models, *Environment and Planning A*, **10**, 371–88.
All reproduced by permission of Pion Ltd., London, and in the case of Figures 9.3, 9.4 and 9.18, Dr J. C. Amson.

Figure 9.12 is based on Figure 2.23 from M. W. Hirsch and S. Smale (1974) *Differential Equations, Dynamical Systems and Linear Algebra*, Academic Press Inc., New York.
Reproduced by permission of Academic Press Inc., New York, and the authors.

Figure 9.14 is based on Figure 8 from J. Maynard Smith (1974) *Models in Ecology*, Cambridge University Press, Cambridge.
Reproduced by permission of Cambridge University Press, Cambridge.

Contents

Preface xi

1. Optimization in locational and transport analysis 1
1.1 Introduction 1
1.2 Perspectives and starting points 1
1.3 Criticisms and research problems 3
1.4 Themes 4
1.4.1 Introduction 4
1.4.2 Mathematical programming methods 5
1.4.3 Behavioural approaches 5
1.4.4 Dynamical systems theory 5
1.4.5 Applications in particular systems of interest 6
1.4.6 Omissions 6
1.5 Main results 6

2. Mathematical programming methods 8
2.1 Introduction 8
2.2 An overview of mathematical programming theory 10
2.2.1 Unconstrained optimization 10
2.2.2 Lagrangian theory 15
2.2.3 Inequality constraints and the Kuhn–Tucker conditions. 16
2.2.4 Duality 19
2.3 The central features of linear programming 21
2.3.1 Primal–dual relations and optimality conditions 21
2.3.2 An example: the transportation problem of linear programming 23
2.4 Examples of model formation from non-linear programs 26
2.4.1 Introduction 26
2.4.2 The fully constrained gravity model: its formulation and dual 26
2.4.3 Relation of the fully constrained spatial-interaction model to the transportation problem of linear programming 29
2.4.4 The dual of the spatial-interaction program 30
2.4.5 The congested assignment problem 32
2.4.6 The combined distribution–assignment model: an example of model integration within a programming framework 35

2.5 Computational methods 38
2.5.1 Introduction 38
2.5.2 General approaches to the unconstrained problem 38
2.5.2.1 Direct search methods 39
2.5.2.2 Gradient methods 39
2.5.3 Constrained optimization 41
2.5.3.1 Penalty-function methods 41
2.5.3.2 Feasible directions algorithms 43

3. Random utility theory and probabilistic choice models 46
3.1 Introduction 46
3.2 The formation of random utility models 47
3.2.1 The choice context 47
3.2.2 The generation of random utility models 47
3.2.3 Independent random utility models 49
3.3 Utility distributions and their properties 50
3.4 The entropy maximizing and group-surplus maximizing approaches 55
3.5 Attribute correlation and model structures 57
3.5.1 The variance–covariance matrix 57
3.5.2 Simultaneous and sequential model structures 59
3.5.3 The generation of share models 63
3.5.4 The derivation of the nested logit function 66
3.6 More general choice model structures 69
3.6.1 Some initial considerations 69
3.6.2 The cross-correlated logit model 70
3.6.3 A general extreme-value class of models 72
3.6.4 The multinominal probit model 73
3.6.5 Generation of choice models by simulation 74
3.7 Examples of the formation of travel-related demand models and economic evaluation measures 75
3.7.1 Introduction 75
3.7.2 Multi-mode travel-choice models 77
3.7.3 Transport planning models 79
3.7.3.1 The dispersion parameters β and λ 80
3.7.3.2 The composite functions $\tilde{c}_{ij}^{n}$ and $\tilde{u}_{i}*$ 80
3.7.3.3 Consumer surplus user benefit measures 81
3.7.3.4 The location model 81
3.7.3.5 A programming derivation of the location model 82
3.8 Conclusion 83

4. Programming approaches to activity location and land-use evaluation 85
4.1 Introduction 85
4.2 The Lowry model: its analytic structure 86
4.2.1 Theoretical foundations 86

4.2.2 The incorporation of planning constraints 89
4.2.3 The spatial characteristics of the interaction variables .. 90
4.2.4 The interdependence of the stock variables........... 91
4.2.5 Problems and limitations 91
4.3 An activity location model based on probabilistic choice theory 93
4.3.1 The probabilistic choice model (PCM).............. 93
4.3.2 Its relationship with the Lowry model 96
4.3.3 A land-use evaluation index—the group surplus measure 97
4.4 The group surplus maximization model for activity allocation . 99
4.4.1 The construction of, and rationale for, the GSM 99
4.4.2 Kuhn–Tucker conditions.......................... 102
4.4.3 Relationship to the Lowry model 104
4.4.4 Relationship to the probabilistic choice model 104
4.5 Duality and the GSM 105
4.5.1 The dual program............................... 105
4.5.2 Duality and the allocation process 106
4.5.3 Duality and evaluation 106
4.6 A variant: the group entropy maximizing model (GEM) 107
4.7 Disaggregation of the mathematical programming activity allocation model.............................. 108
4.8 The calibration process 109

5. Entropy and economic accounts: some new models 111
5.1 Introduction.. 111
5.2 A review of the existing models........................ 112
5.3 The development of an alternative approach............... 117
5.4 More alternative rectangular input–output models 121
5.5 Discussion and further developments 125
5.6 Spatial models....................................... 129
5.7 Concluding remarks................................... 132

6. An extended Lowry model as an input–output model........... 134
6.1 Introduction.. 134
6.2 The Lowry model as an input–output model 135
6.2.1 Summary of the Lowry model 135
6.2.2 Spatially aggregated input–output representation of the Lowry model mechanism 136
6.2.3 Discussion of the input–output representation 140
6.2.4 Spatially disaggregated input–output Lowry model representation.............................. 143
6.2.5 Relation to Garin's matrix representation............ 148
6.3 Extension of the Lowry model to incorporate full inter-activity relations.. 149
6.3.1 The matrix or algebraic representation 149

6.3.2 A maximum-entropy (mathematical programming) representation . . . 152
6.3.3 Further developments . . . 154
6.4 Closing remarks . . . 156

7. Embedding theorems and applications . . . 158
7.1 Introduction . . . 158
7.2 Integration of spatial-interaction submodels within overall mathematical programming frameworks . . . 158
7.2.1 Introduction . . . 158
7.2.2 Equivalence lemma: adding constraints into the objective function . . . 159
7.2.3 A theorem for embedding spatial-interaction submodels in an optimization program . . . 160
7.2.4 Discussion . . . 163
7.2.5 An example: activity location . . . 164
7.2.6 The dual program for the general problem . . . 165
7.3 A first application: an elementary model of residential and employment activities' allocation . . . 167
7.4 A second application: the optimum location and size of shopping centres . . . 169
7.4.1 Introduction . . . 169
7.4.2 A consumers' welfare maximization model . . . 169
7.4.3 Embedding of the Lakshmanan and Hansen shopping model in an optimization framework . . . 170
7.4.4 Disaggregation and dynamics . . . 172
7.5 Concluding considerations . . . 175

8. Optimal design and activity location . . . 177
8.1 Introduction . . . 177
8.2 Spatial-interaction models, locational surplus, and plan design . 180
8.2.1 The benefit functions . . . 180
8.2.2 The locational surplus maximization model: an overview 180
8.3 A family of design problems and models . . . 181
8.3.1 The general design model framework . . . 181
8.3.2 The long-run equilibrium design problem . . . 181
8.3.3 A housing allocation model . . . 183
8.3.4 A service activity reallocation model . . . 185
8.3.5 A housing and service activity redistribution model . . . 185
8.3.6 An employment redistribution model . . . 186
8.4 Some further extensions and comments . . . 186
8.4.1 Introduction . . . 186
8.4.2 A general supply/demand spatial equilibrium framework 187
8.4.3 The locational surplus criterion in the real-world context 188
8.4.4 A note on the β-parameters . . . 188

8.4.5 The dual variables and the generalization of the Hansen accessibility index . . . 189
8.4.6 Some comments on industrial location, accessibility, and plan design . . . 190
8.4.7 The linear programming limiting case . . . 191
8.4.8 Additional remarks on the programming representation. 192
8.5 A brief note on multiple goals . . . 192
8.6 Concluding comments . . . 193

9. Optimization and dynamical systems theory . . . 194
9.1 Introduction. . . . 194
9.2 Catastrophe theory . . . 196
9.3 Bifurcation and differential equations . . . 204
9.4 Summary: Zeeman's six steps for modelling . . . 210
9.5 Examples in locational analysis: retail centres and urban structure 212
9.5.1 The model to be used . . . 212
9.5.2 Hypotheses for centre size: embedding a static model into a dynamic framework . . . 213
9.5.3 A dynamical analysis of equilibrium points. . . . 214
9.5.4 Differential equations and bifurcation for this example . 219
9.5.5 Zonal criticality and retail structure . . . 220
9.5.6 Simultaneous zonal variation . . . 222
9.5.7 Interacting fields: the addition of residential location . . . 224
9.5.8 Order from fluctuations: the work of the Brussels school 227
9.6 Examples in transport analysis: modal choice and network structures . . . 229
9.6.1 Modal choice and structural dynamics . . . 229
9.6.2 Modal choice and hysteresis . . . 231
9.6.3 The evolution of transport network structure. . . . 232

10. Model comparisons: equivalences, similarities, and differences . . . 234
10.1 Phenomena, methods, models . . . 234
10.1.1 Introduction . . . 234
10.1.2 Phenomena of interest . . . 234
10.1.3 Building-block methods . . . 234
10.1.4 A classification of models . . . 235
10.2 Some circumstances in which equivalences arise . . . 235
10.2.1 Introduction . . . 235
10.2.2 Equivalences in the simpler models. . . . 235
10.2.3 Large systems of algebraic equations . . . 240
10.2.4 Nested and linked mathematical programs . . . 241
10.2.5 Equilibrium-point dynamics . . . 241
10.2.6 Applications of these analyses. . . . 242
10.3 Example 1: retailing flows and centre structures . . . 243
10.3.1 Basic models . . . 243

10.3.2 A mechanism to determine $\{W_j\}$ 243
10.3.3 A mathematical programming formulation 244
10.3.4 $\{W_i\}$ in a dynamic framework 246
10.4 Example 2: Lowry models 246
10.4.1 The original model 246
10.4.2 The straightforwardly 'improved' model 247
10.4.3 Matrix inverse formulations 248
10.4.4 Mathematical programming versions of the Lowry model 249
10.4.5 Dynamic models 254
10.5 Example 3: Spatially disaggregated economic account-based models 256
10.5.1 The Leontief matrix inverse form 256
10.5.2 An entropy-maximizing formulation 257
10.5.3 Two alternative mathematical programming models . . 259
10.5.4 Dynamic models 260
10.6 Concluding remarks 262

Bibliography 263

Index 274

Preface

Urban and regional modelling is a field which has developed rapidly over a period of twenty years. By general consent, the development still has a long way to go. Ten years ago, most of the product of the field was to be found in academic journals; there were very few books. Today, there is a wide range of textbooks for the student and planner to choose from, each giving an account of the range of models and their application. This book is of a different kind and the fact that it can be written and assembled is a mark of a stage of advancement of the field. Its focus is on *methods* for modelling, and those concerned with different aspects of optimization in particular. It is argued that a study of optimization gives both a deeper understanding of a wide range of existing models and also provides the tools for the development of many new ones.

Although we concentrate on methods in organizing the book, we do use a wide range of existing and new models as examples. Thus, the reader will be introduced to a large number of new models. We base models on a range of methods—entropy maximizing and random utility techniques, for example—and we are particularly concerned with the relationships between methods. Thus, it should be possible for the reader to site any model in which he or she is interested into a broad context. We also identify, particularly with respect to constructing evaluation indicators and the use of models in design, a number of new ways in which models and optimization methods can be used in planning.

We hope that the book will be of value to a number of different kinds of readers: to geographers, regional scientists, economists, and other social scientists concerned with models of, and theories about, cities and regions; and to planners and engineers seeking tools to help deal with the complexity of the interacting areas they are trying to plan. The methods presented may also be of interest to systems analysts, operational researchers, and applied mathematicians. A background in urban and regional modelling would be useful for the reader, though most of the discussions of particular examples are self-contained. Also, we assume a mathematical background provided by a book such as *Mathematics for Geographers and Planners* by A. G. Wilson and M. J. Kirkby (Oxford University Press, second edition 1980) or an equivalent text for economists or engineers. Again, however, we have tried to make the presentation essentially self-contained by offering, in Chapter 2, an account of the mathematical programming methods which underpin, in various ways, most of the rest of the book. The only additional methods required are some of those of dynamical systems theory which are presented, as needed, in Chapter 9.

In a multi-author book, it is useful to indicate which chapters and areas different authors are responsible for. Alan Wilson has organized aspects of the whole field of work in terms of initiating research grant applications, seeking graduate students and contributing to discussions of what constitute suitable research topics. His own main fields of interest have been aspects of spatial interaction modelling and applications of dynamical systems theory. He has written Chapters 1 and 9 and co-authored Chapters 7 and 10. José Coelho spent three years in Leeds on leave of absence from the University of Lisbon. His main field of interest was the construction of optimizing models which could be used in planning design and associated aspects of modelling theory. He worked with Alan Wilson on embedding theorems and co-authored Chapter 7; and he worked with Huw Williams on mathematical programming versions of the Lowry model and they co-authored Chapters 4 and 8. Sally Macgill worked on new versions of economic-accounting models using entropy-maximizing principles and showed how an extended Lowry model could be seen as an input–output model. She wrote Chapters 5 and 6 on these topics. She also worked with Alan Wilson on equivalences and similarities between apparently different versions of models and they co-authored Chapter 10. Huw Williams has provided the general introduction to mathematical programming in Chapter 2, and Chapter 3, which is a record of much of his own research in random utility theory. Also, as noted earlier, he has co-authored Chapters 4 and 8 with José Coelho.

The work for this book has been carried out over the last five years or so, and we owe acknowledgements to a number of people and organizations. We are all indebted to our colleagues in the University of Leeds for many helpful discussions in seminars and elsewhere. Some of the research was carried out under research programmes funded, in turn, by the Science Research Council and the Social Science Research Council and we are grateful to them for their support. Huw Williams has been directly supported by S.R.C. and S.S.R.C. and Sally Macgill held an S.R.C. studentship during the early part of her research. José Coelho is grateful to the Instituto de Alto Cultura in Lisbon for providing financial support. Chapters of the book have been typed largely by Pamela Talbot, Rosanna Whitehead and Mandy Walton and we are grateful for their help. Gordon Bryant and John Dixon have our thanks for drawing the figures.

Leeds, December 1979

A. G. Wilson
J. D. Coelho
S. M. Macgill
H. C. W. L. Williams

1. Optimization in locational and transport analysis

1.1 INTRODUCTION

Two points should be noted at the outset about the title of this book. First, 'optimization' is used in two senses—as a generic name for certain techniques in mathematics and to denote some notion of 'search for an ideal' in relation to various aspects of the real world. The second sense can be further subdivided into concerns of, say, a government with societal or collective ideals on the one hand, and an individual's search for an optimum *within* a society on the other. We will be concerned with 'optimization' in all these senses, but it is sometimes important to distinguish the sense intended in particular cases. Secondly, 'locational analysis' is used to refer to the location of physical stock such as buildings, and people and organizations and their activities within cities and regions. The reference to transport covers related interactions between locations—for example, the movement of people between homes and workplaces. Some of the methods discussed may also be relevant to other fields of locational analysis if appropriate adjustments are made.

It should also be emphasized at the beginning that the main aim of the book is the presentation and discussion of *methods* of optimization in locational and transport analysis, though this is almost always in the context of applications to particular systems of interest. We ask the reader always to be prepared to look beyond particular examples so that he can use the methods presented in different contexts.

1.2 PERSPECTIVES AND STARTING POINTS

Our perspective is based on the school of urban modelling which has its style and foundations based on the transport models of the late 1950s and early 1960s and the developments following the publication of Lowry's *Model of metropolis* in 1964. In subsequent years, the models have been provided with a variety of theoretical bases, much empirical and planning experience has been gained and the specialism (which exists at an intersection of several disciplines) has become a highly developed one. All this work is taken as read and as a starting point. A large number of books describe the field, including two by Leeds' authors (Wilson, 1974; Wilson, Rees and Leigh, 1977).

As the specialism has become established, so also has it come under attack. The existence of transport models, so one argument runs, was directly responsible for now-unfashionable motorways; models failed to take into account the social problems of high-rise flats; the whole style of planning supported by urban models is outmoded; and so on. Many of the criticisms are based on misunderstanding; many have a grain of truth in them; many more imply substantial challenges in the future of urban and regional science. The present book attempts to take up a few of the challenges. It does rest on the belief that this sort of scientific activity is worth while; in particular, that deeper understanding and analysis can be a force for good in all kinds of circumstances. But perhaps one of the most important tasks of a book like this is to show that the field continues to develop, often in a remarkable way. This in itself is important as much present criticism is based on models that are already outmoded. Many of the results presented here suggest that new models are available for the old in planning practice, where 'old' often means '1977 or later'. But it is also recognized that, as in most sciences, an analyst's utopia remains far away in time and that it will be necessary and important for new methods and models to be subject to the severest scrutiny and criticism, and there will lie the germs of yet another generation of models.

Some of the main features of the school of modelling represented here are sometimes taken as too obvious to be worth remarking on, and yet one at least is a crucial distinguishing mark: that is the way in which the state of an urban system is described algebraically. The spatial system is defined by dividing a study area into zones, labelled by indices such as i and j. Categories of system components are similarly labelled by indices. Thus, T_{ij}^{kn} may be the number of people in car ownership group n who travel to work by mode k from a residence in zone i to a job in zone j. Examples of this kind of representation are well known, and many will be introduced in later chapters. It is immediately apparent, however, that this representation has considerable advantages over others. The 'uniform plain' assumptions typical of much geographical theory, or the monocentric workplaces of much urban economics, are simply unnecessary at the outset. It is also relatively easy to identify and construct (in terms of algebraic description) levels of resolution which are appropriate for particular problems—which balance the information needed for problem solving against the cost of manipulation or of obtaining data.

The particular starting points for this book can be identified reasonably specifically. A substantial family of spatial interaction and activity location models exists at varying degrees of disaggregation and can be used for a variety of purposes. Many of them are based on entropy-maximizing principles and have proved robust and useful (Wilson, 1970). This particular perspective became less all-embracing when links were discovered between linear programming models and entropy-maximizing models (Evans, 1973b; Wilson and Senior, 1974) which enabled the 'family of models' to be broadened and the entropy term to be seen as generating suboptimal *dispersion* in non-linear mathematical programming models which had some behavioural optimizing underpinnings. A large number of ways had been explored of combining various submodels into comprehensive models, with developments based on the Lowry model tending to be the most important.

1.3 CRITICISMS AND RESEARCH PROBLEMS

Four criticisms can be identified which motivated much of the work reported in this book. Firstly, and a favourite one from many social scientists, the 'behavioural' basis of existing models was considered inadequate. Secondly, in spite of the presentation of many models as 'systems theory', interdependence was not always well treated. Thirdly, the dynamical structure of most of the models was very elementary; this deficiency was particularly important since the explicit aim was often to use the models for conditional forecasting in planning. Fourthly, the use of the models in planning followed a number of well-trodden routes with some success, but somehow the initial promise was not fulfilled. The research problems connected with these criticisms are discussed briefly in turn and then two more points, not arising directly from criticism, are added.

Much of the behavioural criticism arises because of the reductionist nature of much social science which involves the assumption that the only worthwhile scale of study is the individual person or organization. This is, of course, a legitimate field of study. However, this viewpoint fails to recognize that there are often phenomena which are of interest in their own right at more aggregate scales. Further, when this point is recognized, there is often a jump to a very aggregated scale (as is common in economics, for example). A distinctive feature of locational and transport analysis is that the scale of interest usually sits between micro and macro perspectives and can usefully be called 'meso'. It is to be expected that methods of analysis distinctive to this scale will exist. So the balance to be struck in our research is to seek to improve the behavioural basis of meso-scale models, for example by connecting back to the classical theories of consumers' behaviour and the theory of the firm but to be able to tackle very difficult *aggregation* problems in passing from micro basis to meso results. This is a well-known problem, of course.

The problems of representing interdependence are also substantial. The main difficulties are of two kinds. Firstly, we become accustomed to using the best available submodels in partial analyses. When such models are stitched together, the resulting total model is too large to handle—and hence the longstanding popularity of the Lowry model as such. Secondly, the interest in compreshensive model-based analysis arises from the possible existence of systemic effects—'behaviour of whole greater than sum of parts'—arising from subsystem interdependence. Typically, this is not the sort of thing we have data on, or at least data we understand, and so model development does not take place in a well-informed empirical environment. (Though it should be said that there is general intuitive knowledge of such phenomena, such as land-use transport interaction, which is often not adequately represented in models.)

Inadequate dynamic structures also arose in part out of lack of time-series data, but more out of theoretical poverty and lack of ambition. At least the ambition was spurred by the publication in 1969 of Forrester's *Urban Dynamics*, which was much disliked but could, in a sense, only be challenged by better theory. Part of the difficulty was the lack of familiarity of urban modellers with the appropriate techniques and an initial task was to put this right.

The most well-trodden path for the use of models in planning was conditional forecasting. If incomes rise by so much, car ownership by something else, and we do this to the road network, what happens? The models have shown quite effectively how such very complex questions can be answered. (There may, of course, then be much argument about the assumptions, but we are beginning to tackle this by sensitivity analysis, as in Bonsall, Champernowne, Mason, and Wilson, 1978.) The much more difficult question was the so-called *combinatorial problem of design*: how to use models to help test a very large number of alternative plans. Mathematical programming provided an obvious technique but it did not seem to mesh well with the underlying set of urban models. We will see later that new methods of dynamical analysis offer approaches in planning which are alternatives to conditional forecasting. This involves the use of the concepts of 'criticality' and 'stability' and planning in terms of resilience and robustness.

The additional points which form the basis of our approach to research problems are of two kinds. Firstly, there was a not-always comfortable commitment to *eclecticism*. Perhaps the most common approach to research is to develop a particular model (or whatever) and then to defend it passionately against all-comers. Much of the research described in this book arises from attempts to put together the best features of different approaches (based on analyses of strengths and weaknesses), though to achieve this often involves the solution of difficult mathematical problems. Fortunately, it sometimes turns out that various approaches are not always as different as appears at first sight.

Secondly, there was a realization that the theory of mathematical programming (and to a lesser extent the closely related topic of dynamical systems theory) underpinned a wide range of urban models, and so a study of the techniques would lead at least to polishing and perhaps to new results.

1.4 THEMES

1.4.1 Introduction

A number of themes, which sometimes overlap and relate to each other, run through the book in various ways. Some will already be evident from the previous section. Here they are presented in a different order (appropriate for following chapters) and more explicitly. In this way, they should serve as a checklist for the reader in reading the rest of the book.

1.4.2 Mathematical programming methods

A very brief overview of the main results we use from mathematical programming theory is presented in Chapter 2. These include developments in entropy-maximizing (EM) theory which arise from recognizing the EM formulation as non-linear mathematical programming (NLP) and the linking of EM approaches with certain mathematical programs as limiting cases. The notion of entropy as introducing *dispersion*—and a degree of sub-optimality—is then a theme which recurs. A number of results concerned with non-linear duals are introduced and again this is a recurring theme. Another major technique theme concerns combination of mathematical programs, either as *embedding*, with one program initially inside another and then combined into one (hierarchical nesting) or the combination of originally unconnected, though related programs—*superproblems*. Embedding is treated in detail in Chapter 7.

It is also useful to draw attention to the role of *constraints* in mathematical programming and in the style of modelling we have been discussing. There is an important affinity. The notion that the system state in phase space follows a trajectory confined to a manifold defined by constraints which contain much information about the system—the constraints largely *are* the model in some cases—is a powerful one, and also relates closely to dynamical systems theory (see section 1.4.4 below).

A special branch of mathematical programming theory is concerned with discrete problems: location–allocation problems for example, or particularly important, networks. Since travel costs appear in most of our models and are based, often implicitly, on underlying network models, these problems are discussed from time to time. This also connects to the notion of superproblems, as the most important combination is of spatial interaction submodels with network assignment models which handles system non-linearities arising from competition. This example is presented at the end of Chapter 2. A wide range of examples of programming methods is presented in Chapters 4–8 and the main results are summarized in Section 1.5 below.

1.4.3 Behavioural approaches

The main contribution under this heading is the wide-ranging application of *random utility theory*, the machinery of which provides a way of solving the aggregation problem mentioned earlier. This overlaps substantially with the previous section in that when the meso-scale results are drawn out from this work, new families of NLP models result. The methods are outlined in Chapter 3.

A quite different and cruder approach involves the study of *change* at a micro scale using the techniques of dynamical systems theory; but this taken up in a wide context in the next subsection below.

1.4.4 Dynamical systems theory

In studying change, it is important to be explicit in distinguishing exogenous and endogenous variables within models. Changes can thus be seen to be driven by the exogenous variables—either model parameters, or variables forecast exogenously, or using other submodels. The main techniques of dynamical systems theory are then: mathematical programming, particularly studying the changing shape of constraint manifolds, given changes in exogenous variables and associated changes of system state (which may involve discrete jumps); differential equations, with particular attention to bifurcation; and catastrophe theory, which can be seen as a special case of mathematical programming where the systems are gradient (or potential) systems and the state non-unique for some (critical) regions of the model parameter space (which can again lead to jumps). So, discrete change is one theme which emerges here. As noted earlier, some of the analyses are at the micro scale, though the main results are at the more usual meso scale.

1.4.5 Applications in particular systems of interest

We often apply different methods to particular systems which recur throughout the book. Of particular interest are models concerned with shopping, transport, spatially disaggregated input–output relations, and the Lowry model. On this basis, we leave to the reader the combination of different innovations relating to any one system of interest, and their wider application, since mostly the particular examples we use are simply examples of a wide class.

1.4.6 Omissions

It is also appropriate to note briefly a number of themes and topics we might have considered but have not. There are two substantial fields of work we do not consider here, mainly because they are as yet rather separate from the integrated themes considered here, but also because they are already treated in books elsewhere. The first is the problem of building discrete facility location models based on the so-called location-allocation procedures (see, for example, Scott, 1971). The second is 'classical' location theory, often based on continous-variable representations of space, and in particular those concerned with spatial pricing (see Takayama and Judge, 1972).

1.5 MAIN RESULTS

In Chapter 2 on techniques, a number of useful extensions (of models or procedures) to previous work are derived—for example, on the conceptual use of non-linear duals and on the integration of assignment procedures with location and interaction models.

The main results of random utility theory are derived in Chapter 3 and applied to a wide range of transport problems. This also lays the foundation for much further work. A detailed understanding is achieved of the differences at the micro scale between entropy-maximizing models and group surplus models (GS) (based on random utility-maximizing principles at the micro scale) and these results are applied in Chapter 4. Models derived from the different bases are the same in aggregate cases, but differ in disaggregated cases. The GS models satisfy Hotelling's integrability conditions while disaggregated EM models do not. Chapter 4 includes a mathematical programming (MP) version of the Lowry models. This can be used, for example, to optimize the spatial distribution of non-basic employment. An important innovation here is the discovery of how to write MP constraints which represent the interdependence of residential and retail submodels within a simple mathematical program. This is an example of combining two separate programmings in an ingenious way and has the consequence that residential location is then explicitly a function of access to services as well as to jobs. Numerical work with the model has shown that this does affect results considerably.

In Chapter 5, a Kullback entropy formulation of an NLP is used to produce a new kind of rectangular spatially disaggregated input–output model, and in Chapter 6, a long-standing problem is solved with the incorporation of a fully explicit input–output model into the Lowry model. Mathematical programming and matrix formulations of this model are explored.

Different contributions to design are presented in Chapters 7 and 8. First, embedding theorems are presented and applied to spatial shopping-centre design. In Chapter 8, the results of Chapters 2, 3, 4, and 7 are applied to the task of optimal design in planning.

Optimization is related to dynamical systems theory in Chapter 9. Various approaches are explored and applied to shopping centre evolution, network infrastructure, transport mode choice, and to urban spatial structure as a whole. In this last application, links are established to central place theory.

Finally, in Chapter 10, various approaches to particular models are compared and the concepts of model equifinality (with respect to approach) and homeo-finality are explored. Some surprising similarities turn up from apparently very different approaches.

2. *Mathematical programming methods*

2.1 INTRODUCTION

The central problem of mathematical programming may be stated with deceptive simplicity as the optimization (in the mathematical sense of maximization or minimization) of a function $f(x_1, \ldots, x_N)$ of several variables $x_1, \ldots, x_N$ subject to a set of constraints. Formally, the structure of most of the problems we shall be concerned with in this book may be expressed as

$$\max_{\{x_1, \ldots, x_N\}} f(x_1, \ldots, x_N) \tag{2.1}$$

subject to

$$\left.\begin{aligned} g_1(x_1, \ldots, x_N) &\leqslant b_1 \\ &\vdots \\ g_M(x_1, \ldots, x_N) &\leqslant b_M \end{aligned}\right\} \tag{2.2}$$

$$x_1 \geqslant 0, \ldots, x_N \geqslant 0 \tag{2.3}$$

where $b_1, \ldots, b_M$ are constants. (This formulation also covers minimization, given the possibility of replacing f by $-f$). Defining the following vectors:

$$\mathbf{x} = \begin{pmatrix} x_1 \\ x_2 \\ \vdots \\ x_N \end{pmatrix}; \quad \mathbf{b} = \begin{pmatrix} b_1 \\ b_2 \\ \vdots \\ b_M \end{pmatrix}$$

$$\mathbf{g}(\mathbf{x}) = \begin{pmatrix} g_1(\mathbf{x}) \\ \vdots \\ g_M(\mathbf{x}) \end{pmatrix}$$

the above program (2.1)–(2.3) may be written

$$\max_{\{\mathbf{x}\}} f(\mathbf{x}) \tag{2.4}$$

subject to

$$\mathbf{g}(\mathbf{x}) \leqslant \mathbf{b} \tag{2.5}$$

$$\mathbf{x} \geqslant 0 \tag{2.6}$$

The problem, simply stated, is one of determining the vector $\mathbf{x}^*$ (i.e. particular values of $\mathbf{x}$, known as **instrument** or **decision** variables) in the **feasible region** (or **opportunity set**) defined by the constraints (2.5) and (2.6) which maximizes the **objective function** $f(\mathbf{x})$.

From a theoretical and a practical viewpoint it is useful to regard the problem (2.4)–(2.6) as successive generalizations of the **unconstrained** multivariate optimization problem

$$\max_{\{\mathbf{x}\}} f(\mathbf{x}) \tag{2.7}$$

and the classical Lagrangian problem

$$\max_{\{\mathbf{x}\}} f(\mathbf{x}) \tag{2.8}$$

subject to the **equality** constraints

$$\mathbf{g}(\mathbf{x}) = \mathbf{b} \tag{2.9}$$

It is not surprising therefore, that the Kuhn–Tucker conditions which characterize the solution of the general non-linear programming problem contain as special cases the Lagrangian conditions for constrained optimization, and the classical conditions for unconstrained optimization over several variables. We shall present these conditions in this context, below.

While general theorems may be used to characterize the solution of problems, they are seldom invoked **directly** in the search for that solution. In numerical work it is often possible to exploit the special structure of a problem, and indeed the study of mathematical programming, in practice, more often than not involves the study of special cases for which particular solution algorithms or approaches have been developed. Of particular importance, from a viewpoint of computing and identifying the solution(s) of a program, are the functional characteristics of $f(\mathbf{x})$ and $\mathbf{g}(\mathbf{x})$.

A mathematical program will, in general, have many solutions for which the theoretically stated conditions for maximization of $f(\mathbf{x})$ will be satisfied. It is important to distinguish **local** optima at which $f(\mathbf{x})$ attains a **relative** maximum in particular limited regions of the feasible space, and a **global** optimum which is the required solution over the whole of the feasible region. Only in special circumstances relating to the functional characteristics of $f(\mathbf{x})$ and $\mathbf{g}(\mathbf{x})$ may we be assured that a local optimum also attains the status of a global optimum. Firstly, it is necessary that the constraints form a convex set (bounded by convex functions†); and secondly, in a maximization (minimization) problem the objective function $f(\mathbf{x})$ must be a concave (convex) function of the instrument variables.

† A function $h(\mathbf{x})$ is termed convex, over a region R, if for any two points $\mathbf{x}_1$ and $\mathbf{x}_2$ in R,

$$h(\alpha\mathbf{x}_1 + (1-\alpha)\mathbf{x}_2) \leqslant \alpha h(\mathbf{x}_1) + (1-\alpha)h(\mathbf{x}_2) \qquad 0 \leqslant \alpha \leqslant 1$$

The inequality sign is reversed for a concave function.

In Figures 2.1–2.3 some of the stationary points (maxima and minima) of a variety of optimization problems are illustrated. From an examination of these two-dimensional contexts, it is easy to imagine the complexity and proliferation of local maxima or minima in high-dimensional problems when $f(\mathbf{x})$ and $\mathbf{g}(\mathbf{x})$ are general functions of $\mathbf{x}$. We are very interested, of course, in the special cases where the optimum is unique. We will also see in Chapter 9 that multi-valuedness plays an important role in aspects of dynamical systems theory.

One very special program structure is obtained when the objective function and constraints are *linear*. As a linear function is a limiting case of both concave and convex forms, the non-existence of local optima in linear programs is assured. The contour diagram for a simple linear program is shown in Figure 2.4. This serves to illustrate the well-known property of linear programs, namely that the solution (if it exists) will be on the boundary (and in general at a vertex or corner) of the feasible region formed by the constraints.

Having examined a few special structures, we return to the problem of characterizing the optimal solution of a non-linear program.

2.2 AN OVERVIEW OF MATHEMATICAL PROGRAMMING THEORY

2.2.1 Unconstrained optimization

It is useful to recall the optimality conditions for the unconstrained multivariate problem (2.7), not only, as we have emphasized above, because the Lagrangian and Kuhn–Tucker conditions are generalizations of this basic form but also, as we shall see later in Section 2.5, because the computational methods for solving the constrained program often rely heavily on techniques developed for unconstrained and equality-constrained problems.

The optimality conditions for the simple unconstrained problem

$$\max_{\{\mathbf{x}\}} f(\mathbf{x})$$

may readily be derived by examining the variation of $f(\mathbf{x})$ about the solution $\mathbf{x}^*$. In the neighbourhood of $\mathbf{x}^*$ we can, if $\mathbf{x}$ is continuous, expand $f(\mathbf{x}^* + \Delta\mathbf{x})$ in a Taylor series as follows:

$$f(\mathbf{x}^* + \Delta\mathbf{x}) = f(\mathbf{x}^*) + \mathbf{G}'\Delta\mathbf{x} + \tfrac{1}{2}\Delta\mathbf{x}'\mathbf{H}\Delta\mathbf{x} + \dots \tag{2.10}$$

in which $\mathbf{G}'$ is the transpose of the gradient vector

$$\mathbf{G} = \frac{\partial f}{\partial \mathbf{x}} = \begin{pmatrix} \dfrac{\partial f}{\partial x_1} \\ \vdots \\ \dfrac{\partial f}{\partial x_N} \end{pmatrix} \tag{2.11}$$

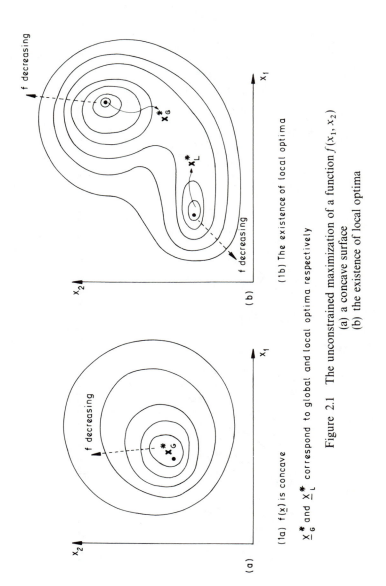

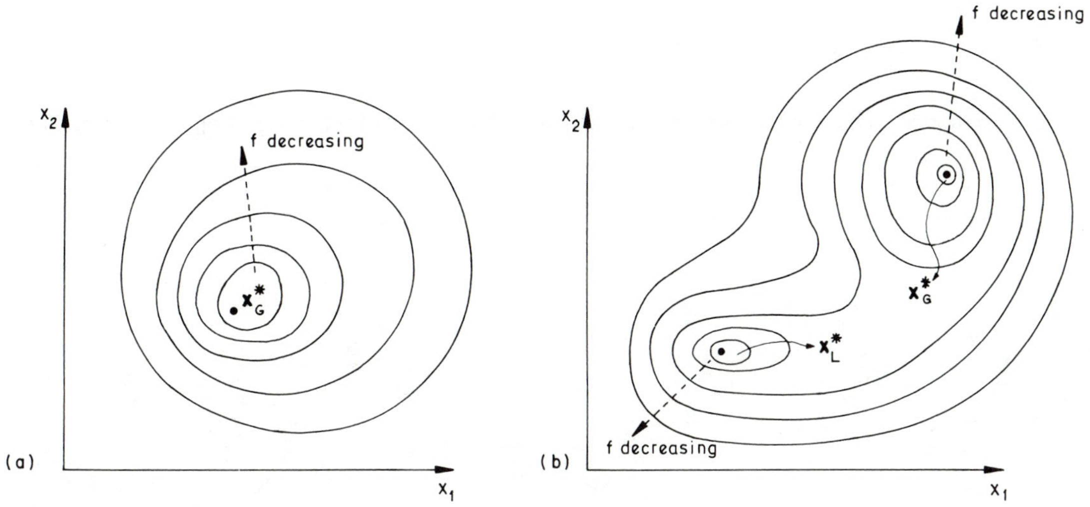

(1a) $f(\underline{x})$ is concave

(1b) The existence of local optima

$\underline{X}_G^*$ and $\underline{X}_L^*$ correspond to global and local optima respectively

Figure 2.1 The unconstrained maximization of a function $f(x_1, x_2)$
(a) a concave surface
(b) the existence of local optima

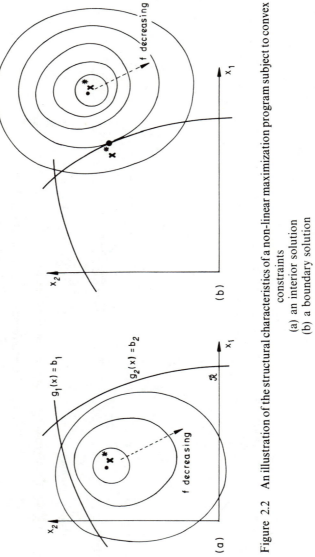

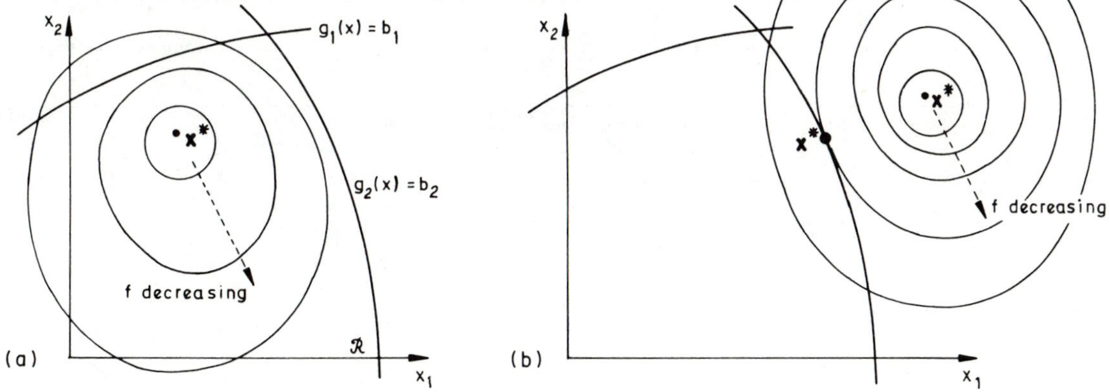

Figure 2.2 An illustration of the structural characteristics of a non-linear maximization program subject to convex constraints
(a) an interior solution
(b) a boundary solution

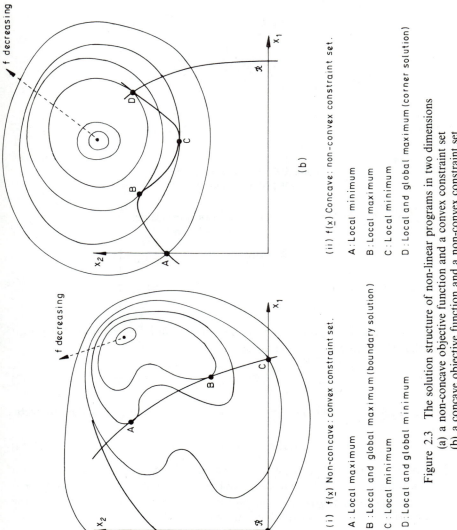

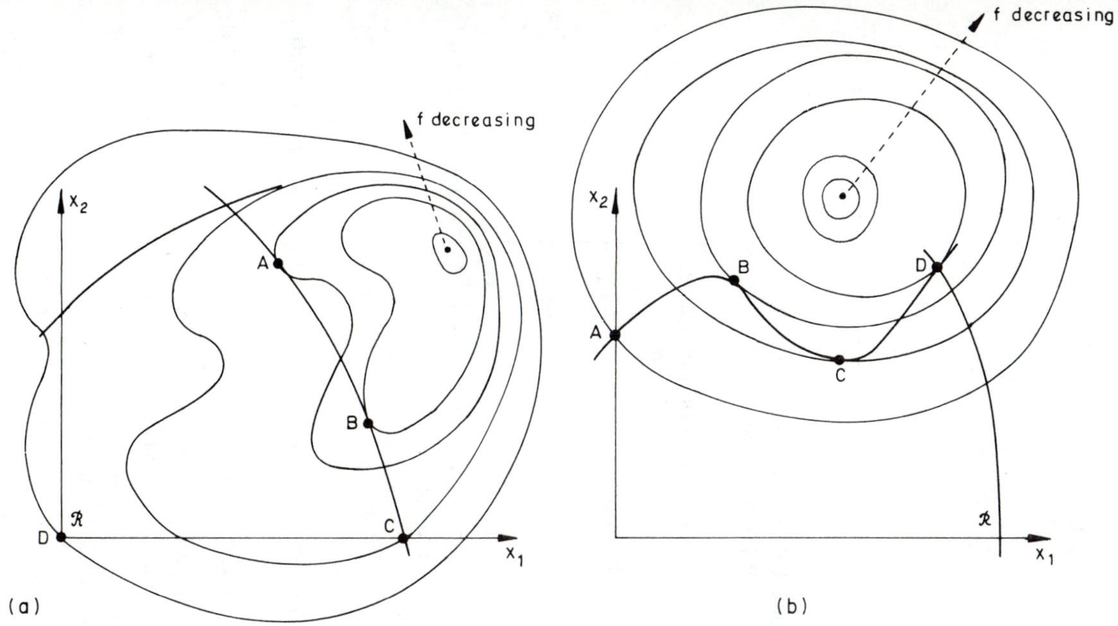

(i) f($\underline{x}$) Non-concave: convex constraint set.

A: Local maximum

B: Local and global maximum (boundary solution)

C: Local minimum

D: Local and global minimum

(ii) f($\underline{x}$) Concave: non-convex constraint set.

A: Local minimum

B: Local maximum

C: Local minimum

D: Local and global maximum (corner solution)

Figure 2.3 The solution structure of non-linear programs in two dimensions
(a) a non-concave objective function and a convex constraint set
(b) a concave objective function and a non-convex constraint set

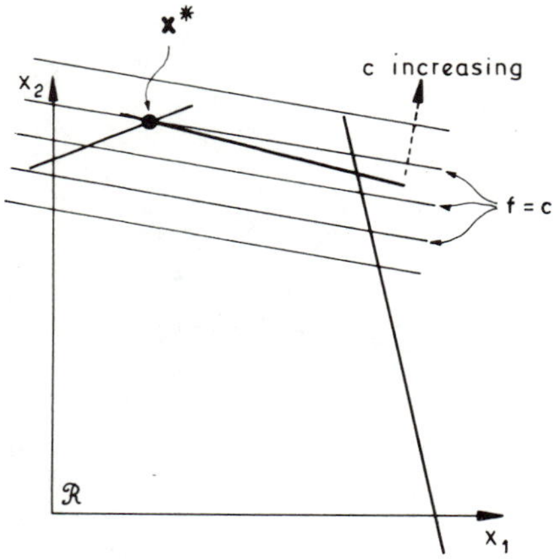

Figure 2.4 Optimum solution of a linear programming problem

and $\mathbf{H}$ is the $N \times N$ Hessian matrix of second derivatives

$$\mathbf{H} = \begin{bmatrix} \dfrac{\partial^2 f}{\partial x_1^2} & \dfrac{\partial^2 f}{\partial x_1 \partial x_2} & \cdots & \dfrac{\partial^2 f}{\partial x_1 \partial x_N} \\ \vdots & & & \\ \dfrac{\partial^2 f}{\partial x_N \partial x_1} & \cdots & \cdots & \dfrac{\partial^2 f}{\partial x_N^2} \end{bmatrix} \tag{2.12}$$

The first-order necessary conditions for a local maximum (or minimum) of $f(\mathbf{x})$ at $\mathbf{x}^*$ is that the gradient vector vanishes at $\mathbf{x}^*$. That is,

$$\frac{\partial f}{\partial x_1}(\mathbf{x}^*) = \frac{\partial f}{\partial x_2}(\mathbf{x}^*) = \ldots = \frac{\partial f}{\partial x_N}(\mathbf{x}^*) = 0 \tag{2.13}$$

Second-order necessary conditions may be derived by examining the quadratic form

$$Q(\mathbf{x}^*) = \Delta\mathbf{x}'\mathbf{H}(\mathbf{x}^*)\Delta\mathbf{x} \tag{2.14}$$

at the point $\mathbf{x}^*$. If $Q(\mathbf{x}^*)$ is negative definite for all $\Delta\mathbf{x}$, the stationary value is a maximum, and if positive a minimum is obtained. A set of necessary and sufficient conditions that Q is negative definite may be expressed in terms of the signs of the leading principal minors of the Hessian matrix,

$$f_{11} < 0, \quad \begin{vmatrix} f_{11} & f_{12} \\ f_{21} & f_{22} \end{vmatrix} > 0, \quad \begin{vmatrix} f_{11} & f_{12} & f_{13} \\ f_{21} & f_{22} & f_{23} \\ f_{31} & f_{32} & f_{33} \end{vmatrix} < 0 \ldots. \tag{2.15}$$

in which f_{ij} denotes $\dfrac{\partial^2 f}{\partial x_i \partial x_j}$. These conditions are therefore sufficient for a maximum. A set of necessary and sufficient conditions that Q is positive definite is

$$f_{11} > 0, \quad \begin{vmatrix} f_{11} & f_{12} \\ f_{21} & f_{22} \end{vmatrix} > 0, \quad \begin{vmatrix} f_{11} & f_{12} & f_{13} \\ f_{21} & f_{22} & f_{23} \\ f_{31} & f_{32} & f_{33} \end{vmatrix} > 0 \ldots . \tag{2.16}$$

and these are therefore sufficient for a minimum of $f(\mathbf{x})$ at $\mathbf{x}^*$. If the Hessian matrix is singular, then the corresponding point $\mathbf{x}^*$ is singular, and this is often of special interest. This topic also has implications for dynamical systems theory and will be pursued further in Chapter 9.

We next consider the necessary modifications to these results in the presence of the equality constraints (2.9).

2.2.2 Lagrangian theory

The effect of introducing constraints

$$\mathbf{g}(\mathbf{x}) = \mathbf{b} \tag{2.17}$$

into the optimization problem is to introduce dependency between the N variables $x_1, \ldots, x_N$. The method of Lagrange multipliers may be used to remove **explicit** consideration of this dependency, at the expense of introducing additional variables into the problem. The solution of the constrained problem may be obtained from the stationary points of the Lagrangian

$$L(\mathbf{x}, \boldsymbol{\lambda}) = f(\mathbf{x}) + \sum_{m=1}^{M} \lambda_m (b_m - g_m(\mathbf{x})) \tag{2.18}$$

$$= f(\mathbf{x}) + \boldsymbol{\lambda}'(\mathbf{b} - \mathbf{g}(\mathbf{x})) \tag{2.19}$$

where $\boldsymbol{\lambda}$ is a vector of the multipliers, and $\boldsymbol{\lambda}'$ its transpose (see, for example, Gillespie (1960)).

The stationary points of $L(\mathbf{x}, \boldsymbol{\lambda})$ are found in the traditional way as for the unconstrained multivariate optimization problem. The $N + M$ first-order conditions are as follows:

$$\frac{\partial L}{\partial x_j} = \frac{\partial f}{\partial x_j} - \sum_m \lambda_m \frac{\partial g_m}{\partial x_j} = 0 \qquad j = 1, \ldots, N \tag{2.20}$$

$$\frac{\partial L}{\partial \lambda_m} = b_m - g_m(\mathbf{x}) = 0 \qquad m = 1, \ldots, M \tag{2.21}$$

The second-order conditions are expressed in terms of the bordered Hessian defined by

$$\hat{\mathbf{H}} = \left[\begin{array}{c|c} \mathbf{0} & \dfrac{\partial \mathbf{g}}{\partial \mathbf{x}} \\ \hline \dfrac{\partial \mathbf{g}'}{\partial \mathbf{x}} & \dfrac{\partial^2 L}{\partial \mathbf{x}^2} \end{array}\right] \tag{2.22}$$

in which $\frac{\partial \mathbf{g}}{\partial \mathbf{x}}$ denotes a matrix with elements $\frac{\partial \mathbf{g}_m}{\partial x_j}$. The conditions for a local maximum are that the last $N-M$ leading principal minors of this bordered Hessian alternate in sign, the sign of the first being $(-1)^{M+1}$

It is immediately apparent from the first-order conditions of the unconstrained problem, in which the values $\mathbf{x}^*$, $\boldsymbol{\lambda}^*$ are sought which maximize $L(\mathbf{x}, \boldsymbol{\lambda})$, are such that the constraints of the original problem are satisfied and the value of the Lagrangian attains the maximum of $f(\mathbf{x})$. That is,

$$L(\mathbf{x}^*, \boldsymbol{\lambda}^*) = f(\mathbf{x}^*) \tag{2.23}$$

2.2.3 Inequality constraints and the Kuhn–Tucker conditions

We now consider the extension of the classical Lagrangian problem to include inequality constraints and outline the (Kuhn–Tucker) conditions which characterize the optimal solution of the general non-linear program. To this end the inequality constraints

$$\mathbf{g}(\mathbf{x}) \leqslant \mathbf{b}$$

are transformed into equality constraints

$$\begin{aligned} g_1(\mathbf{x}) + x_1^s &= b_1 \\ \vdots \qquad & \quad \vdots \\ g_M(\mathbf{x}) + x_M^s &= b_M \end{aligned} \tag{2.24}$$

by introducing the M 'slack' variables $x_1^s \ldots x_M^s$. The non-linear programming problem now becomes one of solving the classical Lagrangian problem

$$\max_{\{\mathbf{x}, \mathbf{x}^s\}} f(\mathbf{x}) \tag{2.25}$$

subject to

$$\mathbf{g}(\mathbf{x}) + \mathbf{x}^s = \mathbf{b} \tag{2.26}$$

and the non-negativity conditions on the instrument **and** slack variables, that is

$$\mathbf{x} \geqslant 0, \qquad \mathbf{x}^s \geqslant 0 \tag{2.27}$$

If a solution to (2.25)–(2.27) exists then it will be found either in the **interior** of the feasible region with $\mathbf{x} > 0$, $\mathbf{x}^s > 0$, or on its **boundary** at which one or more of the variables $(x_1, \ldots, x_N; x_1^s, \ldots, x_M^s)$ are zero. It therefore requires that we examine the solution properties both in the interior and at the boundary of the feasible region. That is, it is necessary to examine the properties of the Lagrangian $L(\mathbf{x}, \mathbf{x}^s, \boldsymbol{\lambda})$, defined by

$$L(\mathbf{x}, \mathbf{x}^s, \boldsymbol{\lambda}) = f(\mathbf{x}) + \boldsymbol{\lambda}'(\mathbf{b} - \mathbf{g}(\mathbf{x}) - \mathbf{x}^s) \tag{2.28}$$

in the neighbourhood of the optimal solution $\mathbf{x}^*$, $\mathbf{x}^{s*}$, $\boldsymbol{\lambda}^*$. Holding all other variables constant at their optimal values, consider the variation of $L(\mathbf{x}, \mathbf{x}^s, \boldsymbol{\lambda})$

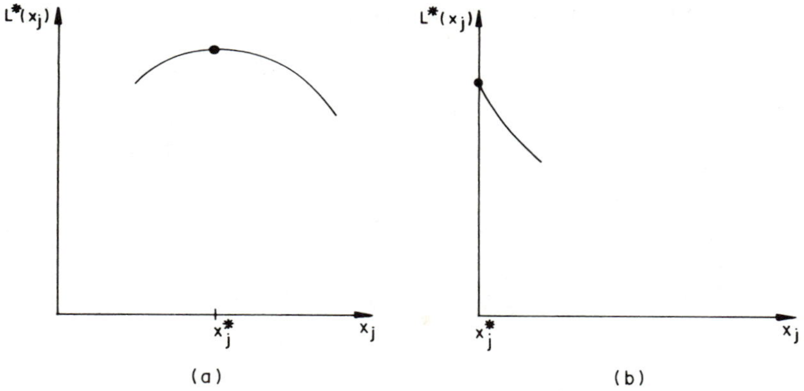

Figure 2.5 The variation of the Lagrangian $L^*(x_j)$ with x_j, holding all other variables constant at their optimal values (a) an interior solution for x_j^*; (b) boundary solution for x_j^*

with the single variable x_j in the neighbourhood of $\mathbf{x}^*, \mathbf{x}^{s*}, \boldsymbol{\lambda}^*$. The two possibilities which exist, in which x_j^* corresponds to an interior and a boundary solution, are illustrated in Figures 2.5(a) and (b) respectively. If x_j^* is an interior solution then the classical conditions for optimality hold at the optimal, namely

$$x_j^* > 0 \qquad \frac{\partial L}{\partial x_j}(\mathbf{x}^*, \mathbf{x}^{s*}, \boldsymbol{\lambda}^*) = 0 \tag{2.29}$$

If, on the other hand, $x_j^* = 0$ (at the boundary) then for a maximum we require that the slope of the Lagrangian function at $x_j^* = 0$ is negative. That is,

$$x_j^* = 0 \qquad \frac{\partial L}{\partial x_j}(\mathbf{x}^*, \mathbf{x}^{s*}, \boldsymbol{\lambda}^*) \leqslant 0 \tag{2.30}$$

Equations (2.29) and (2.30) may be combined to give

$$x_j^* \frac{\partial L}{\partial x_j}(\mathbf{x}^*, \mathbf{x}^{s*}, \boldsymbol{\lambda}^*) = 0, \quad \frac{\partial L}{\partial x_j}(\mathbf{x}^*, \mathbf{x}^{s*}, \boldsymbol{\lambda}^*) \leqslant 0, \quad x_j^* \geqslant 0 \tag{2.31}$$

Grouping similar relations together for all instrument and slack variables, the necessary conditions for a maximum become

$$x_j = 0, \ \frac{\partial L^*}{\partial x_j} \leqslant 0; \qquad x_j > 0, \ \frac{\partial L^*}{\partial x_j} = 0 \qquad j = 1, \ldots, N \tag{2.32}$$

$$x_m^s = 0, \ \frac{\partial L^*}{\partial x_m^s} \leqslant 0; \qquad x_m^s > 0, \ \frac{\partial L^*}{\partial x_m^s} = 0 \qquad m = 1, \ldots, M \tag{2.33}$$

$$\frac{\partial L^*}{\partial \lambda_m} = 0 \qquad m = 1, \ldots, M \tag{2.34}$$

in which the derivatives are evaluated at $(\mathbf{x}^*, \mathbf{x}^{s*}, \boldsymbol{\lambda}^*)$.

For the Lagrangian (2.28), the conditions (2.33) and (2.34) may be written

$$-\boldsymbol{\lambda} \leqslant 0 \qquad -\boldsymbol{\lambda}' . \mathbf{x}^s = 0 \tag{2.35}$$

$$\mathbf{x}^s \geqslant 0. \tag{2.36}$$

The slack variables may be eliminated to yield the Kuhn–Tucker conditions

$$\begin{aligned} x_j = 0, \quad & \frac{\partial f}{\partial x_j} - \sum_m \lambda_m \frac{\partial g_m}{\partial x_j} \leqslant 0 \\ x_j > 0, \quad & \frac{\partial f}{\partial x_j} - \sum_m \lambda_m \frac{\partial g_m}{\partial x_j} = 0 \end{aligned} \tag{2.37}$$

$$\begin{aligned} \lambda_m = 0, \quad & b_m - g_m(\mathbf{x}) \geqslant 0 \\ \lambda_m > 0, \quad & b_m - g_m(\mathbf{x}) = 0 \end{aligned} \tag{2.38}$$

These conditions may also be obtained from the Lagrangian

$$L(\mathbf{x}, \boldsymbol{\lambda}) = f(\mathbf{x}) + \boldsymbol{\lambda}'(\mathbf{b} - \mathbf{g}(\mathbf{x})) \tag{2.39}$$

in terms of which the Kuhn–Tucker conditions (2.37)–(2.38) may be expressed

$$\mathbf{x}^* \frac{\partial L}{\partial \mathbf{x}}(\mathbf{x}^*, \boldsymbol{\lambda}^*) = 0, \quad \frac{\partial L}{\partial \mathbf{x}}(\mathbf{x}^*, \boldsymbol{\lambda}^*) \leqslant 0, \quad \mathbf{x}^* \geqslant 0 \tag{2.40}$$

$$\boldsymbol{\lambda}^* \frac{\partial L}{\partial \boldsymbol{\lambda}}(\mathbf{x}^*, \boldsymbol{\lambda}^*) = 0, \quad \frac{\partial L}{\partial \boldsymbol{\lambda}}(\mathbf{x}^*, \boldsymbol{\lambda}^*) \geqslant 0, \quad \boldsymbol{\lambda}^* \geqslant 0 \tag{2.41}$$

The Kuhn–Tucker theorem establishes the equivalence between the non-linear programming problem and the so-called saddle-point problem. Specifically, $\mathbf{x}^*$ solves the non-linear program if $(\mathbf{x}^*, \boldsymbol{\lambda}^*)$, is a saddle point of the Lagrangian function $L(\mathbf{x}, \boldsymbol{\lambda})$, and under certain conditions $\mathbf{x}^*$ solves the programming problem only if a $\boldsymbol{\lambda}^*$ can be found for which $(\mathbf{x}^*, \boldsymbol{\lambda}^*)$ is the saddle point of $L(\mathbf{x}, \boldsymbol{\lambda})$. The conditions referred to are: firstly, that $f(\mathbf{x})$, in a maximization problem, is concave, and $\mathbf{g}(\mathbf{x})$ are convex functions; secondly, that there exists some vector which satisfies the constraint set

$$\mathbf{g}(\mathbf{x}) \leqslant \mathbf{b}$$

in terms of strict inequalities.

The conditions (2.40)–(2.41) are necessary conditions for $(\mathbf{x}^*, \boldsymbol{\lambda}^*)$ to be a saddle point of $L(\mathbf{x}, \boldsymbol{\lambda})$ and the sufficient conditions are in fact

$$L(\mathbf{x}^* + \Delta\mathbf{x}, \boldsymbol{\lambda}^*) < L(\mathbf{x}^*, \boldsymbol{\lambda}^*) + \left(\frac{\partial L}{\partial \mathbf{x}}(\mathbf{x}^*, \boldsymbol{\lambda}^*)\right)' \Delta\mathbf{x} \tag{2.42}$$

$$L(\mathbf{x}^*, \boldsymbol{\lambda}^* + \Delta\boldsymbol{\lambda}) > L(\mathbf{x}^*, \boldsymbol{\lambda}^*) + \left(\frac{\partial L}{\partial \boldsymbol{\lambda}}(\mathbf{x}^*, \boldsymbol{\lambda}^*)\right)' \Delta\boldsymbol{\lambda} \tag{2.43}$$

for non-negative $\mathbf{x}$ and $\boldsymbol{\lambda}$, in which the transpose vectors are defined by

$$\left(\frac{\partial L}{\partial \mathbf{x}}\right)' = \left(\frac{\partial L}{\partial x_1}, \ldots, \frac{\partial L}{\partial x_N}\right) \tag{2.44}$$

$$\left(\frac{\partial L}{\partial \boldsymbol{\lambda}}\right)' = \left(\frac{\partial L}{\partial \lambda_1}, \ldots, \frac{\partial L}{\partial \lambda_M}\right) \tag{2.45}$$

Before turning to the topic of duality, it is important to comment on the significance of $\boldsymbol{\lambda}^*$. In the classical Lagrangian problem, it is readily shown by differentiating the Lagrangian with respect to $\mathbf{b}$ and evaluating at the program solution, that

$$\boldsymbol{\lambda}^* = \frac{\partial L}{\partial \mathbf{b}}(\mathbf{x}^*, \boldsymbol{\lambda}^*) = \frac{\partial f}{\partial \mathbf{b}}(\mathbf{x}^*) \tag{2.46}$$

or

$$\lambda_m^* = \frac{\partial f}{\partial b_m}(\mathbf{x}^*) \qquad m = 1, \ldots, M \tag{2.47}$$

This condition may be shown to hold also for the non-linear programming problem. From (2.47) the multipliers can be seen to yield important information about the sensitivity of the objective function to variations in the constraint constants. In economic problems $\boldsymbol{\lambda}$ also has a special significance in terms of the 'value' of relaxing constraints and is often referred to as a shadow price vector. Returning to the 'complementary slackness' conditions,

$$x_m^s \lambda_m = 0 \qquad \lambda_m \geqslant 0, \quad x_m^s \geqslant 0 \tag{2.48}$$

which are equivalent to

$$\begin{matrix} x_m^s = 0 & \lambda_m \geqslant 0 \\ x_m^s \geqslant 0 & \lambda_m = 0, \end{matrix} \tag{2.49}$$

it can be seen that if $x_m^s > 0$ then the objective function will be insensitive to the marginal variation in b_m. If, on the other hand, the corresponding constraint is binding, λ_m measures the increase in f per unit increase in b_m.

2.2.4 Duality

For a wide class of mathematical programs it is possible to define a dual program. The notion of duality and the relationship between these programs (primal and dual) is of profound theoretical and practical significance. Here we shall give a brief insight into the nature of duality and its relationship to the properties of Lagrangians. To do appropriate justice and give full significance to this topic we would need to consider in detail the convex analysis theory of Rockafellar (1967, 1969), to which the interested reader is referred for further details.

Consider again the (primal) problem with **equality** constraints

$$\max_{\{\mathbf{x}\}} f(\mathbf{x})$$

subject to

$$\mathbf{g}(\mathbf{x}) = \mathbf{b}$$

When the problem is locally concave in the sense that the Hessian of the Lagrangian evaluated at the constrained maximum is negative definite, the local dual problem is defined as the minimization of the function

$$\min L(\mathbf{x}, \boldsymbol{\lambda}) = f(\mathbf{x}) + \boldsymbol{\lambda}'(\mathbf{b} - \mathbf{g}(\mathbf{x})) \tag{2.50}$$

subject to

$$\frac{\partial f}{\partial \mathbf{x}} - \boldsymbol{\lambda}'\frac{\partial \mathbf{g}}{\partial \mathbf{x}} = 0 \tag{2.51}$$

It can be shown that:

(i) if a solution to one of these primal or dual problems exists, then so does the solution to the other;

(ii) the maximum of the primal program $f(\mathbf{x})$ is equal to the minimum of the dual;

(iii) the value of $\mathbf{x} = \mathbf{x}^*$ which maximizes $f(\mathbf{x})$ subject to the primal constraints is the same as the value of $\mathbf{x}$ which minimizes the dual.

We now consider the dual of the non-linear program (2.4)–(2.6), characterized by the inequality constraints. For this primal program the dual has been defined as the minimization of

$$L(\mathbf{x}, \boldsymbol{\lambda}) - \mathbf{x}'\frac{\partial L(\mathbf{x}, \boldsymbol{\lambda})}{\partial \mathbf{x}} \tag{2.52}$$

subject to

$$\frac{\partial L(\mathbf{x}, \boldsymbol{\lambda})}{\partial \mathbf{x}} \leqslant 0 \tag{2.53}$$

$$\boldsymbol{\lambda} \geqslant 0 \tag{2.54}$$

in which

$$L(\mathbf{x}, \boldsymbol{\lambda}) = f(\mathbf{x}) + \boldsymbol{\lambda}'(\mathbf{b} - \mathbf{g}(\mathbf{x}))$$

A further discussion of non-linear duality may be found in Wolfe (1961) and an interpretation of the terms in (2.52)–(2.54) is given by Balinski and Baumol (1968). The programs (2.50)–(2.51) and (2.52)–(2.54) both contain the mutually dual variables $\boldsymbol{\lambda}$ and $\mathbf{x}$, and will not **in general** be confined to the set of dual variables $\boldsymbol{\lambda}$ alone. At optimality the solutions of the primal and dual programs are mutually complementary and satisfy the three conditions outlined above.

For an important class of spatial interaction and planning programs to be considered below we shall show that the dual may in fact be expressed as a function of the dual variables $\boldsymbol{\lambda}$ alone. In these cases it often proves more convenient for numerical purposes to consider the dual program, and from the optimality conditions obtain the primal variables. In later sections we shall often exploit the 'dimensional reduction' which may be a consequence of forming and solving dual programs.

2.3 THE CENTRAL FEATURES OF LINEAR PROGRAMMING

2.3.1 Primal–dual relations and optimality conditions

In terms of scope of application and theoretical richness, linear programs—in which $f(\mathbf{x})$ and $\mathbf{g}(\mathbf{x})$ are linear functions of the variables $x_1, \ldots, x_N$— are by far the most important special case of the general programming problem. We can appeal directly to the Kuhn–Tucker conditions to obtain the optimality conditions of the linear problem, but as will be seen below these do not help directly in the problem solution.

In canonical form a linear program is expressed as follows:

$$\max_{\{\mathbf{x}\}} \sum_j c_j x_j \tag{2.55}$$

subject to

$$\sum_j a_{mj} x_j \leqslant b_m \qquad m = 1, \ldots, M \tag{2.56}$$

$$x_1, \ldots, x_N \geqslant 0 \tag{2.57}$$

for which the Lagrangian $L(\mathbf{x}, \boldsymbol{\lambda})$ may be written

$$L(\mathbf{x}, \boldsymbol{\lambda}) = \sum_j c_j x_j + \sum_m \lambda_m \left(b_m - \sum_j a_{mj} x_j \right). \tag{2.58}$$

According to the Kuhn–Tucker theory, $\mathbf{x}^* = (x_1^*, \ldots, x_N^*)$ solves the linear program (2.55)–(2.57) if there exists an M-component vector $\boldsymbol{\lambda}^* = (\lambda_1^*, \ldots, \lambda_M^*)$ such that

$$\begin{aligned} \frac{\partial L}{\partial x_j} &= c_j - \sum_m \lambda_m a_{mj} \leqslant 0 \\ x_j \frac{\partial L}{\partial x_j} &= x_j \left(c_j - \sum_m \lambda_m a_{mj} \right) = 0 \end{aligned} \tag{2.59}$$

$$\begin{aligned} \frac{\partial L}{\partial \lambda_m} &= b_m - \sum_j a_{mj} x_j \geqslant 0 \\ \lambda_m \frac{\partial L}{\partial \lambda_m} &= \lambda_m \left(b_m - \sum_j a_{mj} x_j \right) = 0 \end{aligned} \tag{2.60}$$

for $$\mathbf{x} \geqslant 0 \qquad \boldsymbol{\lambda} \geqslant 0.$$

If the problem were to minimize the objective function (2.55) the inequality constraints in (2.59) would be reversed.

The dual of the linear program defined above is written

$$\min_{\{\lambda\}} D(\boldsymbol{\lambda}) = \sum_m \lambda_m b_m \tag{2.61}$$

subject to

$$\sum_m a_{jm} \lambda_m \geqslant c_j \tag{2.62}$$

$$\lambda_1, \ldots, \lambda_m \geqslant 0 \tag{2.63}$$

and it is readily shown that the Lagrangians and Kuhn–Tucker conditions of the primal and dual problems are identical and that the dual of the dual is the original primal problem. It may also readily be checked that this dual is consistent with the general dual form (2.52)–(2.54) (although, in terms of the chronological development of the subject, it is perhaps more accurate to remark that the non-linear dual formulation is consistent with the relations (2.55)–(2.63) in the special case of linear objective functions and constraints).

The following properties characterize the primal and dual programs:

(i) A necessary and sufficient condition for the existence of a solution to a linear program is that there exist vectors $\mathbf{x}$ and $\boldsymbol{\lambda}$ which satisfy the primal and dual constraints respectively.

(ii) If the primal problem involves maximization, the value of the objective function $f(\mathbf{x})$ cannot exceed the value of the objective function in the dual, when evaluated with feasible vectors. At optimality these objective functions are equal, that is

$$\sum_j c_j x_j^* = \sum_m \lambda_m^* b_m \tag{2.64}$$

(iii) A necessary and sufficient condition for $\mathbf{x}^*$ to solve the linear program (2.55)–(2.57) is that there exists a feasible vector $\boldsymbol{\lambda}^*$ for the dual problem for which the condition (2.64) holds.

(iv) Necessary and sufficient conditions for the feasible vectors $\mathbf{x}^*$ and $\boldsymbol{\lambda}^*$ to solve the mutually dual problems are that the complementary slackness conditions (2.59)–(2.60) be satisfied.

From equation (2.47) we have at the optimal solution the marginal relations

$$\lambda_m^* = \frac{\partial f}{\partial b_m}(\mathbf{x}^*) \geqslant 0 \qquad m = 1, \ldots, M \tag{2.65}$$

$$x_j^* = \frac{\partial D}{\partial c_j}(\boldsymbol{\lambda}^*) \geqslant 0 \qquad j = 1, \ldots, N \tag{2.66}$$

which allow the primal and dual variables to be given additional significance both in a sensitivity-testing context, and in a planning context where f represents an economic 'value' and **b** a set of 'resources'. The shadow prices, as described above, represent scarcity values for the corresponding resources.

It is a well-known result from linear algebra and the theory of convex sets that the optimal solution of a linear program will occur at an extreme point of the convex set of feasible solutions, at which point no more than M of the instrument variables will be non-zero. In two dimensions a possible graphical solution was illustrated in Figure 2.4. The extreme points of the feasible region are termed **basic feasible solutions**, and it is required to find that particular solution (or solutions where alternative optima exist) which maximizes $f(\mathbf{x})$. As there are (at most) ${}^{N}C_{M}$ such basic solutions, it is in general necessary to involve an automated search of the extreme points. This may be achieved by using the simplex procedure, which allows efficient transformations between adjacent extreme points in the feasible region. The most efficient codes allow very large programs to be solved.

2.3.2 An example: the transportation problem of linear programming

The transportation problem of linear programming, originally formulated by Hitchcock in 1941, is of great significance in location and spatial-interaction theory. The problem's special structure, reflected in the form of the matrix elements $\{a_{mi}, m = 1, \ldots, M; i = 1, \ldots, N\}$ which are either 0 or 1, permits a simple and efficient algorithm for its solution.

If we define x_{ij} as the number of units (commodities or people) which flow between zones i and j with unit cost c_{ij}, the transportation problem is one of determining the non-negative allocation x_{ij}^*, which minimizes the total transport cost

$$C = \sum_{ij} x_{ij} c_{ij} \tag{2.67}$$

subject to constraints on supply and demand

$$\sum_{j} x_{ij} = O_i \qquad i = 1, \ldots, N \tag{2.68}$$

$$\sum_{i} x_{ij} = D_j \qquad j = 1, \ldots, M \tag{2.69}$$

$$x_{ij} \geqslant 0 \qquad i = 1, \ldots, N; j = 1, \ldots, M$$

Here O_i and D_j are the number of units supplied in zone i and demanded in zone j respectively.

If total supply $\sum_i O_i$ exceeds the total demand $\sum_j D_j$, a dummy destination is added with demand

$$D_{M+1} = \sum_i O_i - \sum_j D_j \tag{2.70}$$

and

$$c_{i,M+1} = 0 \quad \text{for} \quad i = 1, \ldots, N.$$

A dummy supplier is introduced in the converse situation in which total demand exceeds supposed total supply.

We shall assume hereafter that the system is 'balanced', in which case total demand is equal to total supply.

Now, the allocation

$$x_{ij} = \frac{O_i D_j}{T} \tag{2.71}$$

with

$$T = \sum_i O_i = \sum_j D_j \tag{2.72}$$

clearly represents a feasible solution as it satisfies the constraints (2.68)–(2.69), and as we know that the solution is bounded

$$0 \leqslant x_{ij} \leqslant \min\{O_i, D_j\} \tag{2.73}$$

the program possesses an optimal solution. Efficient methods of generating an initial basic feasible solution and the subsequent steps taken to achieve the optimal allocation are described in most books on linear programming. Our intention here is to introduce the program structure together with its dual so they may be related to the spatial interaction model developed below.

Introducing the vectors $\boldsymbol{\alpha}$ and $\boldsymbol{\gamma}$ which have N and M components respectively, the dual of the program (2.67)–(2.69) can be written as follows:

$$\max_{\{\boldsymbol{\alpha}, \boldsymbol{\gamma}\}} \sum_i O_i \alpha_i + \sum_j D_j \gamma_j \tag{2.74}$$

subject to

$$\alpha_i + \gamma_j \leqslant c_{ij} \tag{2.75}$$

Because the constraints (2.68) and (2.69) are **equalities**, all the components $\{\ldots \alpha_i \ldots\}$ and $\{\ldots \gamma_j \ldots\}$ are otherwise unrestricted.

Note that the primal program has $N \times M$ variables and $N + M$ constraints while in the dual these numbers are interchanged. The balancing condition (2.72) may be used to eliminate one of the constraints, so that only $N + M - 1$ of the primal constraints are independent. From the fundamental theorems of linear programming it is thus apparent that in general only $N + M - 1$ out of the NM variables will be non-zero.

The effect of the balancing condition is to leave the dual variables determined up to an arbitrary constant—it is readily apparent from equations (2.74) and (2.75) that if the dual is solved by $\{\boldsymbol{\alpha}^*, \boldsymbol{\gamma}^*\}$ then it is also solved by $\{\boldsymbol{\alpha}^* + \sigma, \boldsymbol{\gamma}^* - \sigma\}$

for an arbitrary constant σ. This indeterminacy may be removed by arbitrarily specifying the value of any **one** of the dual variables.

While the transportation program is most conveniently introduced by considering the minimum cost flow of goods or people, it may be adapted to many allocation processes governed by extremal (optimization) principles. One very important example encountered in urban research is in the resolution of competitive equilibrium processes in, for example, land and housing markets, in which demand and supply are segmented or categorized. The Herbert–Stevens model (Herbert and Stevens, 1960), which may be expressed in transportation program format, was designed as an algorithmic solution to the market clearing problem proposed by Alonso (1964).

This market clearing model generates an allocation T_{ij}^{kw} of workers of type w who work in zone j to houses of type k in zone i, through the following program:

$$\max_{\{\mathbf{T}\}} \sum_{ij} \sum_{kw} T_{ij}^{kw} (b_{ij}^{kw} - c_{ij}) \tag{2.76}$$

subject to

$$\sum_{ik} T_{ij}^{kw} = E_j^w \tag{2.77}$$

$$\sum_{jw} T_{ij}^{kw} \leqslant H_i^k \tag{2.78}$$

$$\mathbf{T} \geqslant 0.$$

The remaining variables are defined as follows:

E_j^w: the number of workers of type w in zone j, which is considered fixed.
H_i^k: the housing stock of type k in zone i.
b_{ij}^{kw}: money bids by workers of type w in zone j for houses of type k in zone i.
c_{ij}: journey-to-work costs between zones i and j.

The worker groups are regarded as competitors in a perfectly competitive spatial market, and their bidding (the travel costs are subtracted from b_{ij}^{kw} to yield a bid for the house alone) generates an upward push of prices for housing services. This effect is reflected in the objective function (2.76) which maximizes total bids for housing services. The constraints (2.77) and (2.78) are accounting relations to ensure that, firstly, the allocation is consistent with the number in each district worker group, and, secondly, that the total number of workers allocated to houses of type k in zone i cannot exceed the available supply.

The dual program may be expressed as follows:

$$\min_{\{\boldsymbol{\alpha}, \boldsymbol{\Upsilon}\}} \sum_{ik} \alpha_i^k H_i^k + \sum_{jw} \gamma_j^w E_j^w \tag{2.79}$$

subject to

$$\alpha_i^k + \gamma_j^w \geqslant b_{ij}^{kw} - c_{ij} \tag{2.80}$$

$$\boldsymbol{\alpha} \geqslant 0; \quad \boldsymbol{\Upsilon} \text{ unbounded}$$

The dual variables $\boldsymbol{\alpha}$ evaluated at optimality are interpreted as actual prices paid in the market—the prices which 'clear the market'. The dual objective function whose first term involves the minimization of total rent (or prices) **paid**, reflects the competition between suppliers who want to avoid unsold or unrented accommodation. As Senior (1977a) remarks, the primal and dual programs reflect both sides of a market confrontation. The set of dual variables $\boldsymbol{\Upsilon} = \{\ldots \gamma_j^w \ldots .\}$ measure the differences between bids by the various consumer groups and the prices actually paid in the market. These consumer surpluses clearly have an important role in the evaluation problem and are further discussed by Williams and Senior (1978).

For a consideration of the many extensions and refinements of the above competitive market model, the reader is referred to the work of Senior (1977a, 1978). We shall return to reconsider the above interpretation of the allocation mechanism in the context of the generation and interpretation of spatial-interaction model programs in Section 2.4.

2.4 EXAMPLES OF MODEL FORMATION FROM NON-LINEAR PROGRAMS

2.4.1 Introduction

In this section we give examples of the formation of models from non-linear programs, and show how the programming framework may be exploited for the investigation of model properties. Firstly, we consider the formation of the fully constrained gravity model which plays a central role in transport and many land-use studies for the purpose of distributing trips. The significance of the primal and dual relations associated with the model is emphasized, and the connection with the transportation problem of linear programming noted.

In the second example we examine a multicommodity network flow problem—the congested assignment problem—which again is of central importance in transport and location models, as the interaction costs are generated by the assignment of trips to appropriate modal networks. Here a network equilibrium is derived in which there is no incentive for drivers to change from their perceived minimum-cost routes between any pair of zones.

As a final example we discuss the integration of the above problems and the generation of distribution and assignment patterns which are mutually consistent. This is a good example of a 'superproblem' in which two models are interfaced within a mathematical programming framework.

2.4.2 The fully constrained gravity model: its formulation and dual

The spatial-interaction model has the following familiar form:

$$T_{ij} = A_i B_j O_i D_j F_{ij} \tag{2.81}$$

in which we denote

T_{ij}: the number of trips between zones i and j
O_i: the number of trips generated in zone i
D_j: the number of trips attracted to zone j
F_{ij}: a factor reflecting the spatial impedance characterizing separation between zones i and j. Typically F_{ij} has the exponential form $e^{-\beta c_{ij}}$, in which c_{ij} is the generalized cost of travel between i and j and β is a dispersion parameter.

We shall hereafter restrict attention to this exponential form.

The balancing factors A_i and B_j are computed to satisfy the trip end constraints

$$\sum_j T_{ij} = O_i \tag{2.82}$$

$$\sum_i T_{ij} = D_j \tag{2.83}$$

and by substituting (2.81) in (2.82) and (2.83), the factors are readily seen to satisfy

$$A_i = \left\{\sum_j B_j D_j e^{-\beta c_{ij}}\right\}^{-1} \tag{2.84}$$

$$B_j = \left\{\sum_i A_i O_i e^{-\beta c_{ij}}\right\}^{-1} \tag{2.85}$$

That the model (2.81)–(2.85) could be derived from an optimization problem was first noted by Murchland (1966), and by Wilson (1967) who provided the well-known entropy-maximizing (or maximum probability) interpretation of the trip pattern. That is, the model was formed as the Kuhn–Tucker relations associated with the mathematical program.

The program considered by Wilson was of the following form:

$$\max_{\{T\}} S = -\sum_{ij} T_{ij} \ln T_{ij} \tag{2.86}$$

subject to

$$\sum_j T_{ij} = O_i$$

$$\sum_i T_{ij} = D_j$$

$$\sum_{ij} T_{ij} c_{ij} = C \tag{2.87}$$

$$T \geqslant O$$

in which equation (2.87) is an expression for the total *observed* travel cost expended in the travel pattern $\{T\}$ The optimal solution has an interpretation as the most probable distribution of trips consistent with all known information associated with the cost and trip-end constraints.

The Lagrangian associated with this program is written

$$L(\mathbf{T}, \boldsymbol{\alpha}, \boldsymbol{\gamma}, \beta) = S(\mathbf{T}) + \sum_i \alpha_i (O_i - \sum_j T_{ij}) + \sum_j \gamma_j (D_j - \sum_i T_{ij})$$

$$+ \beta (C - \sum_{ij} T_{ij} C_{ij}) \tag{2.88}$$

and L is to be maximized over non-negative values of the interaction matrix $\{T_{ij}\}$. It is known (see for example Evans, 1973a,b) that the maximization problem results in non-zero values for each T_{ij} for finite values of β. The non-negativity conditions are in this case redundant, and the optimal solution

$$T_{ij}^* = \exp[-\alpha_i^* - \gamma_j^* - \beta^* c_{ij}] \tag{2.89}$$

results. (An asterisk is used to denote the value of a quantity at optimality.) By defining A_i and B_j as

$$A_i = \frac{\exp[-\alpha_i^*]}{O_i} \tag{2.90}$$

$$B_j = \frac{\exp[-\gamma_j^*]}{D_j} \tag{2.91}$$

the doubly constrained gravity model (2.81) results. Because $-x \ln x$ is strictly concave over the positive x-axis, the program has a unique solution.

The entropy-maximizing framework may readily be extended to incorporate *prior information* (Kullback, 1959; Snickars and Weibull, 1977) by replacing S by

$$S' = -\sum_{ij} T_{ij} \ln \frac{T_{ij}}{\bar{T}_{ij}} \tag{2.92}$$

in which $\bar{T}_{ij}$, the prior distribution, may for example, in a forecasting context, be a known trip matrix, or of the form (2.71),

$$\bar{T}_{ij} = \frac{O_i D_j}{T} \tag{2.93}$$

which represents the expected distribution in the absence of a trip cost constraint. In this case maximization of S' conditional on (2.82), (2.83) and (2.87) generates the model

$$T_{ij}^* = \frac{O_i D_j}{T} \exp[-\bar{\alpha}_i^* - \bar{\gamma}_j^* - \beta c_{ij}] \tag{2.94}$$

in which the duals $\bar{\alpha}^*$, and $\bar{\gamma}^*$ are related to α^* and γ^* by

$$\alpha_i^* = \bar{\alpha}_i^* - \ln\left(\frac{O_i}{\sqrt{T}}\right) \qquad i = 1, \ldots, N \tag{2.95}$$

$$\gamma_j^* = \bar{\gamma}_j^* - \ln\left(\frac{D_j}{\sqrt{T}}\right) \qquad j = 1, \ldots, N \tag{2.96}$$

The prior distribution (2.93) allows, in a sense, compensation for different zone sizes and releases the *explicit* dependence of the dual variables on the trip end capacities.

Throughout this book it is understood that any appropriate prior information may be added to interaction models in this way.

2.4.3 Relation of the fully constrained spatial-interaction model to the transportation problem of linear programming

Although long suspected, it has only recently been proved by Evans (1973b) that, as the parameter β tends to infinity, the model (2.81), with an exponential deterrence function, tends to the solution of the transportation programming model derived from (2.67)–(2.69).

Introducing the inequality constraint $\mathbf{T} \geqslant 0$, the appropriate first-order conditions for the Lagrangian (2.88) are given by

$$T_{ij}\frac{\partial L}{\partial T_{ij}} = 0, \quad \frac{\partial L}{\partial T_{ij}} \leqslant 0, \quad T_{ij} \geqslant 0 \tag{2.97}$$

and it is necessary to consider the possibility of boundary solutions. Because $\dfrac{\partial L}{\partial T_{ij}}$ is not defined **at** the boundary $T_{ij} = 0$, an appropriate limiting process must be invoked. The reader is refered to the article by Evans (1973b) for a rigorous consideration of procedure. Here we shall provide an intuitive demonstration of the limiting process by first noting the equivalence of the models generated by the program (2.86)–(2.87) and the following program:

$$\max_{\{\mathbf{T}\}} -\frac{1}{\beta^*}\sum_{ij} T_{ij} \ln T_{ij} - \sum_{ij} T_{ij} c_{ij} \tag{2.98}$$

subject to

$$\sum_j T_{ij} = O_i$$

$$\sum_i T_{ij} = D_j$$

$$\mathbf{T} \geqslant 0$$

for the value of β^* which satisfies the trip cost constraint (2.87) (see Chapter 7 below). The optimal Lagrange multipliers $\hat{\boldsymbol{\alpha}}^*$, $\hat{\boldsymbol{\gamma}}^*$ of the resultant model

$$T_{ij} = \exp[-\beta^*(\hat{\alpha}_i^* + \hat{\gamma}_j^* + c_{ij})] \tag{2.99}$$

are related to those associated with the entropy-maximizing program by

$$\begin{aligned} \beta^* \hat{\boldsymbol{\alpha}}^* &= \boldsymbol{\alpha}^* \\ \beta^* \hat{\boldsymbol{\gamma}}^* &= \boldsymbol{\gamma}^* \end{aligned} \tag{2.100}$$

The program (2.98) is now seen as a member of a family, parameterized by β^* (variation in β^* in the entropy-maximizing model corresponds to the variation of the total travel cost C), and as β^* becomes very large the relative contribution of the dispersion term

$$-\frac{1}{\beta^*}\sum_{ij} T_{ij} \ln T_{ij}$$

becomes small compared with interaction cost considerations. In the limit as β^* tends to infinity, the structure of and the solution of the non-linear program tends to that associated with the transportation problem discussed above in Section 2.3.

2.4.4 The dual of the spatial-interaction program

Considerations of the dual of the program generating the fully constrained gravity model were first given by Wilson and Senior (1974) using the Lagrangian approach outlined in section 2.2.4 and by Evans (1973a) using Rockafellar theory. We shall follow the Wilson–Senior Lagrangian formulation, in which the dual is derived from equations (2.52)–(2.54). For the program (2.86)–(2.87) the dual can be written

$$\min_{\{\mathbf{T}, \boldsymbol{\alpha}, \boldsymbol{\Upsilon}, \beta\}} D(\mathbf{T}, \boldsymbol{\alpha}, \boldsymbol{\Upsilon}, \beta) = \sum_{ij} T_{ij} + \sum_i O_i \alpha_i + \sum_j D_j \gamma_j + \beta C \tag{2.101}$$

subject to

$$-\ln T_{ij} + \alpha_i + \gamma_j + \beta c_{ij} \leqslant 0 \tag{2.102}$$

for all i and j.

Invoking the property that the non-negativity condition in the primal program is redundant for finite β (or equivalently for values of C greater than C', the least cost corresponding to the linear programming problem), the dual becomes

$$\min_{\{\mathbf{T}, \boldsymbol{\alpha}, \boldsymbol{\Upsilon}, \beta\}} D(\mathbf{T}, \boldsymbol{\alpha}, \boldsymbol{\Upsilon}, \beta) = \sum_{ij} T_{ij} + \sum_i O_i \alpha_i + \sum_j D_j \gamma_j + \beta C$$

subject to

$$-\ln T_{ij} + \alpha_i + \gamma_j + \beta c_{ij} = 0 \tag{2.103}$$

Eliminating T_{ij} in the objective function by substitution from the constraint (2.103), the dual becomes expressible in terms of dual variables alone:

$$\min_{\{\boldsymbol{\alpha}, \boldsymbol{\Upsilon}, \beta\}} D(\boldsymbol{\alpha}, \boldsymbol{\Upsilon}, \beta) = \sum_{ij} \exp[-\alpha_i - \gamma_j - \beta c_{ij}] + \sum_i O_i \alpha_i + \sum_j D_j \gamma_j + \beta C \tag{2.104}$$

From the Kuhn–Tucker theory we can say that $(\boldsymbol{\alpha}^*, \boldsymbol{\Upsilon}^*, \beta^*)$ solves the dual if T_{ij}^* defined by

$$T_{ij}^* = \exp[-\alpha_i^* - \gamma_j^* - \beta^* c_{ij}]$$

solves the primal (entropy-maximizing) problem.

It can immediately be seen from the dual objective function that if the problem is balanced with $\sum_i O_i = \sum_j D_j$ then for any value δ

$$D(\boldsymbol{\alpha}^*, \boldsymbol{\Upsilon}^*, \beta^*) = D(\boldsymbol{\alpha}^* + \delta, \boldsymbol{\Upsilon}^* - \delta, \beta^*) \tag{2.105}$$

That is, if the vectors $\boldsymbol{\alpha}^*$, $\boldsymbol{\Upsilon}^*$ solve the dual, then so do the vectors $\alpha^* + \delta$, $\Upsilon^* - \delta$. This property corresponds to the well-known observation that the balancing factors A_i and B_j may be transformed by a multiplicative constant k,

$$\begin{aligned} A_i &\to kA_i \\ B_i &\to \frac{1}{k}B_j \qquad \text{for all } i \text{ and } j \end{aligned} \tag{2.106}$$

to give the same spatial-interaction pattern. Thus, in common with the linear programming transportation problem, the solution of the primal program is unique, while there exist an infinite number of solutions to the dual.

It is interesting to note the formulation of dual spatial-interaction programs achievable from geometric programming, as, for example, discussed by Nijkamp and Paelinck (1974), Charnes *et al.* (1976), Dinkel *et al.* (1977), and others. The interest in geometric programming formulations is derived from the observation that the program (2.86)–(2.87) which generates the spatial-interaction model (2.89) corresponds to the *dual* of a (primal) geometric program. The latter, once formulated, may then be identified as the dual of the spatial-interaction program.

Following Nijkamp (1979) we write the program (2.86)–(2.87) in terms of the trip interaction probabilities

$$p_{ij} = \frac{T_{ij}}{T} \tag{2.107}$$

as follows:

$$\max_{\{\mathbf{P}\}} S = -\sum_{ij} p_{ij} \ln p_{ij} \tag{2.108}$$

subject to

$$\sum_{ij} p_{ij} = 1 \tag{2.109}$$

$$-\frac{O_i}{T}\sum_{ij} p_{ij} + \sum_j p_{ij} = 0 \qquad \text{for all } i \tag{2.110}$$

$$-\frac{D_j}{T}\sum_{ij} p_{ij} + \sum_i p_{ij} = 0 \qquad \text{for all } j \tag{2.111}$$

$$-\sum_{ij} p_{ij} c_{ij} + \frac{C}{T}\sum_{ij} p_{ij} = 0 \tag{2.112}$$

$$\mathbf{P} \geqslant 0 \tag{2.113}$$

Now it may be shown that this model is a specific member of the class of dual geometric programming models (see, for example, Nijkamp, 1972). The primal program associated with this dual, equations (2.108)–(2.113), may be written in a number of forms, including the following unconstrained problem (Nijkamp, 1979):

$$\min_{\{\mathbf{x},\,\mathbf{y},\,\mathbf{z}\}} D'(\mathbf{x}, \mathbf{y}, \mathbf{z}) = \left\{ \prod_i \prod_j x_i^{O_i/T} y_j^{D_j/T} z^{-C/T} \right\}^{-1} \left(\sum_{ij} x_i y_j z^{-c_{ij}} \right) \tag{2.114}$$

By invoking the transformation

$$x_i = e^{-\alpha_i} \tag{2.115}$$

$$y_j = e^{-\gamma_j} \tag{2.116}$$

$$z = e^{\beta} \tag{2.117}$$

and minimizing the function

$$D = \log D' \tag{2.118}$$

it may readily be verified (after allowing for the different formulations in terms of $\{p_{ij}\}$ and $\{T_{ij}\}$) that the geometric program (2.114) is equivalent to the given in equation (2.104).

2.4.5 The congested assignment problem

The assignment submodel in a transport system planning model is designed for the purpose of forecasting the traffic on each link of a highway and/or public transport network. This problem is interpreted mathematically as one of computing a network equilibrium of demand and supply in which the level of service (times and costs) on each link is consistent with the route demand assumed to be generated by these costs, and also corresponds to a behavioural principle of route choice.

Our interests in the assignment program in this introductory chapter on mathematical programming and its applications are threefold: firstly, it is another excellent example of the Kuhn–Tucker conditions providing a formal mathematical statement of a particular model, in this case the boundary conditions are an essential component of the solution. Secondly, the structure of the program constraints are of considerable interest and provide the basis for further comments on the structure of models considered in Chapter 10. Thirdly, in the context of the book as a whole, many of the activity–spatial interaction models to be considered will include transport (generalized) costs which are given *exogenously*. From the results of this section, and that following, which combines the distribution and assignment models, it will be apparent how an endogenous determination of interaction costs and network flows may be achieved, if required, within a programming framework from which rigorous solution procedures may be formulated. In other words, route demand (with associated congestion) and activity location (distribution) patterns may be made mutually dependent and simultaneously determined.

To avoid an excessively lengthy presentation of the congested assignment problem we shall sketch the derivation and interpretation of the Kuhn–Tucker conditions, corresponding to the program formulated by Beckmann, *et al.* (1956). A detailed and rigorous treatment is given by Murchland (1969) and by Evans (1973a). We must start with a brief description of Wardrop's (1952) principle and some general features of trip assignment.

If each traveller between any pair of zones in a city or region linked by a network L is assumed to select the shortest path (measured in terms of times or generalized costs) the resultant network equilibrium of flows is characterized by Wardrop's principle, which states that the cost c_{ij}^r on all selected routes $R(ij)$ used for travel between nodes i and j are equal and less than or equal to those for non-used routes. Throughout the 1960s several heuristic methods such as the incremental and iterative loading capacity restraint procedures (see for example Van Vliet, 1976) were developed to treat the allocation of trips to routes under congested conditions, which are a result of the finite capacity of the network. By making route choice sensitive to congestion on the network, these methods attempted in various ways to generated an aggregate flow pattern which was characterized by Wardrop's principle. In common with many heuristic approaches, however, little if anything was known of the properties of the resultant flow pattern, whether iterative methods converged, and indeed the extent to which they approximated the Wardrop equilibrium flow pattern.

Recently, much interest has been shown in the implementation of assignment algorithms based on rigorous mathematical programming principles. The formulation of a 'multicommodity' network flow program which generated the Wardrop relations embodied in the Kuhn–Tucker conditions was first given by Beckmann, McGuire and Winsten (1956).

To introduce this program and the equivalent 'existence problem' we define the following:

T_{ij}: the number of trips between zones i and j

L: a network formed from a set of links $L(l, l')$ connecting pairs of nodes l and l'

J: the set of nodes identified with zone centroids

$x_{ll'}^i$: the flow on link $L(l, l')$ in the network L which originates from zone i

$x_{ll'}$: the total flow on link $L(l, l')$

$c_{ll'}(x_{ll'})$: The (generalized) cost of travel along the link $L(l, l')$ when the total flow carried is $x_{ll'}$. It shall be assumed that $c_{ll'}(x)$ is a strictly increasing function of x.

Any feasible pattern of flows $\{x_{ll'}^i\}$ must satisfy the following continuity conditions at each node of the network:

$$\sum_{l'} x_{l'l}^i - \sum_{l'} x_{ll'}^i = \begin{cases} 0 & \text{for all } l \notin J \\ T_{ij} & \text{for } l = j, j \in J \end{cases} \tag{2.119}$$

That is, at 'intermediate' nodes in the network, no net accumulation of flow can occur while at all nodes corresponding to zone centroids $(j \in J)$ the flow into the node is T_{ij}, the trip matrix element associated with i and j.

Clearly, the total flow on any link may be written as the sum over separate contributions from different zones

$$x_{ll'} = \sum_i x^i_{ll'} \tag{2.120}$$

It may be shown (see, for example, Evans, 1973a) that the following conditions are necessary and sufficient for the existence of an assignment consistent with Wardrop's principle, namely that quantities $\{x^i_{ll'}\}$, $\{x_{ll'}\}$, and $\{\xi^i_l\}$ be found such that for a set of feasible flows, satisfying equations (2.119)–(2.120) the following relations hold:

$$\text{If } x^i_{ll'} > 0 \quad \text{then} \quad \xi^i_l + c_{ll'}(x_{ll'}) = \xi^i_{l'} \tag{2.121}$$

$$\text{If } x^i_{ll'} = 0 \quad \text{then} \quad \xi^i_l + c_{ll'}(x_{ll'}) > \xi^i_{l'} \tag{2.122}$$

for all $\{i, l, l'\}$, where ξ^i_l is the shortest path from i to any node l. The conditions (2.121)–(2.122) indicate that $x^i_{ll'}$ can only be non-zero if the shortest path to l' from centroid i is equal to the shortest path to node l, plus the link cost $c_{ll'}$, evaluated with the total flows $\{x_{ll'}\}$.

It was shown by Beckmann *et al.* (1956) that there could be found an analytically equivalent optimization problem to this 'existence' problem which could be written:

$$\min_{\{x_{ll'}\},\{x^i_{ll'}\}} C = \sum_{L(l,l')\in L} \int_0^{x_{ll'}} \mathrm{d}x\, c_{ll'}(x) \tag{2.123}$$

subject to

$$\sum_i x^i_{ll'} - x_{ll'} = 0 \qquad \text{for all} \quad L(l, l') \in L \tag{2.124}$$

$$\sum_{l'} x^i_{l'l} - \sum_{l'} x^i_{ll'} = \begin{cases} 0 & \text{if } l \notin J \quad (2.125) \\ T_{ij} & \text{if } l = j \in J \quad (2.126) \end{cases}$$

$$\{x^i_{ll'}\} \geqslant 0 \quad \{x_{ll'}\} \geqslant 0 \text{ for all link flows} \tag{2.127}$$

Introducing the multipliers $\{\eta_{ll'}\}$ associated with the set of equations (2.124) and $\{\xi^i_l\}$ associated with the continuity conditions (2.125)–(2.126), the Lagrangian for the problem may be written

$$\begin{aligned} L = C &+ \sum_{L(l,l')\in L} \eta_{ll'} \left(x_{ll'} - \sum_i x^i_{ll'} \right) \\ &+ \sum_i \sum_{l\notin J} \xi^i_l \left(\sum_{l'} x^i_{ll'} - \sum_{l'} x^i_{l'l} \right) \\ &+ \sum_i \sum_{j\in J} \xi^i_j \left(T_{ij} + \sum_{l'} x^i_{jl'} - \sum_{l'} x^i_{l'j} \right) \end{aligned} \tag{2.128}$$

By invoking the Kuhn–Tucker conditions

$$x_{ll'}\frac{\partial L}{\partial x_{ll'}}=0;\qquad \frac{\partial L}{\partial x_{ll'}}\geqslant 0,\qquad x_{ll'}\geqslant 0 \text{ for all } L(l,l')\in L \tag{2.129}$$

$$x^i_{ll'}\frac{\partial L}{\partial x^i_{ll'}}=0;\qquad \frac{\partial L}{\partial x^i_{ll'}}\geqslant 0,\qquad x^i_{ll'}\geqslant 0 \text{ for all } L(l,l')\in L \text{ and origin nodes } i \tag{2.130}$$

$$\frac{\partial L}{\partial \eta_{ll'}}=0\qquad \text{for all } L(l,l')\in L \tag{2.131}$$

$$\frac{\partial L}{\partial \xi^i_l}=0\qquad \text{for all centroids } i, \text{ and nodes } l \tag{2.132}$$

the above existence conditions, necessary and sufficient for a Wardrop equilibrium assignment, may readily be derived.

The achievement of an equivalent optimization problem may now be exploited in examining general questions of solution existence, uniqueness, and the convergence of solution algorithms. These topics are extensively treated by Murchland (1969), Evans (1973a), Nguyen (1974), Wigan (1977), among others.

In the above problem we have considered the matrix $\{T_{ij}\}$ to be fixed and given exogenously. We now briefly consider the unification of the distribution and assignment programs to achieve consistency between the models.

2.4.6 The combined distribution–assignment model: an example of model integration within a programming framework

One of the criticisms of land-use transportation models has been that the separate development of activity location/spatial interaction models and assignment models in conventional approaches has led to the inconsistent determination of the equilibrium which is assumed to exist between the demand for travel and supply of transportation service expressed through generalized link costs. Thus the distribution of trips would produce through a following assignment stage interzonal route costs which were not consistent with the original trip pattern. Attempts have been made in several of the more recent transport studies to remedy this situation by **ad hoc** iterative procedures involving iterative implementation of the distribution and assignment stages with successive updating of the travel costs. As in the conventional treatments of the assignment problem, little was known of the convergence properties of such methods, if indeed they did converge.

The combination of a spatial interaction model and assignment model and its generation within a mathematical program constitutes an excellent example of the formation of a superproblem. This strategy will be generalized and applied in later sections. We shall here comment only briefly of the combined distribution and assignment problem. The interested reader may again find a much fuller discussion in the work of Murchland (1969), Evans (1973a), Nguyen (1974) and Wigan (1977).

The (existence) problem at hand is one of finding quantities $\{x^i_{ll'}\}$, $\{x_{ll'}\}$, $\{\xi^i_l\}$, and $\{T_{ij}\}$ which simultaneously satisfy:

(i) Wardrop's assignment conditions (2.119)–(2.122);

(ii) A specified activity location/trip interaction pattern for given link cost functions

$$c_{ll'} = c_{ll'}(x_{ll'}) \tag{2.133}$$

In the case for which the activity-location/interaction model is of doubly constrained (exponential) gravity form, the existence problem is specified through the following set of equations:

$$x_{ll'} = \sum_i x^i_{ll'} \qquad \text{(equation 2.120)}$$

$$\sum_{l'} x^i_{l'l} - \sum_{l'} x^i_{ll'} = \begin{cases} 0 & \text{if } l \notin J \\ T_{ij} & \text{if } l = j \in J \end{cases} \qquad \text{(equation 2.119)}$$

and

$$\text{if } x^i_{ll'} > 0 \text{ then } \xi^i_l + c_{ll'}(x_{ll'}) = \xi^i_{l'} \qquad \text{(equation 2.121)}$$

$$\text{if } x^i_{ll'} = 0 \text{ then } \xi^i_l + c_{ll'}(x_{ll'}) > \xi^i_{l'} \qquad \text{(equation 2.122)}$$

$$\text{for } T_{ij} = A_i B_j O_i D_j e^{-\beta \xi^i_j} \tag{2.134}$$

$$\text{with } \sum_j T_{ij} = O_i \qquad \text{(equation 2.82)}$$

$$\sum_i T_{ij} = D_j \qquad \text{(equation 2.83)}$$

It may readily be shown (see for example the discussion in Wigan, 1977, and Evans, 1973a) that the equivalent optimization problem P may be written

$$\min \sum_{L(l,l') \in L} \int_0^{x_{ll'}} \mathrm{d}x\, c_{ll'}(x) + \frac{1}{\beta} \sum_{ij} T_{ij} \log T_{ij} \tag{2.135}$$

subject to equations (2.119)–(2.120) and (2.82)–(2.83), and the non-negativity conditions on the flow and trip matrix elements. This equivalence may be demonstrated through the straightforward application of the Kuhn–Tucker optimality conditions as outlined in section 2.2.3.

Again the properties of the solution to the combined problem may be sought in the structure and characteristics of the generating program. Here we wish to note a general feature of this combination process which will be encountered a number of times in subsequent chapters. The above problem as we noted involved the achievement of spatial interaction and flow patterns which were mutually consistent, and this was obtained through an overall extremal or optimization process in which the network flow and trip interaction variables were 'interfaced' by a particular conservation or accounting constraint (equation 2.119). The accounting or constraint relations may be represented by the nodes in Figure

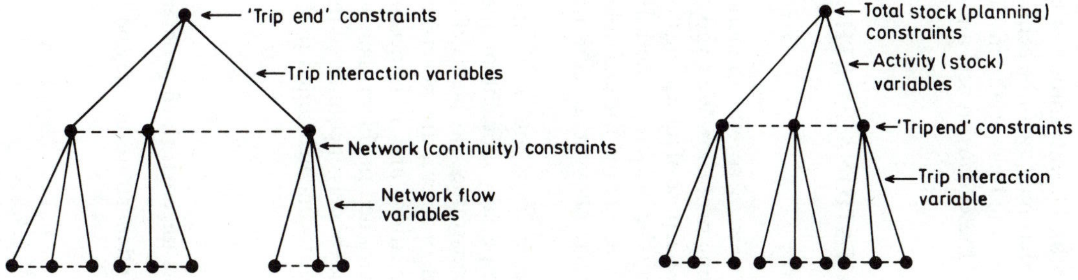

(a) (b)

Figure 2.6 (a) A spatial interaction–network flow problem (b) An activity location–spatial interaction problem

2.6(a) in which the branches correspond to either trip interaction or network flow variables.

In the land-use design models considered in a later chapter, we shall be similarly concerned with the derivation of mutually consistent activity and trip interaction variables again achievable within a 'superprogram' with the pictorial representation shown in figure 2.6(b). The integration of land-use design, spatial interaction, and network flow models may again be achieved within a program, the constraints of which are represented pictorially in Figure 2.6(b).

A more formal examination of these 'consistency' and 'embedding' relations will be presented in Chapter 7.

2.5 COMPUTATIONAL METHODS

2.5.1 Introduction

In this section we give an overview of several numerical approaches to the solution of non-linear programming problems. We stress that this survey is non-exhaustive and is primarily for the purpose of indicating how models encountered in the course of this book may be solved numerically. Further, we omit considerations of those programs including, for example, linear and quadratic objective forms for which particular idiosyncracies in their program structure may be exploited.

Many different solution strategies exist and it should be emphasized that particular methods may prove especially efficient or inefficient for particular problems. All approaches to the constrained problems, if not direct extensions of the corresponding unconstrained problem, are nevertheless highly reliant on methods for multivariate optimization, and it is necessary first to consider solution strategies for the unconstrained problem. This will in fact prove to be especially relevant as several of the models encountered in later chapters may be expressed in terms of the solution to unconstrained problems.

2.5.2 General approaches to the unconstrained problem

Two broad, but not totally independent strategies may be adopted for solution of the problem

$$\max_{x} f(x_1, \ldots, x_N) \qquad \text{(equation 2.7)}$$

In direct *search processes*, use is made of function evaluation alone, while in *gradient* methods, additional information is required of the vector $\mathbf{G}$ defined in equation (2.11). It has been found that if the function $f(\mathbf{x})$ has continuous derivatives and the gradients can be evaluated analytically, then the latter methods are superior.

2.5.2.1 *Direct search methods*

Direct search methods generally determine through function evaluation a direction of search in which a maximum is expected to be. The maximum is then found either by executing a linear search in the direction determined, or approached by taking a fixed step towards it. Since the direction chosen is not necessarily correct, the process must be repeated. After each maximization search step, further searches are carried out until a convergence criterion is satisfied.

One obvious strategy is to adopt searches parallel to the coordinate axes for each variable in turn. To improve the efficiency of this 'one at a time' approach, a development involves movement along directions which are changing during the search process. One such method is the so-called conjugate directions search method due to Powell (1964) in which a succession of searches are made 'one at a time' in each of n sets of independent directions. The iterative process is initiated with the coordinate axes as search directions and updated as follows:

$$\{\mathbf{n}_1^{k+1}, \mathbf{n}_2^{k+1}, \ldots, \mathbf{n}_n^{k+1}\} = \{\mathbf{n}_2^k, \mathbf{n}_3^k, \ldots, \mathbf{n}_n^k, (\mathbf{x}^{k+1} - \mathbf{x}^k)\} \qquad (2.136)$$

in which $\{\mathbf{n}_1^k, \ldots, \mathbf{n}_n^k\}$ are the n directional vectors at the kth iteration when the current argument of the function is $\mathbf{x}^k$. $\mathbf{x}^{k+1}$ is the point produced after linear searches in each direction in turn.

2.5.2.2 *Gradient methods*

Procedures based on gradient search may be further categorized into first or second orders according to whether the gradient vector $\mathbf{G}$ alone is required, or whether additionally the Hessian $\mathbf{H}$ or an approximation to it is involved.

Because the gradient direction is locally the direction of fastest increase or decrease, one strategy for optimization involves performing unidirectional searches along the gradient direction until no further increase in $f(\mathbf{x})$ is achieved. The method of steepest ascent is defined by the recursive equation

$$\mathbf{x}^{k+1} = \mathbf{x}^k + \mu_k \mathbf{G}^k \quad k = 0, 1, \ldots \qquad (2.137)$$

in which the parameter μ_k which determines the length of step at the kth iteration is such as to maximize the increase in $f(\mathbf{x})$ in the direction $\mathbf{G}^k$. That is,

$$f(\mathbf{x}^{k+1}) = \max_{\mu} f(\mathbf{x}^k + \mu \mathbf{G}^k) \qquad (2.138)$$

which is achieved in a unidirectional search. An illustration of the method of steepest ascent is given in Figure 2.7.

Gradient methods may be improved, particularly in the vicinity of an optimum, if the Hessian or an estimate of it is available. If we expand $f(\mathbf{x}^*)$ in a Taylor series about the solution $\mathbf{x}^*$,

$$f(\mathbf{x}^*) \simeq f(\mathbf{x}) + \mathbf{G}'.\Delta\mathbf{x} + \tfrac{1}{2}\Delta\mathbf{x}'\mathbf{H}\Delta\mathbf{x} \qquad (2.139)$$

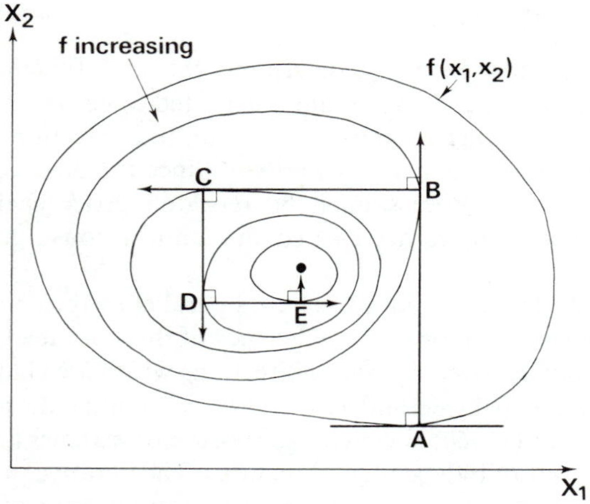

Figure 2.7 An illustration of the method of steepest ascent

and ignore terms higher than the second, then an estimate of the optimal value $\mathbf{x}^*$ from a point $\mathbf{x}$ where

$$\mathbf{x}^* = \mathbf{x} + \Delta\mathbf{x}$$

may be achieved by differentiating (2.139) with respect to the components of $\Delta\mathbf{x}$ assuming $\mathbf{G}$ and $\mathbf{H}$ to be held constant. The first-order condition for maximizing the right-hand side of (2.139) is

$$\Delta\mathbf{x} = -\mathbf{H}^{-1}.\mathbf{G} \tag{2.140}$$

which gives the optimal move from point $\mathbf{x}$.

In a generalized version of the method the recursive equation

$$\mathbf{x}^{k+1} = \mathbf{x}^k - \mathbf{H}^{-1}.\mathbf{G} \tag{2.141}$$

obtained from (2.140), is replaced by

$$\mathbf{x}^{k+1} = \mathbf{x}^k - \mu_k\mathbf{H}^{-1}.\mathbf{G} \tag{2.142}$$

in which μ_k is determined by linear search.

A disadvantage of this Newton–Raphson approach is that the matrix $\mathbf{H}$ must be computed and inverted at each iteration. To avoid this inconvenience, a number of variants have been proposed, including the class of so-called quasi-Newton or variable matrix methods in which the iterative formula (2.142) is replaced by

$$\mathbf{x}^{k+1} = \mathbf{x}^k - \mu_k\mathbf{F}^k.\mathbf{G} \tag{2.143}$$

in which $\mathbf{F}^k$ is a sequence of matrices which will approximate $\mathbf{H}^{-1}$ and μ^k is determined through unidirectional search along $\mathbf{F}^k\mathbf{G}$. (See, for example, Adby and Dempster, 1974.) The sequence is updated according to the changes in the gradient vector. In this way there is no need to compute and invert the Hessian at each iteration.

In place of an approximation to the inverse Hessian matrix, the Fletcher–Reeves conjugate gradient algorithm (Fletcher and Reeves, 1964) exploits the properties of mutual conjugate directions† defined with respect to **H** in the maximization of quadratic functions. Unidirectional searches are performed along vectors $\mathbf{u}^k$ generated recursively at each iteration by

$$\mathbf{u}^k = -\mathbf{G}^{k-1} + \frac{\mathbf{G}'^{k-1}.\mathbf{G}^{k-1}}{\mathbf{G}'^{k-2}.\mathbf{G}^{k-2}}.\mathbf{u}^{k-1} \qquad k = 1, 2, \ldots \qquad (2.144)$$

with $\qquad \mathbf{u}^1 = -\mathbf{G}^0$

When the algorithm is used for general unconstrained functions a new cycle is started after each set of n iterations with the steepest-ascent search direction in the current point. The algorithm requires only two vectors to be stored—a search direction and a gradient, at any point of the iterative process.

The Fletcher–Reeves method was employed by Champernowne *et al.* (1976) for the calibration of a model related to the doubly constrained gravity model through the direct minimization of the dual program (2.104). Such a method was employed because the conventional approach which involves the assymetric treatment of the duals (α, γ) and β was inappropriate for the model considered.

2.5.3 Constrained optimization

The constrained optimization problem may be approaches by two distinct strategies involving *implicit* and *explicit* treatment of constraints. The former includes elimination methods, in which constraints may be removed by variable transformation, together with the important class of penalty-function methods in which the constraints are transformed into additional terms in the objective function using penalty or barrier functions, thereby allowing unconstrained optimization methods to be employed. In addition to a description of the penalty-function approach we shall consider a number of methods based on the second general strategy for the constrained problem which involves the explicit treatment of constraints in the general class known as feasible directions algorithms.

2.5.3.1 Penalty-function methods

The rationale for penalty-function methods is very appealing. One of the most direct ways of modifying the unconstrained problem to incorporate constraints is to view them as obstacles to the search path and impose large penalties if the 'barriers' are transgressed.

† Two vectors **u** and **v** are said to be mutually conjugate with respect to a matrix **A** if **u** and **Av** are mutually orthogonal, that is if $\mathbf{u}'\mathbf{A}\mathbf{v} = 0$.

A typical penalty-function formulation for the constrained problem

$$\max f(\mathbf{x}) \tag{2.145}$$

subject to

$$g_i(\mathbf{x}) \leqslant 0 \qquad i = 1, 2, \ldots \tag{2.146}$$

is to maximize the unconstrained function

$$f(\mathbf{x}) - \sum_i \eta_i (g_i(\mathbf{x}))^2 \tag{2.147}$$

where summation is taken over the violated constraints and the weights η_i, specified by the user, represent how strongly are penalized constraints which are not satisfied. By solving a sequence of unconstrained maximization problems of the form (2.147) with increasing values for the weights, the solution of the corresponding constrained problem may be obtained in the limit.

In more general terms, the composite function (2.147) may be replaced by

$$\psi(\mathbf{x}, \eta) = f(\mathbf{x}) - \eta \sum_i \phi_i(g_i(\mathbf{x})) \tag{2.148}$$

in which η is a controlling factor and ϕ_i is any function of one variable u which gives a positive contribution only when the constraint is violated, that is,

$$\phi_i(u) \begin{cases} = 0 & \text{for } u < 0 \\ > 0 & \text{for } u > 0 \end{cases} \tag{2.149}$$

This type of penalty function, often called a 'loss function', generates a sequence of points which proceed to the maximum from outside the feasible region.

Another class of penalty-function methods are those which proceed to the optimal point from inside the feasible region, and involve a composite function of the form

$$\psi(\mathbf{x}, \eta) = f(\mathbf{x}) - \eta \sum_i \phi_i(g_i(\mathbf{x})) \tag{2.150}$$

in which η is the positive controlling factor and ϕ is a function of a variable u with a positive singularity at the constraint boundary $g_i(\mathbf{x}) = 0$, that is,

$$\begin{aligned} \phi_i(u) &\quad > 0 \quad \text{for } u < 0 \\ \phi_i(0-) &= +\infty \end{aligned} \tag{2.151}$$

Penalty functions of the latter barrier type require an initial point strictly inside the feasible region due to the singularity of $\psi(\mathbf{x}, \eta)$ at the boundary, but they have the advantage of preventing constraint violation during computation.

A typical example of this 'barrier function' approach is provided by the sequential unconstrained maximization technique developed by Fiacco and McCormick (1968).

Several authors have proposed the use of mixed penalty methods in order to exploit the advantages and reduce the difficulties arising from each formulation, in addition to improving the numerical efficiency of the optimization procedure. A good general discussion of these and other aspects of the approach may be found in Adby and Dempster (1974).

Finally, we note that certain composite functions may be viewed as an approximation to the Lagrangian of the constrained problem, which allows a link to be established between the method and Lagrangian theory.

2.5.3.2 Feasible directions algorithms

In the gradient methods described above (section 2.5.2.2) the general strategy was one of successively determining the direction of greatest ascent (or descent) and, at each stage, using linear search to arrive at the appropriate step length. In the constrained problem a boundary of the feasible region will usually be encountered as the search progresses, and it is necessary to determine the new direction if constraints are not to be violated. This poses an obvious problem: what is the direction of maximum rate of increase in $f(\mathbf{x})$ which does not involve violation of the problem constraints?

It can be shown (see Hadley, 1964) that this direction may be determined by solving a quadratic program. Because this is generally a non-trivial non-linear programme itself and must be repeated whenever a boundary is confronted, it is usual to sacrifice the aim of moving along the direction of maximum rate of increase for a path which can be shown to lead to an improvement in the current solution. This latter strategy is embodied in Zoutendijk's method of feasible directions in which the following steps are considered (Zoutendijk, 1960):

(i) given a feasible solution x';
(ii) test for the optimal solution;
(iii) determine a feasible direction, i.e. a direction $\mathbf{d}^k$ pointing towards the feasible region such that

$$\mathbf{G}^k(\mathbf{x}^k) \, . \, \mathbf{d}^k > 0; \tag{2.152}$$

(iv) determine a step length μ^k and move to the point

$$\mathbf{x}^{k+1} = \mathbf{x}^k + \mu^k \mathbf{d}^k; \tag{2.153}$$

(v) Return to step (ii).

The condition (2.152) ensures that a local move along the feasible direction $\mathbf{d}^k$ will produce an improvement in the current solution. This general scheme becomes a distinct algorithm whenever a precise direction-generation procedure and step-length calculation method are chosen.

We shall summarize three such schemes:
(a) The Frant–Wolfe method (Frank and Wolfe, 1956)
(b) The reduced-gradient method (Wolfe, 1965; Faure and Huard, 1965)
(c) The gradient-projection method (Rosen, 1960).

(a) *The Frank–Wolfe method.* This method involves a linear approximation of the objective function and may be constrained to linearly constrained problems of the form

$$\max_{\mathbf{x}} f(\mathbf{x}) \tag{2.154}$$

subject to

$$\mathbf{Ax} \leqslant \mathbf{b} \tag{2.155}$$

$$\mathbf{x} \geqslant 0 \tag{2.156}$$

The direction-generation procedure consists of solving the linear programming problem

$$\max_{\mathbf{x}} \mathbf{G}(\mathbf{x}^k) \, . \, \mathbf{x} \tag{2.157}$$

subject to

$$\mathbf{Ax} \leqslant \mathbf{b} \tag{2.158}$$

$$\mathbf{x} \geqslant 0 \tag{2.159}$$

in which $\mathbf{G}(\mathbf{x}^k)$ is the gradient of the objective function evaluated at the current point $\mathbf{x}^k$. The new direction is defined by

$$\mathbf{d}^k = \mathbf{x}^* - \mathbf{x}^k \tag{2.160}$$

where $\mathbf{x}^*$ is the optimal solution of the linear program defined above. Having determined the step direction, the step length is calculated through univariate search along $\mathbf{d}^k$.

Since the method requires the solution of a linear program at each iteration, it may exhibit very poor performance. This is mainly due to the fact that the solution of a linear program is at a vertex of the convex feasible region, while the optimum of a non-linear programming problem may of course be at an interior point *or* at a boundary point. This tends to create oscillations in the search direction when the Frank–Wolfe method is adopted unless additional restrictions to the linear programming problem are included.

(b) *The reduced-gradient method.* In this method, also designed for linear constraints, the direction of move is given by the gradient of the objective function suitably 'reduced' to accommodate the constraints of the problem. Whereas in the Frank–Wolfe method the optimal solution of (2.157)–(2.159) is sought to define the 'best' feasible direction, in the reduced-gradient approach, a procedure similar to an iteration of the simplex method is used to generate a feasible direction.

The search path in the reduced-gradient algorithm is not bound to go from vertex to vertex. There is therefore the possibility that the feasible region will be crossed in just one iteration, even when the algorithm is applied to linear programming problems. For the class of problems encountered in this book we have found this scheme to be very successful.

A variant of the above approach is the convex simplex method (Zangwill, 1967) which differs from the reduced-gradient algorithm in the way the feasible direction $\mathbf{d}$ is defined. In its updating procedure this algorithm follows the simplex method very closely. Moreover, when the objective function is linear the convex simplex method reduces itself to the simplex method.

(c) *The gradient-projection method.* In the gradient-projection method the search direction is given by the orthogonal projection of the gradient $\mathbf{G}(\mathbf{x}^k)$ over the intersection subspace of the tangent hyperplanes to the active constraints. This strategy is most easily explained with reference to a geometric interpretation which is given in Figure 2.8. At point $\mathbf{x}^k$ no further move can be made in the direction of the gradient without violating the linear boundary constraint AB. The algorithm essentially obtains the perpendicular projection $\mathbf{d}_p$ onto the set of feasible solutions, as shown. At the next stage the move proceeds along the boundary until the extreme point A is reached, at which point the procedure terminates.

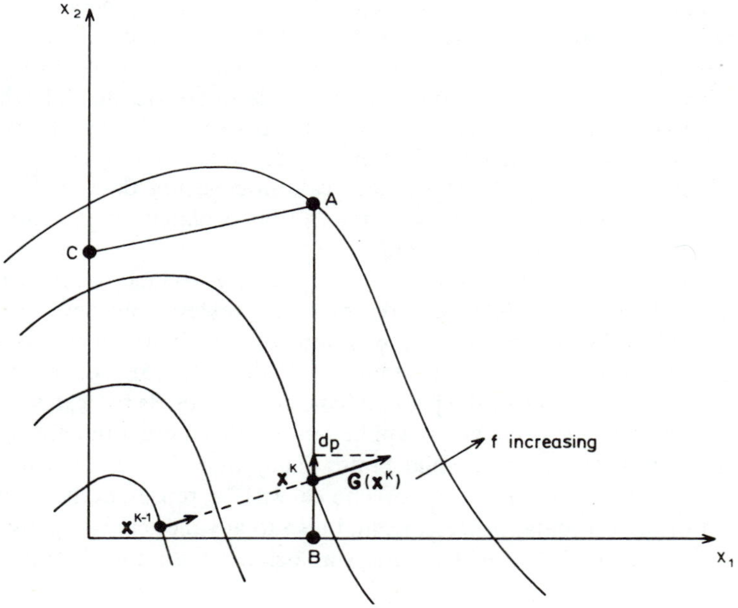

Figure 2.8 A geometric representation of the gradient-projection method

Clearly this method can be used with linear and non-linear constraints. However, when the latter are present the search along the direction tangential to the active constraints can lead to an infeasible point, and an additional step towards the feasible region may then be required. For a fuller discussion of the approach, the book by Hadley (1970) should be consulted.

3. *Random utility theory and probabilistic choice models*

3.1 INTRODUCTION

In this section we describe and develop a theory of choice between discrete alternatives which forms the basis for spatial interaction (location) models; mode choice and route choice models; and for models which embrace combinations of these choice 'dimensions' (e.g. location–mode, mode and route, etc.). The theory is essentially concerned with an explanation of the dispersion or variability which characterizes the observed actions of different individuals confronted by a choice. The models generated by the theory are calibrated using data pertaining to individuals, and the general strategy of description, explanation and forecasting is, for this reason, termed a 'micro approach'.

The theory to be described—random utility theory—which has its origins in the work of Thurstone (1927), has recently been extensively developed in a traveller- and location- choice context by, among others: Charles River Associates (1972), Ben-Akiva (1973); Domencich and McFadden (1975); Cochrane (1975); and Williams (1977). Some of the probabilistic choice models to be described (in Section 3.2 below) have been widely applied in the 1970s predominantly, but by no means exclusively, in the United States. They owe their popularity to the marriage of a behavioural theory of choice with a representation allowing efficient information (data) processing, and also to advances in sampling theory and econometric methods of estimation associated with the discrete choice processes.

In Section 3.2 the choice context and formation of probabilistic choice models is outlined with emphasis on the important class of independent random utility models. The utility distributions and their properties are then explored in Section 3.3. In Section 3.4 the framework developed is compared and contrasted with the widely applied entropy-maximizing approach. The alternative theoretical derivations of the logit model are noted, and the concept of group-surplus maximization is introduced.

The limitations associated with simple 'independent' choice models are confronted in Section 3.5 when the phenomenon of attribute correlation is introduced into the basic theory. A detailed explanation of model structure is presented, and shown to relate very intimately to the nature of perceived similarity between choice substitutes. The assumption of optimal choice behaviour which underpins the modelling framework implies important restrictions on the form of so-called 'partial share' models and on the relative sizes of certain parameters embedded in them. These conditions are shown to have important practical implications.

The full incorporation of correlation between the utility distributions associated with different choice alternatives allows the formation of more general model structures which are consistent with utility-maximizing behaviour at the micro level. These models are explored in Section 3.6. Finally, in Section 3.7, we present a range of examples in which travel-demand and location models are generated in conjunction with their associated user-benefit measures.

3.2 THE FORMATION OF RANDOM UTILITY MODELS

3.2.1 The choice context

We consider a population of individuals Π confronted by a choice between a discrete set of alternatives $\mathbf{A} = \{A_1, \ldots, A_\rho, \ldots, A_N\}$ which may correspond to a collection of modes, routes, locations, or a combination of these. Each individual in Π is considered to select one and only one option from **A.** If it is noted that over a particular population individuals select non-identical alternatives, the modeller will, in a behavioural theory of choice, attribute this dispersion to one or more facets of the choice context. In random utility theory it is assumed that individuals behave rationally, each selecting that option from **A** which offers maximum net utility or surplus. If each alternative were perceived by members of Π to be endowed with the same utility, individuals confronted by the same set of choices would presumably select the same option. However, in general the perceived benefits and costs derivable from a given member of the set **A** will vary over the population Π. Because some of the attributes associated both with the individuals and with the available alternatives, which are relevant to the choice process, will be *unobserved* by the modellers, and furthermore because the valuation of observable attributes which characterize alternatives may be non-uniform, it is not certain which alternative will be selected by that individual. Probabilistic concepts are thus introduced. The uncertainty here is associated with the observer, or modeller, who attributes random components to the utility or suplus functions—while each member of Π is considered to choose optimally and consistently within his own frame of reference.

3.2.2 The generation of random utility models

We shall examine the choice behaviour of members of Π, and shall derive an expression for the probability P_ρ that an individual will select A_ρ from the set of options $\{A_1, \ldots, A_\rho, \ldots, A_N\}$, the members of which are characterized by the vectors of *observable* attributes $\{\mathbf{Z}_1, \ldots, \mathbf{Z}_\rho, \ldots \mathbf{Z}_N\}$. The vector $\mathbf{Z}_\rho$, for example, will include components such as time, cost, etc., associated with the choice A_ρ.

The probability that an individual selects a given option may be computed if a functional form is assigned to the random component of the surplus s_ρ (utility minus cost) associated with each alternative. The essence of the approach is embodied in the formal expression

$$P_\rho = \int_{R_\rho} \mathrm{d}sH\{s_1, \ldots, s_\rho, \ldots, s_N\} \equiv \int_{R_\rho} \mathrm{d}sH(s) \tag{3.1}$$

in which $H\{s_1, \ldots, s_\rho, \ldots, s_N\}$ is the joint density function of the surplus variables $s_1, \ldots, s_\rho, \ldots, s_N$. d$s$ is the volume element $\prod_\rho \mathrm{d}s_\rho$, and R_ρ is the region of integration defined by

$$s_\rho = \max_{\rho'} s_{\rho'}, \text{ and } s_\rho \geqslant 0 \tag{3.2}$$

Over the region R_ρ, A_ρ is therefore the preferred option.

Two methods of generating random utility models have been widely used. In the random-coefficient approach, the net utilities are expressed as functions of a vector of random parameters α, with mean value $\bar{\alpha}$, such that

$$s_\rho = s_\rho(\alpha, Z) \qquad \rho = 1, \ldots, N \tag{3.3}$$

Equation (3.1) is then transformed into α-space to give

$$P_\rho(\bar{\alpha}, Z) = \int_{R_\rho} \mathrm{d}\alpha H^0(\alpha) \tag{3.4}$$

and integration is over the portion of that space equivalent to R_ρ defined by equations (3.2) and (3.3). Such an approach has been adopted by Quandt (1968), Blackburn (1970), Golob and Beckmann (1971), Harris and Tanner (1974), and others.

The alternative approach is to specify a functional form for each random variable s_ρ directly. The net utility is written

$$s_\rho = \bar{s}_\rho(\bar{\alpha}, Z) + \varepsilon_\rho \qquad \rho = 1, \ldots, N \tag{3.5}$$

in terms of a 'representative' component $\bar{s}_\rho(\bar{\alpha}, Z)$ which is equal for all members of Π, and a stochastic component ε_ρ *for which a distribution is assumed.* All the variation in tastes represented by the random coefficients α in equation (3.3), together with the variation arising from unobserved factors, is condensed into the random variable ε_ρ. Such an approach has been adopted by Charles River Associates (1972), McFadden (1974), Domencich and McFadden (1975), Cochrane (1975), and Williams (1977), and will be used in this chapter.

The computation of a consumer surplus measure of benefit which is fully consistent with the model generated within the random utility approach may now be expressed (Harris and Tanner, 1974) in terms of the change in the expected value of surplus $\langle s \rangle$ which, for given values of **Z**, is defined by

$$\langle s \rangle = \sum_\rho \int_{R_\rho} \mathrm{d}\mathbf{s}\, s_\rho H(\mathbf{s}, \mathbf{Z}) \tag{3.6}$$

3.2.3 Independent random utility models

An important class of problems is that for which the distribution of the random variables ε_ρ ($\rho = 1, \ldots, N$) (and therefore s_ρ) are independent. If $\Gamma_\rho(s, \bar{s}_\rho)\,ds_\rho$ is the probability that the perceived surplus for alternative X_ρ lies in the range $[s_\rho, s_\rho + ds_\rho]$, while $\theta_\rho(s)$ is the probability that all other alternative have surplus less than s, then we may write

$$P_\rho = \text{prob}\,\{s_\rho \geqslant s_{\rho'},\ \text{for all}\ X_{\rho'} \in X;\ s_\rho \geqslant 0\} \tag{3.7}$$

$$= \int_0^\infty ds\Gamma_\rho(s, \bar{s}_\rho)\,\theta_\rho(s) \tag{3.8}$$

$$= \int_0^\infty ds\Gamma_\rho(s, \bar{s}_\rho) \prod_{\rho' \neq \rho} \int_{-\infty}^{s} ds'\Gamma_{\rho'}(s', \bar{s}_{\rho'}) \tag{3.9}$$

We shall now assume that the net benefit associated with all alternatives is considerably greater than zero so that the lower limit of integration may be extended from zero to minus infinity with negligible approximation for bell-shaped distributions, to give:

$$P_\rho = \int_{-\infty}^\infty ds\Gamma_\rho(s, \bar{s}_\rho) \prod_{\rho' \neq \rho} \int_{-\infty}^{s} ds'\Gamma_{\rho'}(s', \bar{s}_{\rho'}) \tag{3.10}$$

in which case

$$\sum_\rho P_\rho = 1 \tag{3.11}$$

It may be shown (Charles River Associates, 1972, Cochrane, 1975) that the multinomial logit model

$$P_\rho = \frac{\exp(\lambda \bar{s}_\rho)}{\sum_\rho \exp(\lambda \bar{s}_\rho)} \tag{3.12}$$

is generated if the random component ε_ρ varies according to a Weibull distribution

$$\Gamma(\varepsilon_\rho, \lambda) = \Gamma(s - \bar{s}_\rho, \lambda) \tag{3.13}$$

$$= \lambda \exp(-\lambda\varepsilon_\rho)\exp[-\exp(-\lambda\varepsilon_\rho)] \tag{3.14}$$

which has standard deviation σ, given by†

$$\sigma = \pi(\lambda\sqrt{6})^{-1} \tag{3.15}$$

† The mean value of the distribution (3.14) is not in fact $\bar{s}_\rho$ but $\bar{s}_\rho + \gamma/\lambda$ where γ, Euler's constant, is approximately 0.577. Since the distributions $\Gamma_1, \ldots, \Gamma_\rho, \ldots, \Gamma_N$ and the resultant demand model are (translationally) invariant with respect to addition or subtraction of constants to the quantities $\bar{s}_1, \ldots, \bar{s}_\rho, \ldots, \bar{s}_N$, the distinction between $\bar{s}_\rho$ and the mean value may effectively be ignored.

as noted, for example, by Cochrane (1975). The relation (3.14) confirms, as would be expected, that the dispersion parameter λ is inversely related to the standard deviation of the underlying distributions of net utility.

The linear model

$$P_\rho = \frac{\exp\left(\lambda \sum_\eta \alpha_\eta Z_\rho^\eta\right)}{\sum_\rho \exp\left(\lambda \sum_\eta \alpha_\eta Z_\rho^\eta\right)} \tag{3.16}$$

with

$$\bar{s}_\rho = \sum_\eta \alpha_\eta Z_\rho^\eta \tag{3.17}$$

is much favoured by demand analysts due to its ease of calibration and manipulation. A very full discussion of its applications to modelling choice behaviour has been given by McFadden (1973, 1976). Alternative models may be generated by taking different functional forms for the values $\bar{s}_\rho$ in terms of the attributes Z, or of the form for the residual ε_ρ.

3.3 UTILITY DISTRIBUTIONS AND THEIR PROPERTIES

Consider the choice context in which the net utility or surplus for the alternatives in **A** is defined by

$$s_\rho = u - c_\rho \qquad \rho = 1, \ldots, N \tag{3.18}$$

in which u is a utility value, constant over all options, and the perceived costs c_ρ are each distributed according to Γ_ρ ($\rho = 1, \ldots, N$). This problem is of considerable importance in the more complex choice models discussed in later sections.

For each individual, the problem of selecting the option with maximum surplus is clearly equivalent to that of determining the minimum perceived cost alternative. It is important now to distinguish between three cost or disutility measures $\bar{c}_\rho$, $\bar{\tilde{c}}_\rho$, and $\tilde{c}_*$ which are defined as follows:

$\bar{c}_\rho$ is the mean value of the distribution of disutility Γ_ρ for alternative A_ρ, as perceived by the choice-making population Π.

$\bar{\tilde{c}}_\rho$ is the mean value of the distribution of disutility Γ_ρ for alternative A_ρ, as perceived by that portion Π_ρ of the population Π which selects A_ρ.

$\bar{\tilde{c}}_*$ is the mean value of the distribution of disutility Λ_* over the selected choices. Λ_* will be described as the distribution of minimum perceived costs. The asterisk is used to denote an operation, defined below, over the set of distributions associated with the choice set **A**.

The distributions $\tilde{\Gamma}_\rho$ and Λ_*, and their respective mean values $\bar{\tilde{c}}_\rho$, $\tilde{c}_*$ are defined (for independent normalized distributions Γ_ρ, $\rho = 1, \ldots N$) as follows:

$$\tilde{\Gamma}_\rho(c) = \Gamma_\rho(c) \prod_{\rho' \neq \rho} \int_c^\infty \mathrm{d}c' \Gamma_{\rho'}(c') \tag{3.19}$$

$$\Lambda_*(c) = \sum_\rho \tilde{\Gamma}_\rho \tag{3.20}$$

$$= \sum_\rho \left[\Gamma_\rho(c) \prod_{\rho' \neq \rho} \int_c^\infty \mathrm{d}c' \Gamma_{\rho'}(c') \right] \tag{3.21}$$

with

$$\bar{\tilde{c}}_\rho = \left[\int_{-\infty}^\infty \mathrm{d}c\, c \tilde{\Gamma}_\rho(c) \right] \Big/ \left[\int_{-\infty}^\infty \mathrm{d}c \tilde{\Gamma}_\rho(c) \right] \tag{3.22}$$

and

$$\tilde{c}_* = \int_{-\infty}^\infty \mathrm{d}c\, c \Lambda_*(c) \tag{3.23}$$

The distribution Λ_* is normalized if the distributions Γ are themselves normalized.

We may write immediately from equations (3.20) and (3.23),

$$\tilde{c}_* = \sum_\rho \left[\int_{-\infty}^\infty \mathrm{d}c\, c \tilde{\Gamma}_\rho(c) \right] \left[\int_{-\infty}^\infty \mathrm{d}c \tilde{\Gamma}_\rho(c) \right] \Big/ \int_{-\infty}^\infty \mathrm{d}c \tilde{\Gamma}_\rho(c) \tag{3.24}$$

$$= \sum_\rho P_\rho \bar{\tilde{c}}_\rho \tag{3.25}$$

on using the relation

$$P_\rho = \int_{-\infty}^\infty \mathrm{d}c \tilde{\Gamma}_\rho(c) \tag{3.26}$$

$\tilde{c}_*$ is thus the weighted mean of N values $\bar{\tilde{c}}_\rho$. The quantity $\tilde{c}_\rho$ will be less than $\bar{c}_\rho$ because the population Π_ρ which selects A_ρ, does so precisely because that option is perceived as the least-cost alternative, while those in the population Π who are involved in the calculation of $\bar{c}_\rho$ do not all select that option.

For later reference, the distribution of minimum costs Λ_*, defined by equation (3.21), will be written symbolically as

$$\Lambda_*(c, \bar{c}_1, \ldots, \bar{c}_\rho, \ldots, \bar{c}_N) = \min[\Gamma_\rho(c, \bar{c}_\rho) \quad \rho = 1, \ldots, N] \tag{3.27}$$

$\tilde{c}_*$, the mean value of Λ_*, is a function of the costs $\bar{c}_1, \ldots, \bar{c}_\rho, \ldots, \bar{c}_N$.

$$\tilde{c}_* = \tilde{c}_*(\bar{c}_1, \ldots, \bar{c}_\rho, \ldots, \bar{c}_N) \tag{3.28}$$

It can be shown (Williams, 1977, appendix 2) that the function $\tilde{c}_*$ has the following properties:

(a) $\tilde{c}_* > 0$ if the distribution $\Gamma_\rho(c)$ vanish for $c < 0$ (3.29)

(b) $\dfrac{\partial \tilde{c}_*}{\partial \bar{c}_\rho} = P_\rho(\bar{c}_1, \ldots, \bar{c}_\rho, \ldots, \bar{c}_N)$† (3.30)

† This condition holds if the distributions Γ are translationally invariant.

(c) $\tilde{c}_* < \min_{\rho} (\bar{c}_1, \ldots, \bar{c}_\rho, \ldots, \bar{c}_N)$ (3.31)

(d) $\lim_{\lambda \to \infty} c_*(\lambda) = \min_{\rho} (\bar{c}_1, \ldots, \bar{c}_\rho, \ldots, \bar{c}_N)$ (3.32)

where the distributions Γ are parameterized by λ.

It is clear that for any model we may similarly define the distributions of surplus $\Gamma_\rho(s)$, $\tilde{\Gamma}_\rho(s)$ ($\rho = 1, \ldots, N$) and $\Lambda_*(s)$ together with their mean values $\bar{s}_\rho$, $\bar{\bar{s}}_\rho$, and $\tilde{s}_*$. In this case

$$\Lambda_*(s, \bar{s}_1, \ldots, \bar{s}_\rho, \ldots, \bar{s}_N) = \max [\Gamma_\rho(s, \bar{s}_\rho) \quad \rho = 1, \ldots, N] \tag{3.33}$$

For a model in which Γ are independent, the benefit ΔS, defined by equation (3.6) reduces to

$$\Delta S = <\tilde{s}^{\mathrm{F}}> - <\tilde{s}^{\mathrm{I}}> = \int_{-\infty}^{\infty} \mathrm{d}s s[\Lambda_*(s, \mathbf{Z}^{\mathrm{F}}) - \Lambda_*(s, \mathbf{Z}^{\mathrm{I}})] \tag{3.34}$$

where $\tilde{s}_*^{\mathrm{F}}$ and $\tilde{s}_*^{\mathrm{I}}$ are computed with the two sets of vectors $\mathbf{Z}^{\mathrm{F}}$ and $\mathbf{Z}^{\mathrm{I}}$ respectively. If the utility u is constant over all choices, then ΔS is given by

$$\Delta S = (u - \tilde{c}_*^{\mathrm{F}}) - (u - \tilde{c}_*^{\mathrm{I}}) = c_*^{\mathrm{I}} - c_*^{\mathrm{F}} \tag{3.35}$$

The computation of benefit now directly involves the mean value of the distribution $\Lambda_*(c)$ evaluated in the initial and final states.

A distinction between the distribution $\Gamma_\rho(c)$, $\tilde{\Gamma}_\rho(c)$, and $\Lambda_*(c)$, and between their mean values $\bar{c}_\rho$, $\bar{\bar{c}}_\rho$, and $\tilde{c}_*$ is illustrated in Figure 3.1 for binary selection. The functions $\Gamma_\rho(c, \bar{c}_\rho)$ are taken as independent normal distributions

$$\Gamma(c, \bar{c}_\rho) = \frac{1}{(2\pi)^{1/2}\sigma} \exp\left[-\frac{(c - \bar{c}_\rho)^2}{2\sigma^2} \right] \quad \rho = 1, 2 \tag{3.36}$$

Figure 3.1(a) corresponds to the condition $\bar{c}_1 = \bar{c}_2$, in which the underlying normal distributions completely overlap, which results in equal choice between

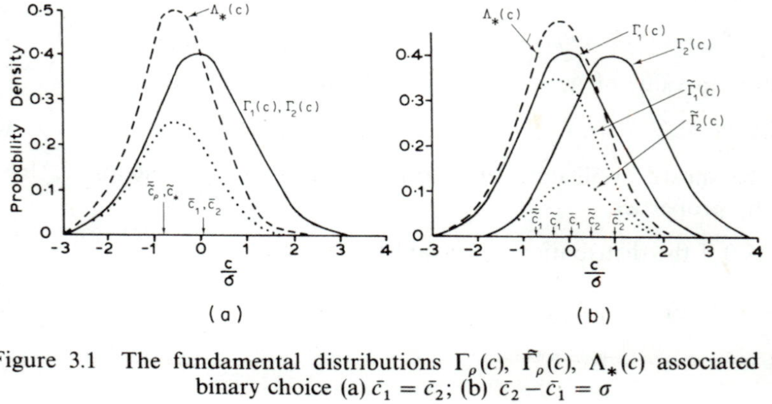

Figure 3.1 The fundamental distributions $\Gamma_\rho(c)$, $\tilde{\Gamma}_\rho(c)$, $\Lambda_*(c)$ associated with binary choice (a) $\bar{c}_1 = \bar{c}_2$; (b) $\bar{c}_2 - \bar{c}_1 = \sigma$

the movements, $P_1 = P_2 = 0.5$, while Figure 3.1(b) illustrates the case in which $(\bar{c}_2 - \bar{c}_1)$ equals σ, the common standard deviation, and corresponds to the share probability $P_1 = 0.76$. The form of the probit (cumulative normal) model

$$P_1(\chi) = \frac{1}{(2\pi)^{\frac{1}{2}}} \int_{-\infty}^{\chi} \mathrm{d}x \exp\left(-\frac{1}{2}x^2\right) \tag{3.37}$$

with

$$\chi = \frac{\bar{c}_2 - \bar{c}_1}{\sigma}$$

which results from the normal distributions (3.36), and the accompanying cost statistics $\bar{c}_\rho$, $\bar{\bar{c}}_\rho$, and $\tilde{c}_*$, appear in Figures 3.2 and 3.3 respectively†. The sigmoid curves are illustrated for a number of σ values. The binary probit model presented

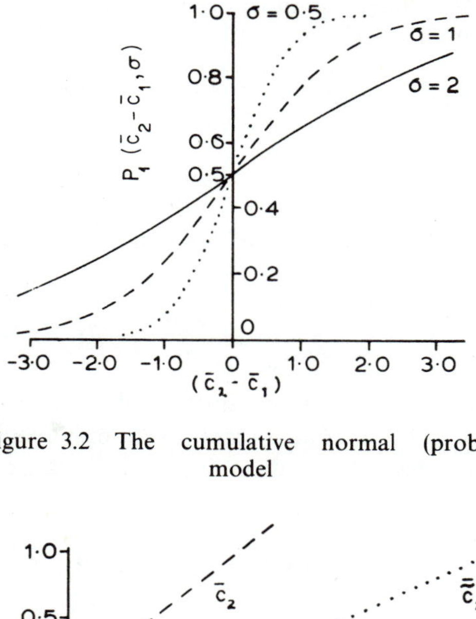

Figure 3.2 The cumulative normal (probit) model

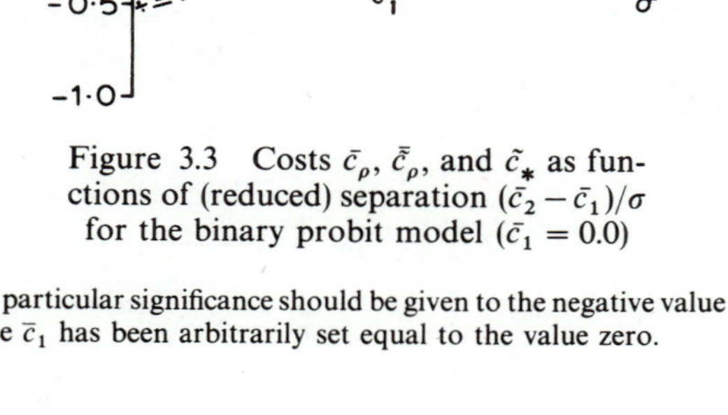

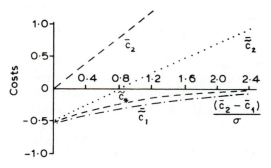

Figure 3.3 Costs $\bar{c}_\rho$, $\bar{\bar{c}}_\rho$, and $\tilde{c}_*$ as functions of (reduced) separation $(\bar{c}_2 - \bar{c}_1)/\sigma$ for the binary probit model $(\bar{c}_1 = 0.0)$

† In this instance no particular significance should be given to the negative value for disutilities (costs) in Figure 3.3 because $\bar{c}_1$ has been arbitrarily set equal to the value zero.

here is virtually indistinguishable from the binary logit model

$$P_\rho(\bar{c}_1, \bar{c}_2) = \frac{\exp(-\lambda \bar{c}_\rho)}{\sum_\rho \exp(-\lambda \bar{c}_\rho)} \tag{3.38}$$

when

$$\lambda = \frac{\pi}{\sigma(\text{Weibull})\sqrt{6}} \simeq \frac{4}{\sigma(\text{Normal})\sqrt{2\pi}} \tag{3.39}$$

a relation which may be derived by examining cross-elasticities at $\bar{c}_1 = \bar{c}_2$.

The property (3.30) may be exploited directly for the estimation of benefit in cases for which the utility values u_ρ are equal. For a differential change ($d\bar{c}_1, \ldots, d\bar{c}_\rho, \ldots, d\bar{c}_N$), the change in the quantity $\tilde{c}_*$, which from equation (3.35) gives the benefit dS, is given immediately by

$$d\tilde{c}_* = \sum_\rho \frac{\partial \tilde{c}_*}{\partial \bar{c}_\rho} d\bar{c}_\rho \tag{3.40}$$

$$= \sum_\rho P_\rho \, d\bar{c}_\rho \tag{3.41}$$

For the finite change $\{\bar{c}^{\mathrm{I}}_\rho\} \to \{\bar{c}^{\mathrm{F}}_\rho\}$ the change in $\tilde{c}_*$ may be computed by line integration as

$$\Delta\tilde{c}_* = \sum_\rho \int_0^1 d\hat{\sigma} \frac{d\bar{c}_\rho(\hat{\sigma})}{d\hat{\sigma}} P_\rho(\hat{\sigma}) \tag{3.42}$$

where the evaluation is along a path parameterized by $\hat{\sigma}$, between the initial and final cost states $\{\bar{c}^{\mathrm{I}}_\rho\}$ and $\{c^{\mathrm{F}}_\rho\}$. The integral (3.42) may readily be shown to be unique and $\Delta\tilde{c}_*$ may be defined unambiguously because the necessary integrability conditions,

$$\frac{\partial P_\rho}{\partial \bar{c}_{\rho'}} = \frac{\partial P_{\rho'}}{\partial \bar{c}_\rho} \qquad \text{for all} \quad \chi_\rho, \chi_{\rho'} \in \chi \quad A_\rho, A_{\rho'} \in A \tag{3.43}$$

follow directly from equation (3.30) when the symmetry of the second derivative $\partial^2 \tilde{c}_* / \partial \bar{c}_\rho \partial \bar{c}_{\rho'}$ is exploited. Thus $\Delta\tilde{c}_*$ may be computed directly from a knowledge of the distributions Γ_ρ through equation (3.23), or (if it is known that the demand function P_ρ is derivable from translationally invariant distributions) directly from (3.42), which is formally equivalent to the generalized surplus integral of Hotelling (1938). For the multinomial logit model (3.12), the path integral (3.42) which is equal to ΔS, is given by

$$\Delta S = -\Delta\tilde{c}_* = \frac{1}{\lambda} \ln \frac{\sum_\rho \exp(-\lambda \bar{c}^{\mathrm{F}}_\rho)}{\sum_\rho \exp(-\lambda \bar{c}^{\mathrm{I}}_\rho)} \tag{3.44}$$

a result which has been derived by Wilson and Kirwan (1969), and by Neuberger (1971).

The absolute value for $\tilde{c}_*$ (given $\bar{c}_\rho$) may be obtained by direct integration from its definition or, by exploiting the property shown in equation (3.30), through path integration by defining initial and final cost states $\mathbf{c}^{\mathrm{I}}$ and $\mathbf{c}^{\mathrm{F}}$ as follows

$$\mathbf{c}^{\mathrm{I}} = (\infty \ldots \infty, \bar{c}_\rho, \infty \ldots \infty),\ \mathbf{c}^{\mathrm{F}} = (\bar{c}_1, \ldots, \bar{c}_\rho, \ldots, \bar{c}_N) \qquad (3.45)$$

For the Weibull distribution we have

$$\tilde{c}^{\mathrm{I}}_* - \tilde{c}^{\mathrm{F}}_* = -\sum_\rho \int_0^1 \mathrm{d}\hat{\sigma} \frac{\mathrm{d}\bar{c}_\rho}{\mathrm{d}\hat{\sigma}} P_\rho(\hat{\sigma}) = \frac{1}{\lambda} \ln \frac{\sum_\rho \exp(-\lambda \bar{c}^{\mathrm{F}}_\rho)}{\sum_\rho \exp(-\lambda \bar{c}^{\mathrm{I}}_\rho)} \qquad (3.46)$$

$$= \frac{1}{\lambda} \ln\left[\sum_\rho \exp(-\lambda \bar{c}_\rho)\right] + \bar{c}_\rho \qquad (3.47)$$

Now, $\tilde{c}^{\mathrm{I}}_*$ is the mean value of the Weibull distribution $\Gamma_\rho(c)$, so finally

$$\tilde{c}_* = \tilde{c}^{\mathrm{F}}_* = -\frac{1}{\lambda} \ln \sum_\rho \exp(-\lambda \bar{c}_\rho) + \frac{\gamma}{\lambda} \qquad (3.48)$$

This result will be used extensively in the models of Section 3.5.

3.4 THE ENTROPY MAXIMIZING AND GROUP-SURPLUS MAXIMIZING APPROACHES

The random utility models described above are underpinned by a rational assessment of choice alternatives in which the probabilistic component of the theory arises because of the different perception of utility associated with any choice alternative. The approach is essentially *behavioural* in nature. An alternative, and widely used, method of generating travel demand and activity location models is that proposed by Wilson (1967, 1970) based on the concept of entropy, in which the most probable distribution of trips is determined subject to any known constraints.

The entropy of the discrete probability distribution $\{P_1, \ldots, P_\rho, \ldots, P_N\}$, defined as

$$W = -\sum_\rho P_\rho \ln P_\rho \qquad (3.49)$$

is maximized subject to the normalization condition

$$\sum_\rho P_\rho = 1 \qquad (3.50)$$

the cost constraint

$$\sum_\rho P_\rho \bar{c}_\rho = C \qquad (3.51)$$

and any other constraints on the distribution. The quantity C is the *observed* expenditure (or perceived cost) for the population of which the 'individual' of interest is a member, with $\bar{c}_\rho$ the mean cost attributed to the option A_ρ by an observer (the model builder). The optimal solution of the program (3.49)–(3.51) is given by

$$P_\rho = \exp(-\tilde{\eta} - \lambda\bar{c}_\rho) \tag{3.52}$$

where $\tilde{\eta}$ and λ are calculated to satisfy the constraints (3.50) and (3.51). $\tilde{\eta}$ is trivially determined from equation (3.50) to be

$$\tilde{\eta} = \ln \sum_\rho \exp(-\lambda\bar{c}_\rho) \tag{3.53}$$

The formal similarity between $\tilde{\eta}$ and $\tilde{c}_*$ expressed in equation (3.48), is no coincidence. From equations (3.53) and (3.48) we have immediately

$$\tilde{\eta} = -\lambda\tilde{c}_* + \gamma \tag{3.54}$$

It can also be seen from equations (3.42), (3.44), (3.49), and (3.53) that, for fixed values of λ, the generalized surplus may be written

$$\Delta S = -\sum \int_0^1 \mathrm{d}\hat{\sigma} \frac{\mathrm{d}\bar{c}_\rho(\hat{\sigma})}{\mathrm{d}\hat{\sigma}} P_\rho(\hat{\sigma}) = \tilde{c}_*^{\mathrm{I}} - \tilde{c}_*^{\mathrm{F}} = \frac{1}{\lambda}(\tilde{\eta}^{\mathrm{F}} - \tilde{\eta}^{\mathrm{I}}) \tag{3.55}$$

$$= C^{\mathrm{I}} - C^{\mathrm{F}} + \frac{1}{\lambda}(\mathrm{W}^{\mathrm{F}} - \mathrm{W}^{\mathrm{I}}) \tag{3.56}$$

where the ^{I} and ^{F} notation again denotes evaluation of a variable in the initial and final cost states corresponding to $\mathbf{c}^{\mathrm{I}}$ and $\mathbf{c}^{\mathrm{F}}$ respectively.

Several comments about the entropy-maximizing method and its application are appropriate at this point. The methodology, which may be applied with equal validity at the macro and micro levels, is characterized by the *lack* of assumptions made concerning the decision-making process at the level of the individual. The probability distribution (3.52) obtained by maximizing entropy is, as Wilson (1970) has pointed out, that which makes the weakest assumptions consistent with what is known and reflected in the constraints. It is not generally within the spirit of the approach to pin down the variability in behaviour to any particular source, as this would, in effect, present additional assumptions or hypotheses which could be exploited in model formation. In the above calculation, which gives the result (3.56), for example, the population must in fact be endowed with economic rationality in order to be able to interpret ΔS as the appropriate measure of benefit. In this case, W/λ in equation (3.56) becomes formally equivalent to a utility function for the group, measured in money units.

Derivation of the relation (3.56) by path integration of the demand function is not restricted to the form of the entropy function (3.49) and may be obtained for *any* program with the following structure:

$$\max_{\{P\}} \frac{1}{\lambda} \hat{W}(\mathrm{P}) - \sum_\rho P_\rho \bar{c}_\rho \tag{3.57}$$

subject to

$$\sum_{\rho} P_{\rho} = 1 \tag{3.58}$$

plus any other constraints on the probability distribution. Here the objective function may be interpreted as the net utility or surplus function for the group Π *in the sense that maximization of this quantity reproduces the variability existing at the individual level.* Where one seeks to explain trip variability in terms of *economic rationality* it is more appropriate to start from a utility-based approach rather than an information-theory approach, as noted by Beckmann and Golob (1971), Neuberger (1971), and others. If the economic implications of a change in the vector of perceived costs are to be determined and interpreted, it is indeed essential that the basis upon which decisions are considered to be made is fully recognized. If, however, the observer is not prepared to attribute the variability at the micro level or macro level to any particular source, or is confronted by several sources of variability, which would overtax theoretical or data capability at the level of resolution considered, the entropy-maximizing method would seem to be an appropriate model-building device.

For the remainder of the chapter we shall assume that the choice models are underpinned by a *rational decision process* for which random utility theory is appropriate. In later discussion it proves very convenient to relax the notational distinction between utility or benefit u, disutility c, and surplus s. In the following section, a single variable u will be used to denote all components, which will be termed utility components. Perceived generalized cost will thus be included implicitly as a negative utility measured in money or time units. The net utility or surplus will, as before, be calculated consistently. Where it proves convenient, a generalized cost notation will be introduced. Note that the dispersion parameter, say λ may be absorbed into the utility function u in an expression such as $\exp(\lambda u)$ which will be written $\exp(\hat{u})$. If u is in money units then λ will be the coefficient of the attribute money (out-of-pocket cost) in the linear expression

$$\hat{u}_{\rho} = \sum_{\eta} \bar{\alpha}_{\eta} Z_{\rho}^{\eta} \tag{3.59}$$

3.5 ATTRIBUTE CORRELATION AND MODEL STRUCTURES

3.5.1 The variance–covariance matrix

The assumption of independence of the utility components associated with different choice alternatives endows the models developed in Section 3.3 with a relatively simple structure. An important property of the multinomial logit (MNL) model, for example, is that the ratio of choice probabilities for any two alternatives,

$$\frac{P_{\rho}}{P_{\rho'}} = \exp[\lambda(\bar{u}_{\rho} - \bar{u}_{\rho'})] \tag{3.60}$$

is independent of the characteristics of *other* members of the choice set **A**.

This property of the MNL model, which is referred to as the 'independence from irrelevant alternatives' (IIA) characteristic, because the ratio of probabilities is invariant with respect to addition or removal of other members to or from the set **A**, is regarded as a mixed blessing. It can be and often has been exploited in the modelling of 'new' alternatives whose market share may be derived simply from a knowledge of its attributes. There are many choice situations, however, in which the IIA property leads to intuitively unreasonable implications, particularly if a subset of the alternative choices have similar characteristics. An extreme example which emphasizes the limitations of these simple choice models, and the MNL model in particular, is epitomized in the red-bus/blue-bus conundrum (Mayberry, 1970). Simply by redefining the available alternatives (in this case painting half the available red buses blue) and uncritically applying the MNL model to an increased number of alternatives, an intuitiviely unacceptable reapportionment of trips occurs.

The problem of apportioning trips here is seen in terms of mode definition, or how to prespecify alternatives which in some sense had equal 'weight' in the choice process. In this example the attribute colour has (presumably) negligible effect on bus patronage, which suggests that the notion of 'weight' has something to do with 'independent' contributions to the cost or utility functions. Here the red and blue buses are endowed with attributes which are perfectly correlated. That is, the times, costs, comfort factors, etc., of the two public transport alternatives are adjudged to be identical by each individual although there will in general be an interpersonal variation in the perception of these attributes.

The phenomenon of attribute correlation and the violence rendered to Luce's axiomatic approach, when choice between 'similar' alternatives has been considered, has long been recognized in mathematical psychology. Luce's (1959) model would suggest that the relative probability of complex choice alternatives could be computed by summing the values of the component attributes to produce a total utility value u_ρ as in equation (3.12). (This is, in effect, assumed in equation (3.12)). That this could yield absurd results when common elements were present in the choices was pointed out by Restle (1961), who argued that the choice between two complex and 'overlapping' alternatives depends not at all on the elements common to the two alternatives, but only on differential elements. Restle developed a set-theoretical approach to accommodate the effect while maintaining the flavour of the axiomatic formulation. The red-bus/blue-bus anomaly here is simply a manifestation of the problems arising from applying Luce's axiom in a condition of substantial 'overlapping' or 'correlation' of attributes.

Within random utility theory the notion of 'similarity' between alternatives has a mathematical interpretation in terms of the correlation between the stochastic residuals in equation (3.5). If we define the variance–covariance matrix $\mathbf{\Sigma}$ the elements of which are given by

$$\Sigma_{\rho\rho'} = E(\varepsilon_\rho \varepsilon_{\rho'}) \tag{3.61}$$

with $E(\cdot)$ denoting an expectation value, it can immediately be seen that the assumption of *identical* and *independently* distributed (IID) residuals is associated with a very simple diagonal structure for the matrix Σ.

For all IID models, and the MNL model in particular, $\Sigma_{\rho\rho'}$ is given by

$$\Sigma_{\rho\rho'} = \sigma^2 \delta_{\rho\rho'} \tag{3.62}$$

in which

$$\sigma^2 = E(\varepsilon_\rho \varepsilon_{\rho'}) \qquad \text{for all } A_\rho \in A \tag{3.63}$$

and

$$\delta_{\rho\rho'} = \begin{cases} 1 & \text{if } \rho = \rho' \\ 0 & \text{otherwise} \end{cases} \tag{3.64}$$

The limitations in the simple model structure may also be seen if we examine the matrix χ of elasticity parameters $\chi_{\rho\rho'}$ defined by

$$\chi_{\rho\rho'} = \frac{\bar{U}_{\rho'}}{P_\rho} \frac{\partial P_\rho}{\partial \bar{U}_{\rho'}} \tag{3.65}$$

For the MN logit model this matrix element is given by

$$\chi_{\rho\rho'} = \lambda \bar{U}_{\rho'}(\delta_{\rho\rho'} - P_{\rho'}) \tag{3.66}$$

The whole matrix χ is characterized by the single parameter λ and the restrictive properties of cross-substitution in the structure of the MNL may be traced directly to the simple structure of the matrix Σ.

In the remainder of this section we shall consider the formation and application of models, conceived within the framework of random utility theory, which are consistent with more general variance–covariance matrices and therefore less restrictive properties of cross-substitution. Before deriving specific models and seeking the more general implications of correlation for model structure, it is important to review the issue of model structure and some of the opinions which have shaped our thinking on it over the past twenty years, and in particular in the early 1970s.

3.5.2 Simultaneous and sequential model structures

The traditional urban-transport planning process involved a decomposition in which a set of submodels was used to correspond to the address sequence shown in Table 3.1. The form of demand model was originally based on a dynamic scenario of the trip decision process in which a person decided whether to make a trip, then whether to take public or private transport, where to perform a given activity, and then finally which route to take. The trip decision was thus conceived within a framework in which successive choices were conditional on the predecessor.

The ambiguity of the relative positions of the modal split and distribution of the sequence was recognised at an early stage. The earliest models which

Table 3.1

Submodel	Address
Urban-regional growth model	What will be the magnitude of urban-regional growth?
Activity location land-use allocation model	Where will the activities be located?
Generation	How many trips will the activities generate?
Modal split	What mode of travel will be used?
Distribution	Where will these trips go?
Assignment	Which route will these trips take?
Evaluation	What is the 'best' transportation system?

considered modal split were expressed in terms of the sequence G/MS/D/A of generation (G), modal split (MS), distribution (D), and assignment (A) submodels. In order to accommodate modal competition in what was considered to be a more satisfactory way, trip interchange models of modal split were developed by the Traffic Research Corporation in the early 1960s within the model structure G/D/MS/A which has dominated applications to date. A person was now considered to choose mode after the location decisions had been settled. More recently the argument that location and mode were simultaneously determined in the decision process has gained prominence and this appears to have influenced the recent British Studies in Coventry and London in which the structure G/D–MS/A has been applied. This embodies a model combining the distribution and modal-split segments.

It has been argued that the 'simultaneous' (or 'joint') and 'recursive' (or 'sequential') model structures are based on different hypotheses about the underlying travel decision-making process. Thus Ben-Akiva (1974) remarks: 'The simultaneous structure is very general and does not require any specific assumptions. The recursive structure represents a specific conditional decision structure, i.e. the traveller is assumed to decompose his trip decision into several stages. Thus, simultaneous and recursive structures represent simultaneous and sequential decision making processes.' We shall see later that this influential view is in fact misleading, and that there is in fact within any **theoretical** framework a more rigorous interpretation of the distinction between alternative model structures.

Four basic approaches to forecasting travel demand in equilibrium are now usually identified: aggregate sequential; disaggregate sequential; aggregate simultaneous; and disaggregate simultaneous. The first descriptor refers to whether the model addresses data pertaining to groups or to individuals (or households), whereas the latter characterizes the representation of the trip decision process itself. A full range of examples of these distinct approaches is given by Charles

River Associates (1972) and by Ruiter (1973). A small number of recent British studies have employed hybrid forms. In the model developed by Bonsall *et al.* (1977) and the models in the Coventry Transportation Study (1973) and Greater London Study (Havers, *et al.*, 1974), for example, the trip-generation stage has been developed at the household level with other submodels addressing aggregate data, and as mentioned above the distribution and modal-split models have been combined in a joint structure similar to that proposed by Wilson (1969). These examples are further discussed by Senior and Williams (1977).

In order to illustrate, and later develop, the implications of these concepts and later rationalize the distinction between model structures, we shall refer now to the work of Ben-Akiva (1973). This work involved the calibration of a multinomial logit model, for choice over frequency, destination, and mode of shopping trips, both directly and indirectly (see Manheim, 1973), and compared the results with those of recursive models based on two possible sequences, each with different composite cost functions. These processes may be illustrated most simply for choice contexts involving the two dimensions $X = \{X_1, \ldots, X_\rho, \ldots, X_N\}$ and $Y = \{Y_1, \ldots, Y_\mu, \ldots, Y_M\}$ for which choice alternatives are of the form $\{X_1 Y_1, \ldots, X_\rho Y_\mu, \ldots, X_N Y_M\}$. $X_\rho Y_\mu$ might for example refer to a particular location and mode combination.

The 'simultaneous' model structure adopted was of multinomial logit form

$$P_{\rho\mu} = \frac{\exp(\bar{\hat{u}}(\rho, \mu))}{\sum_{\rho\mu} \exp(\bar{\hat{\mathrm{u}}}(\rho, \mu))} \tag{3.67}$$

with

$$\bar{\hat{u}}(\rho, \mu) = \bar{\hat{u}}_\rho + \bar{\hat{u}}_\rho + \bar{\hat{u}}_{\rho\mu} \tag{3.68}$$

$$\bar{\hat{u}}_\rho = \sum_\eta \bar{\hat{\alpha}}_\eta^X Z_\rho^\eta \tag{3.69}$$

$$\bar{\hat{u}}_\mu = \sum_\eta \bar{\hat{\alpha}}_\eta^Y Z_\mu^\eta \tag{3.70}$$

$$\bar{\hat{\mathrm{u}}}_{\rho\mu} = \sum_\eta \bar{\hat{\alpha}}^{XY} Z_{\rho\mu}^\eta \tag{3.71}$$

In the multinomial logit model the exponent is of linear form in the components of the attribute vector $\mathbf{Z}$. $\bar{\hat{u}}_\rho$ and $\bar{\hat{u}}_\mu$ denote the components which are constant over the choice dimensions Y and X respectively, and $\bar{\hat{\mathrm{u}}}_{\rho\mu}$ is that component which varies across both dimensions. Recall that $\bar{\hat{\mathrm{u}}}$ indicates that utility components are dimensionless.

Now equation (3.67) may be written in the following indirect form:

$$p_{\rho\mu} = \frac{\exp(\bar{\hat{\mathrm{u}}}_\rho + \bar{\hat{\mathrm{u}}}_{\rho\mu})}{\sum_\rho \exp(\bar{\hat{u}}_\rho + \bar{\hat{\mathrm{u}}}_{\rho\mu})} \frac{\exp(\bar{\hat{u}}_\mu + \bar{\hat{\mu}}_{\mu^*})}{\sum_\mu \exp(\bar{\hat{\mathrm{u}}}_\mu + \bar{\hat{\mathrm{u}}}_{\mu^*})} \tag{3.72}$$

with the composite utility defined by

$$\hat{u}_{\mu*} = \ln \sum_{\rho} \exp(\bar{\hat{u}}_{\rho} + \bar{\hat{u}}_{\rho\mu}) \tag{3.73}$$

The conditional probability $P_{\rho.\mu}$ and marginal probability P_{μ} can be identified as

$$P_{\rho|\mu} = \frac{\exp(\bar{\hat{u}}_{\rho} + \bar{\hat{u}}_{\rho\mu})}{\sum_{\rho} \exp(\bar{\hat{u}}_{\rho} + \bar{\hat{u}}_{\rho\mu})} \tag{3.74}$$

and

$$P_{\mu} = \frac{\exp(\bar{\hat{u}}_{\mu} + \hat{u}_{\mu*})}{\sum_{\mu} \exp(\bar{\hat{u}}_{\mu} + \hat{u}_{\mu*})} \tag{3.75}$$

Equally, we can write

$$P_{\rho\mu} = P_{\mu|\rho} P_{\rho} \tag{3.76}$$

with

$$P_{\mu|\rho} = \frac{\exp(\bar{\hat{u}}_{\mu} + \bar{\hat{u}}_{\rho\mu})}{\sum_{\mu} \exp(\bar{\hat{u}}_{\mu} + \bar{\hat{u}}_{\rho\mu})} \tag{3.77}$$

and

$$P_{\rho} = \frac{\exp(\bar{\hat{u}}_{\rho} + \hat{u}_{\rho*})}{\sum_{\rho} \exp(\bar{\hat{u}}_{\rho} + \hat{u}_{\rho*})} \tag{3.78}$$

with

$$\hat{u}_{\rho*} = \ln \sum_{\mu} \exp(\bar{\hat{u}}_{\mu} + \bar{\hat{u}}_{\rho\mu}) \tag{3.79}$$

The expressions (3.67), (3.72), and (3.76) are mathematically equivalent. The parameters $\bar{\hat{\alpha}}^{X}$, $\bar{\hat{\alpha}}^{Y}$, and $\bar{\hat{\alpha}}^{XY}$ in the 'simultaneous' model may therefore be estimated directly from the multinomial logit model (3.67) or indirectly from either of the forms (3.72) and (3.76). Ben-Akiva has compared this model with two recursive forms, which may be written as follows:

$$P_{\rho\mu} = P_{\mu} P_{\rho|\mu} = \frac{\exp(\bar{\hat{u}}_{\mu} + \hat{u}_{\mu*}\delta)}{\sum_{\mu} \exp(\bar{\hat{u}}_{\mu} + \hat{u}_{\mu*}\delta)} \frac{\exp(\bar{\hat{u}}_{\rho} + \bar{\hat{u}}_{\rho\mu})}{\sum_{\rho} \exp(\bar{\hat{u}}_{\rho} + \bar{\hat{u}}_{\rho\mu})} \tag{3.80}$$

and

$$P_{\rho\mu} = P_{\rho} P_{\mu|\rho} = \frac{\exp(\bar{\hat{u}}_{\rho} + \hat{u}_{\rho*}\delta')}{\sum_{\rho} \exp(\bar{\hat{u}}_{\rho} + \hat{u}_{\rho*}\delta')} \frac{\exp(\bar{\hat{u}}_{\mu} + \bar{\hat{u}}_{\rho\mu})}{\sum_{\mu} \exp(\bar{\hat{u}}_{\mu} + \bar{\hat{u}}_{\rho\mu})} \tag{3.81}$$

where δ and δ' are coefficients of the composite variables. The possible forms of the composite functions (e.g. for $\tilde{u}_{\mu*}$) considered were

$$\text{‘weighted prices’ } \hat{u}_{\mu*}(\eta) = \sum_{\rho} P_{\rho|\mu} Z_{\rho}^{\eta} \tag{3.82}$$

$$\text{‘weighted generalized costs’ } \hat{u}_{\mu*} = \sum_{\rho} P_{\rho|\mu} \bar{\hat{u}}_{\rho\mu} \tag{3.83}$$

$$\text{‘weighted utilities’ } \bar{\hat{u}}_{\mu*} = \sum_{\rho} P_{\rho|\mu} (\bar{\hat{u}}_{\rho} + \bar{\hat{u}}_{\rho\mu}) \tag{3.84}$$

and ‘log of the denominator’ $\hat{u}_{\mu*} = \ln \sum_{\rho} \exp(\bar{\hat{u}}_{\rho} + \bar{\hat{u}}_{\rho\mu})$ (equation (3.73)). The functional form (3.83) was used by Charles River Associates (1972) in their disaggregate shopping model.† Similar forms exist for $\hat{u}_{\rho*}$.

It can be seen that if the unity constraint on the coefficient of $\hat{u}_{\mu*}$ in equation (3.75) is relaxed, a model similar to the recursive structure (3.80) is obtained with the composite cost form (3.73), and corresponds to one of Manheim's special product (MSP) models. As Ben-Akiva notes, ‘Writing the indirect structure of the simultaneous model without the unity constraint corresponds to an alternative formulation of a recursive model . . . An empirical “test of simultaneity” can be made by estimating the unconstrained model (MSP *model*) and performing a statistical test to determine *whether the* δ *parameter* is significantly different from one.’ (The italics have been inserted by the author.)

It is clear from the above discussion on the two approaches that the parameter δ plays an important role in the distinction between the structures, although its significance was not elucidated in the work of Ben-Akiva (1973) or in Manheim's (1973) characterization of demand models. Why does the assumption of sequence destroy the property of independence from irrelevant alternatives of the choice axiom when the recursive structure involves the composite utility (3.73)? Are, in fact, the models (3.67) and (3.80) consistent with the utility functions assumed to have generated them? These issues centre directly on the phenomenon of *attribute correlation* to which we must return.

3.5.3 The generation of share models

The existence of a semantic difficulty in relation to the descriptors ‘simultaneous’ and ‘sequential’ and time ordering of the trip decision process has been noted on several occasions (see for example Kraft, 1974). However, the important point is that both representations of the trip decision process are underpinned by a utility function which, in the theory of rational choice, governs an optimal selection over the set of alternatives. Once this function has been specified, the generation of the demand model becomes a mathematical exercise within a specific theoretical

† A full discussion of the implications of these forms for the marginal rate of substitution between pairs of variables throughout the model is given by Ben-Akiva (1973).

framework. Thus, in the random utility framework, the model may in principle be developed through multiple integration as in equation (3.1). Any classification of the analytic structure of the demand function, insofar as it pertains to the *behavioural* context of the model, is thus determined by the properties and structure of the utility function itself. How, in fact, a traveller identifies his optimal choice in a maze of possibilities and disregards suboptimal alternatives is not particularly relevant in an optimizing cross-sectional approach, although a mental picture of how this *may* occur can be useful in interpreting certain characteristics of the resultant model.

Consider choice in the two dimensions X and Y to be governed by a utility function of the form

$$U(X, Y) = U^X + U^{XY} \tag{3.85}$$

in terms of which the components $u(\rho, \mu)$ may be written

$$u(\rho, \mu) = u_\rho + u_{\rho\mu} \qquad \rho = 1, \ldots, N;\ \mu = 1, \ldots, M \tag{3.86}$$

The subscripts ρ and μ will refer as before to components in the X and Y dimensions respectively.

In the function (3.86), the existence of correlation between the utility distributions for different (ρ, μ) alternatives is explicitly recognized through the commonality of the component u_ρ, for all choices $(\rho|\mu = 1, \ldots, M)$. By writing $u(\rho, \mu)$ in terms of a representative component and stochastic residual as follows:

$$u(\rho, \mu) = \bar{u}_\rho + \bar{u}_{\rho\mu} + \varepsilon_\rho + \varepsilon_{\rho\mu} \tag{3.87}$$

it can immediately be seen that the variance–covariance matrix $\mathbf{\Sigma}$ is defined in terms of the associated elements

$$\Sigma_{\rho\mu, \rho'\mu'} = E(\varepsilon_\rho + \varepsilon_{\rho\mu}, \varepsilon_{\rho'} + \varepsilon_{\rho'\mu'}) \tag{3.88}$$

If the components u_ρ and $u_{\rho\mu}$ are independently distributed and we further impose the requirement that

$$E(\varepsilon_\rho\ \varepsilon_{\rho'}) = \sigma_X^2 \delta_{\rho\rho'} \tag{3.89}$$

$$E(\varepsilon_{\rho\mu}\ \varepsilon_{\rho'\mu'}) = \sigma_{XY}^2 \delta_{\rho\rho'} \delta_{\mu\mu'} \tag{3.90}$$

then the matrix elements of $\mathbf{\Sigma}$ may be written

$$\Sigma_{\rho\mu, \rho'\mu'} = \sigma_X^2 \delta_{\rho\rho'} + \sigma_{XY}^2 \delta_{\rho\rho'} \delta_{\mu\mu'} \tag{3.91}$$

In order to find the model structure which is appropriate to this variance–covariance matrix, we write the probability $P_{\rho\mu}$ that the alternative (ρ, μ) will be selected, as follows:

$$P_{\rho\mu} = \text{prob}\,\{u(\rho, \mu) > u(\rho', \mu') \text{ for all } \rho' \in X, \mu' \in Y\} \tag{3.92}$$

This model will be characterized as follows:

$$P_{\rho\mu} = F_{\rho\mu}(\bar{\mathbf{u}}_X, \bar{\mathbf{u}}_{XY}; \boldsymbol{\sigma}_X, \boldsymbol{\sigma}_{XY}) \tag{3.93}$$

in which the vectors $\bar{\mathbf{u}}_X$, $\bar{\mathbf{u}}_{XY}$ contain all mean values of the utility components in the $\{X, Y\}$ choice complex, with $\boldsymbol{\sigma}_X$ and $\boldsymbol{\sigma}_{XY}$ the corresponding vectors of standard deviations. That is, $\bar{\mathbf{u}}_X = (\bar{u}_1, \ldots \bar{u}_\rho, \ldots, \bar{u}_N)$, $\bar{\mathbf{u}}_{XY} = (\bar{u}_{11}, \ldots, \bar{u}_{\rho\mu}, \ldots, \bar{u}_{NM})$, and $\boldsymbol{\sigma}_X = (\sigma_1, \ldots, \sigma_\rho, \ldots, \sigma_N)$, $\boldsymbol{\sigma}_{XY} = (\sigma_{11}, \ldots, \sigma_{\rho\mu}, \ldots, \sigma_{NM})$. With the additional restrictions (3.89) and (3.90), $F_{\rho\mu}$ may be simplified to

$$P_{\rho\mu} = F_{\rho\mu}(\bar{\mathbf{u}}_X, \bar{\mathbf{u}}_{XY}; \sigma_X, \sigma_{XY}) \tag{3.94}$$

Now the choice process *may* be conceptualized in terms of the way an individual acquires full information in order to act optimally. The additive separable utility function lends itself naturally to a partitioning of the alternatives in the hierarchy shown in Figure 3.4(b). This is essentially a representation of *Strotz's utility tree* (Strotz, 1957). For each alternative, X_ρ, an individual t will determine the maximum value

$$u^t_{\rho *} = \max_{\mu} u^t_{\rho\mu} \tag{3.95}$$

and then select alternative X_ρ if

$$u^t_\rho + u^t_{\rho *} = \max_{\rho'} (u^t_{\rho'} + u^t_{\rho' *}) \tag{3.96}$$

Over the whole choice-making population,

$$P_{\rho\mu} = \text{prob}\{u_\rho + u_{\rho *} > u_{\rho'} + u_{\rho' *}, \text{ for all } \rho' \in X, \text{ and } u_{\rho\mu} > u_{\rho\mu'}, \text{ for all } \mu' \in Y\} \tag{3.97}$$

where $u_{\rho *}$ is a random composite utility variable drawn from the distribution $\Lambda_{\rho *}$ of maximum utility,

$$u_{\rho *} = \max_{\rho} \{u_{\rho 1}, \ldots, u_{\rho\mu}, \ldots, u_{\rho M}\} \tag{3.98}$$

Because of the independence of the distributions in the separate choice dimensions, we may write

$$P_{\rho\mu} = \text{prob}\{u^+_{\rho *} > u^+_{\rho' *}, \text{ for all } \rho' \in X\}\text{prob}\{u_{\rho\mu} > u_{\rho\mu'}, \text{ for all } \mu' \in Y\} \tag{3.99}$$

where $u^+_{\rho *}$ is distributed according to the sum of the independent random variables u_ρ and $u_{\rho *}$. This distribution $\Lambda^+_{\rho *}$ has mean $\bar{u}^+_{\rho *}$ and standard deviation $\sigma^+_{\rho *}$ given by

$$\bar{u}^+_{\rho *} = \bar{u}_\rho + \tilde{u}_{\rho *} \tag{3.100}$$

$$\sigma^+_{\rho *} = (\sigma^2_\rho + \sigma^2_{\rho *})^{1/2} \tag{3.101}$$

$P_{\rho\mu}$ may thus be expressed in the product form

$$\begin{aligned} P_{\rho\mu} &= P_\rho(\bar{u}^+_{1*}, \ldots, \bar{u}^+_{\rho *}, \ldots, \bar{u}^+_{N*}) P_{\mu|\rho}(\bar{u}_{\rho 1}, \ldots, \bar{u}_{\rho\mu}, \ldots, \bar{u}_{\rho M}) \\ &\equiv P_\rho(\bar{u}^+_{X*}) P_{\mu|\rho}(\bar{u}_{\rho Y}) \end{aligned} \tag{3.102}$$

in an obvious vector notation, while the difference in generalized consumer surplus ΔS arising from the change $(\bar{u}^{\mathrm{I}}_{11}, \ldots, \bar{u}^{\mathrm{I}}_{\rho\mu}, \ldots, \bar{u}^{\mathrm{I}}_{NM}) \to (\bar{u}^{\mathrm{F}}_{11}, \ldots,$

$\bar{u}^{\mathrm{F}}_{\rho\mu}, \ldots, \bar{u}^{\mathrm{F}}_{NM})$ becomes

$$\Delta S = \sum_{\rho\mu} \int_0^1 \mathrm{d}\hat{\sigma} \frac{\mathrm{d}\bar{u}_{\rho\mu}}{\mathrm{d}\hat{\sigma}} P_\rho (\bar{u}^+_{X*}) P_{\mu|\rho} (\bar{u}_{\rho Y}) \tag{3.103}$$

$$= \sum_{\rho} \int_0^1 \mathrm{d}\hat{\sigma} P_\rho (\bar{u}^+_{X*}) \sum_{\mu} P_{\mu|\rho} (\bar{u}_{\rho Y}) \frac{\mathrm{d}\bar{u}_{\rho\mu}}{\mathrm{d}\hat{\sigma}} \tag{3.104}$$

For translationally invariant distribution (cf. equation 3.30),

$$\frac{\mathrm{d}\tilde{u}_{\rho*}}{\mathrm{d}\hat{\sigma}} = \sum_{\mu} P_{\mu|\rho} \frac{\mathrm{d}\bar{u}_{\rho\mu}}{\mathrm{d}\hat{\sigma}} \tag{3.105}$$

which reduces equation (3.104) to

$$\Delta S = \sum_{\rho} \int_0^1 \mathrm{d}\hat{\sigma} \frac{\mathrm{d}\bar{u}^+_{\rho*}}{\mathrm{d}\hat{\sigma}} P_\rho (\bar{\mathbf{u}}^+_{X*}) \tag{3.106}$$

on using the independence of $\bar{u}_\rho$ from the path parameter. ΔS may also be written in terms of the difference

$$\Delta S = \tilde{U}^{\mathrm{F}}_{**} - \tilde{U}^{\mathrm{I}}_{**} \tag{3.107}$$

between the mean value of the maximum surplus distributed as Λ_{**}, evaluated in the final and initial states respectively.

In essence, the equations (3.104) and (3.106) indicate tat the surplus change corresponding to the change $(\bar{u}^{\mathrm{I}}_{11}, \ldots, \bar{u}^{\mathrm{I}}_{\rho\mu}, \ldots. \bar{u}^{\mathrm{I}}_{NM}) \rightarrow (\bar{u}^{\mathrm{F}}_{11}, \ldots .\bar{u}^{\mathrm{F}}_{\rho\mu}, \ldots. \bar{u}^{\mathrm{F}}_{NM})$ computed with the choice probabilities $P_{\rho\mu}$ is identical to that obtained for the change $(\bar{u}^{+\mathrm{I}}_{1*}, \ldots. \bar{u}^{+\mathrm{I}}_{\rho*}, \ldots. \bar{u}^{+\mathrm{I}}_{N*}) \rightarrow (\bar{u}^{+\mathrm{F}}_{1*}, \ldots. \bar{u}^{+\mathrm{F}}_{\rho*}, \ldots. \bar{u}^{+\mathrm{F}}_{N*})$ computed with the probabilities P_ρ. This property of vertical aggregation in a utility tree is not surprising, and arises because of the optimality condition underpinning the choice process. All relevant information for optimal choices to be made is passed 'up the tree' as was illustrated in the possible conceptualization of the decision process for the individual. This argument may be immediately generalized (see Williams, 1977) to an arbitrary number of dimensions over which choice is governed by an additive, separable utility function.

We can thus see that the structure of the variance–covariance matrix of the form (3.91) leads naturally to the product share form (3.102) in which the composite utilities play the vital role of transmitting appropriate information for optimal choices to be made and consistent benefit measures to be achievable from (3.106).

3.5.4 The derivation of the nested logit function

To derive a demand function we must assume specific forms for the utility distributions (or those of the stochastic residuals). If $\{u_{\rho\mu}\}$ are Weibull distributed with standard deviation given by

$$\sigma_{\mathrm{XY}} = \frac{\pi}{\sqrt{6}\lambda} \tag{3.108}$$

then $P_{\mu|\rho}$ may be written in the logit form (see equations (3.7)–(3.17))

$$P_{\mu|\rho} = \frac{\exp(\lambda \bar{u}_{\rho\mu})}{\sum_{\mu} \exp(\lambda \bar{u}_{\rho\mu})} \tag{3.109}$$

Now the Weibull distribution, as Domencich and McFadden (1975) note, is the only distribution which retains its functional form under the maximization process (3.98). That is, the distribution of

$$\mu_{\rho*} = \max\{u_{\rho 1}, \ldots, u_{\rho\mu}, \ldots, u_{\rho M}\} \tag{3.110}$$

is also of Weibull form with mean and standard deviation given respectively by

$$\bar{u}_{\rho*} = \frac{1}{\lambda} \log \sum_{\mu} \exp(\lambda \bar{u}_{\rho\mu}) \tag{3.111}$$

$$\sigma_{X*} = \frac{\pi}{\sqrt{6}\lambda} \tag{3.112}$$

In order to derive an expression for the marginal probability P_ρ it is necessary to determine the distribution of the sum of u_ρ and $u_{\rho *}$, that is, the distribution of

$$u_\rho^+ = u_\rho + u_{\rho*} \tag{3.113}$$

As u_ρ and $u_{\rho*}$ are independent variates we may write the variance and mean value of u_ρ^+ as follows:

$$(\sigma_\rho^+)^2 = \sigma_\rho^2 + \frac{\pi^2}{6\lambda^2} \tag{3.114}$$

$$\bar{u}_\rho^+ = \bar{u}_\rho + \frac{1}{\lambda} \log \sum_{\mu} \exp(\lambda \bar{u}_{\rho\mu}) \tag{3.115}$$

The distribution $\Gamma(u_\rho, \bar{u}_\rho)$ is as yet unspecified. By specifying that distribution we immediately determine the distribution of u_ρ^+ and hence the resultant probabilistic choice model, P_ρ. Alternatively, we can require u_ρ^+ to be of a particular functional form and determine (by convolution) that distribution $\Gamma(u, \bar{u}_\rho)$ which generates that form. In particular, we shall require that u_ρ^+ be Weibull distributed with standard deviation σ_X^+ and mean value given by equations (3.114) and (3.115) respectively. We may immediately write down the resulting functional form for $P_{\rho\mu}$ as

$$P_{\rho\mu} = \frac{\exp[\beta(\bar{u}_\rho + \bar{u}_{\rho*})]}{\sum_{\rho} \exp[\beta(\bar{u}_\rho + \bar{u}_{\rho*})]} \cdot \frac{\exp(\lambda \bar{u}_{\rho\mu})}{\sum_{\mu} \exp(\lambda \bar{u}_{\rho\mu})} \tag{3.116}$$

with

$$\bar{u}_{\rho*} = \frac{1}{\lambda} \log \sum_{\mu} \exp(\lambda \bar{u}_{\rho\mu}) \tag{3.117}$$

and

$$\beta = \frac{\pi}{\sqrt{6}\sigma_X^+} \tag{3.118}$$

$$= \frac{\pi}{\sqrt{6}}\left(\sigma_X^2 + \frac{\pi^2}{6\lambda^2}\right)^{-1/2} \tag{3.119}$$

The surplus change ΔS accompanying changes in the vector of utilities $\{\bar{u}_{\rho\mu}\}$ may now be written

$$\Delta S = \sum_{\rho\mu} \int_0^1 \mathrm{d}\hat{\sigma} \frac{\mathrm{d}\bar{u}_{\rho\mu}}{\mathrm{d}\hat{\sigma}} P_{\rho\mu} \tag{3.120}$$

$$= \sum_{\rho} \int_0^1 \mathrm{d}\hat{\sigma} \frac{\exp[\beta(\bar{u}_\rho + \bar{u}_{\rho *})]}{\sum_\rho \exp[\beta(\bar{u}_\rho + \bar{u}_{\rho *})]} \cdot \frac{\mathrm{d}\bar{u}_{\rho *}}{\mathrm{d}\sigma} \tag{3.121}$$

$$= \frac{1}{\beta} \log \frac{\sum_\rho \exp[\beta(\bar{u}_\rho^{\mathrm{I}} + \bar{u}_{\rho *}^{\mathrm{F}})]}{\sum_\rho \exp[\beta(\bar{u}_\rho + \bar{u}_{\rho *}^{\mathrm{I}})]} \tag{3.122}$$

The composite utility U_{**} is from (3.107), given by

$$\bar{U}_{**} = \frac{1}{\beta} \log \sum_\rho \exp[\beta(\bar{u}_\rho + \bar{u}_{\rho *})] \tag{3.123}$$

$$= \frac{1}{\beta} \log \left\{ \sum_\rho \left\{ \sum_\mu \exp[\lambda(\bar{u}_\rho + \bar{u}_{\rho\mu})] \right\}^{\beta/\lambda} \right\} \tag{3.124}$$

It is immediately apparent from the expressions for the choice model $P_{\rho\mu}$ and from the composite utility U_{**}, that when the standard deviation σ_X vanishes the results correspond to those of the multinomial logit model. This is to be expected as the variance–covariance matrix becomes diagonal.

It may further be seen from equation (3.119) that because $\sigma_X \geqslant 0$, the dispersion parameters must satisfy the inequality

$$\beta \leqslant \lambda \tag{3.125}$$

if the choice model (3.116) is to be consistent with utility maximization. This is a most important restriction the violation of which is considered in Section 3.7. The above model may as before be generalized to several choice dimensions and be expressed as a multiple nested logit function (Williams, 1977).

That the model (3.116) violates the 'independence from irrelevant alternatives' axiom (as indeed it should) may readily be shown. The ratio $P_{\rho\mu}/P_{\rho'\mu'}$ for the model is given by

$$\frac{P_{\rho\mu}}{P_{\rho'\mu'}} = \frac{\exp(\beta\bar{u}_\rho + \lambda\bar{u}_{\rho\mu})}{\exp(\beta\bar{u}_{\rho'} + \lambda\bar{u}_{\rho'\mu'})} \left\{ \frac{\sum_\nu \exp(\lambda\bar{u}_{\rho\nu})}{\sum_\nu \exp(\lambda\bar{u}_{\rho\nu'})} \right\}^{(\beta-\lambda)/\lambda} \tag{3.126}$$

which is seen to depend on the utility values of alternatives other than $\{\rho, \mu\}$ and $\{\rho', \mu'\}$. Only if the parameter relation $\beta = \lambda$ holds will the independence axiom be satisfied, and from equation (3.118) it is apparent that this will only be the case if the variance associated with the utility component U^X is zero. In other words, the independence axiom will not hold in the presence of correlation between the alternatives.

Note that it is not necessary to go *constructively* through the stages expressed in equations (3.97)–(3.102) to show that the nested logit model (3.116) is derivable from utility maximizing principles. It is sufficient to know that the model satisfies the integrability conditions (see equation (3.43)) which ensure the uniqueness of the benefit function (3.120).

3.6 MORE GENERAL CHOICE MODEL STRUCTURES

3.6.1 Some initial considerations

The nested logit model, which as we have shown above is consistent with stochastic utility maximzation, embodies more general properties of cross-substitution than the multinomal logit model. The matrix elements of the elasticity parameter matrix for the model (3.116) are, for example, given by

$$\chi_{\rho\mu,\rho'\mu'} = \frac{\bar{U}_{\rho'\mu'}}{P_{\rho\mu}} \frac{\partial P_{\rho\mu}}{\partial \bar{U}_{\rho'\mu'}} \tag{3.127}$$

$$= \{\lambda(\delta_{\mu\mu'} - P_{\mu'|\rho})\delta_{\rho\rho'} + \beta(\delta_{\rho\rho'} P_{\mu'|\rho} - P_{\rho'\mu'})\}\, \bar{U}_{\rho'\mu'} \tag{3.128}$$

When correlation is absent, the parameters β and λ are equal and the multinomial logit results are obtained:

$$\chi_{\rho\mu,\rho'\mu'} = \lambda(\delta_{\mu\mu'}\delta_{\rho\rho'} - P_{\rho'\mu'})\bar{U}_{\rho'\mu'} \tag{3.129}$$

While the more general structure of the nested logit model is considered a virtue, it does introduce an asymmetic structure into the form of the model—precisely because the variance–covariance matrix $\mathbf{\Sigma}$ is assymetric—under the interchange of ρ, ρ' and μ, μ'. In this section we shall consider models which are not subject to such structural restrictions and the methods of generating them which are consistent with the theoretical framework outlined in the above sections.

We shall first describe the cross-correlated logit (CCL) model which is a generalization of the nested and multinomial logit forms. The class of general extreme-value models developed by McFadden (1978) will then be described. Recent progress on the implementation of the general probit model (Domencich and McFadden, 1975) will then be discussed, before we consider the use of simulation for the generation of models, and testing of theoretical propositions which they may embrace.

One task which we must consider below is the formulation of a model appropriate to a utility function of the following (symmetric) form:

$$U(X,\ Y) = U^X + U^{XY} + U^Y \tag{3.130}$$

which has the following components:

$$U_{(\rho,\mu)} = U_\rho + U_{\rho\mu} + U_\mu \tag{3.131}$$

We noted in Secton 3.5 that this is the form of utility function employed by Ben–Akiva (1973), and others, in the consideration of joint choice models.

Writing $U_{\rho\mu}$ in terms of its representative component and stochastic residuals,

$$U_{(\rho,\mu)} = \bar{U}_\rho + \bar{U}_{\rho\mu} + \bar{U}_\mu + \varepsilon_\rho + \varepsilon_{\rho\mu} + \varepsilon_\mu \tag{3.132}$$

and assuming the usual properties of independence of the ε vectors, namely

$$\begin{aligned} E(\varepsilon_\rho\varepsilon_\mu) &= 0 \\ E(\varepsilon_\rho\varepsilon_{\rho'\mu'}) &= 0 \\ E(\varepsilon_\mu\varepsilon_{\rho'\mu'}) &= 0 \quad \text{for all } \rho,\rho' \in X,\ \mu,\mu' \in Y \end{aligned} \tag{3.133}$$

it is readily seen that the matrix elements of the variance–covariance matrix are given by

$$\textstyle\sum_{\rho\mu,\rho'\mu'} = \sigma_X^2\delta_{\rho\rho'} + \sigma_{XY}^2\delta_{\rho\rho'}\,\delta_{\mu\mu'} + \sigma_Y^2\delta_{\mu\mu'} \tag{3.134}$$

in which

$$\begin{aligned} E(\varepsilon_\rho\varepsilon_{\rho'}) &= \delta_{\rho\rho'}\sigma_X^2 \\ E(\varepsilon_\mu\varepsilon_{\mu'}) &= \delta_{\mu\mu'}\sigma_Y^2 \\ E(\varepsilon_{\rho\mu}\varepsilon_{\rho'\mu'}) &= \delta_{\rho\rho'}\delta_{\mu\mu'}\sigma_{XY}^2 \end{aligned} \tag{3.135}$$

Just as the variance–covariance matrix (3.91) endows the nested logit model with a nested or tree structure of the form (3.102), the matrix (3.134) embodies a 'cross-correlated' structure, which must reduce to the nested or multinomial logit form when σ_X or σ_Y, or both, vanish. It is clear from equations (3.132) and (3.134) that use of the representative utility

$$\bar{U}_{(\rho,\ \mu)} = \bar{U}_\rho + \bar{U}_{\rho\mu} + \bar{U}_\mu \tag{3.136}$$

in conjunction with the MNL model is, within this theoretical framework, in general inconsistent, because the latter is formulated from a diagonal variance–covariance matrix.

3.6.2 The cross-correlated logit model

As far as we are aware, no exact closed-form expression has been found for the choice model corresponding to the utility function (3.130) with variance–covariance matrix (3.134). One approximate form—the cross-correlated logit model—has, however, been suggested by Williams (1977). Consider again the nested logit model (3.116) which may be written in the following form:

$$P_{\rho\mu} = \frac{\exp\left[\beta\left(\bar{U}^*_\rho + \frac{\lambda}{\beta}\bar{U}_{\rho\mu}\right)\right]}{\sum_{\rho'\mu'} \exp\left[\beta\left(\bar{U}^*_{\rho'} + \frac{\lambda}{\beta}\bar{U}_{\rho'\mu'}\right)\right]} \tag{3.137}$$

with

$$\bar{U}^*_\rho = \bar{u}_\rho + \frac{\beta-\lambda}{\beta}\tilde{u}_{\rho*} \tag{3.138}$$

and $\tilde{u}_{\rho*}$, β given by equations (3.117) and (3.119), respectively. This has the appearance of a multinomial logit model with *transformed* utility variables and dispersion parameters.

We seek a symmetric form for $P_{\rho\mu}$ under interchange of the X and Y dimensions which reduces to the nested and multinomial logit functions in the appropriate limits adopted by the variance–covariance matrix.

Such a model is given by

$$P_{\rho\mu} = \frac{\exp Q_{\rho\mu}}{\sum_{\rho'\mu'} \exp Q_{\rho'\mu'}} \tag{1.139}$$

in which

$$Q_{\rho\mu} = \beta\bar{U}^*_\rho + \xi\bar{U}^*_\mu + \lambda\bar{U}_{\rho\mu} \tag{3.140}$$

$$\bar{U}^*_\rho = \bar{u}_\rho + \frac{\beta-\lambda}{\beta}\tilde{u}_{\rho*} \tag{3.141}$$

$$\bar{U}^*_\mu = \bar{u}_\mu + \frac{\xi-\lambda}{\xi}\tilde{u}_{\mu*} \tag{3.142}$$

and the composite utilities are given by

$$\tilde{u}_{\rho*} = \frac{1}{\lambda}\log\sum_\mu \exp(\lambda\bar{u}_{\rho\mu} + \xi\bar{u}_\mu) \tag{3.143}$$

$$\tilde{u}_{\mu*} = \frac{1}{\lambda}\log\sum_\rho \exp(\lambda\bar{u}_{\rho\mu} + \beta\bar{u}_\rho) \tag{3.144}$$

with

$$\beta = \frac{\pi}{\sqrt{6}}\left(\sigma_X^2 + \frac{\pi^2}{6\lambda^2}\right)^{-1/2} \tag{3.145}$$

$$\xi = \frac{\pi}{\sqrt{6}}\left(\sigma_Y^2 + \frac{\pi^2}{6\lambda^2}\right)^{-1/2} \tag{3.146}$$

A pictorial representation of this model and the nested and multinomial logit special cases are given in Figure 3.4.

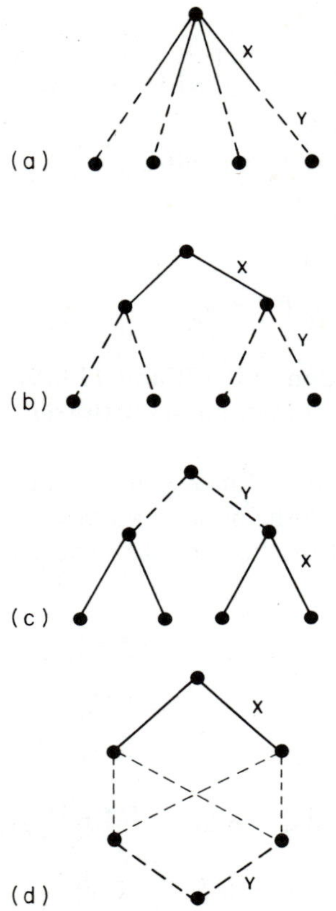

Figure 3.4 *Alternative model structures:*
(*a*) uncorrelated structure ($X - Y$)
(b) nested structure (X/Y)
(c) nested structure (Y/X)
(d) cross-correlated structure ($X * Y$)

The estimating procedure for this model is rather involved (see Williams, 1977) and for this reason does not commend itself. We shall argue below (section 3.6.5) that it may in fact not be necessary to implement such a form in practice.

3.6.3 A general extreme-value class of models

Since the extensions to the logit family were derived, a number of other theoretical innovations have been introduced. One such development is the derivation of a general extreme-value class of models by McFadden (1978). In order to introduce this class, recall that the Weibull distribution has the cumulative distribution given by

$$F(\varepsilon_\rho) = \exp[-\exp(-\varepsilon_\rho)] \tag{3.147}$$

and the joint cumulative distribution of $\varepsilon_1, \ldots, \varepsilon_\rho, \ldots, \varepsilon_N$, from which the multinomial logit model may be derived, is given by

$$F(\varepsilon) = \prod_\rho \exp[-\exp(-\lambda\varepsilon_\rho)] \tag{3.148}$$

$$= \exp[-\sum_\rho \exp(-\lambda\varepsilon_\rho)] \tag{3.149}$$

McFadden (1978) has recently shown that by taking a joint extreme-value distribution for the utility of the alternatives given by

$$F(\varepsilon) = \exp[-G(e^{-\varepsilon_1}, \ldots, e^{-\varepsilon_\rho}, \ldots, e^{-\varepsilon_N})] \tag{3.150}$$

in which $G(Y_1, \ldots, Y_N)$ is a non-negative, linear homogeneous function of $Y_1, \ldots, Y_N$, choice models may be generated which are consistent with stochastic utility maximization. For $F(\varepsilon)$ to define a valid cumulative distribution function, it is sufficient that $G(Y_1, \ldots, Y_N)$ be positive for non-zero $Y_1, \ldots, Y_N$ and that $(-1)^{k-1} \dfrac{\partial^k G}{\partial Y_{i_1}, \ldots, \partial Y_{i_k}}$ be non-negative for $k = 1, \ldots, N$ and any combinations from $(Y_1, \ldots, Y_N)$.

The choice model resulting from this distribution function may be written in the form

$$P_\rho = \frac{V_\rho G_\rho(V_1, \ldots, V_N)}{G(V_1, \ldots, V_N)} \tag{3.151}$$

with

$$G_\rho = \frac{\partial G}{\partial V_\rho} \tag{3.152}$$

and

$$V_\rho = \exp(\hat{u}_\rho) \tag{3.153}$$

As McFadden notes, any parametric function G satisfying the sufficient conditions for F to be a cumulative distribution function yields a practical parametric choice model. This class again includes the nested and multinomial logit models as special cases. As far as we are aware, however, no model appropriate to the utility function (3.130) has been derived and implemented from the general extreme-value system.

3.6.4 The multinomial probit model

An alternative approach to the generation of choice models employs the assumption that the stochastic residuals $(\varepsilon_1, \ldots, \varepsilon_\rho, \ldots, \varepsilon_N)$ are distributed multivariate normal $\mathrm{N}(\mathbf{0}, \Sigma)$ with variance–covariance matrix Σ, such that

$$P_\rho = \int_{R_\rho} \mathrm{d}\boldsymbol{\varepsilon}\, \mathrm{N}(\mathbf{0}, \Sigma) \tag{3.154}$$

R_ρ is the region of integration defined in equation (3.1) over which A_ρ is the preferred option. As $\sum$ may assume an arbitrary form, this approach to the correlation problem is potentially very powerful. In order to provide a practical method of forming choice models, the multiple integration in (3.154) must, however, prove numerically tractable. For a small number of choice alternatives (up to 4 or 5) this is indeed the case by straightforward numerical integration, and has been exploited in different forms in mode choice contexts by Langdon (1976), Hausman and Wise (1978), and others.

Several attempts have been made to extend the applicability of the multinomial probit model through the adoption of various approximations. The most promising to date has been provided by the introduction of the Clark approximation (Clark, 1961) by Daganzo, Bouthelier, and Sheffi (1977). This essentially involves approximating the maximum of two normal variates by a normal distribution. By repeated applications of this approximation, the probit model may be reduced to a univariate integration. At the present time, programs exist for the computation and estimation of choice probabilities for up to 15–20 choice alternatives, although there are still practical and theoretical difficulties associated with the estimation procedure when the Clark approximation is adopted.

3.6.5 Generation of choice models by simulation

The final method for generating choice models which we shall discuss is the direct appeal to numerical integration of equation (3.1) through simulation. The method involves the repeated sampling of utility values for 'test individuals' from prespecified utility distributions and allocating each 'individual' t to an alternative according to the sampled values $\{U_1^t, \ldots, U_\rho^t, \ldots, U_N^t]$. This Monte Carlo approach to the numerical integration of equation (3.1) has been employed by a number of authors, including Manski and Lerman (1978), Robertson and Kennedy (1979), Ortuzar (1978), and Williams and Ortuzar (1979). It is an expensive procedure for implementing models in which it is required to estimate parameters, and emphasizes the value of suitable closed-form expressions. One important use is the testing of particular approximations in model development as exemplified in the tests of the Clark approximation, described above, by Albright, *et al.* (1977), and in the work of Williams and Ortuzar (1979), who have examined the accuracy of the cross-correlated logit model and the conditions under which alternative members of the logit family are theoretically appropriate in a modelling context. They have concluded that the selection of a model from the nested logit and multinomial logit alternatives is not unduly restrictive, which appears to obviate the need to estimate the more complicated cross-correlated logit model.

Williams and Ortuzar (1979) have further employed simulation to implement choice models, which allow the assumption that a complete set of alternatives be scrutisized by each individual in a choice-making population to be relaxed. This involves a choice-set generating process which draws a set of choice alternatives

A^t for each 'test individual' according to a parameterized set size distribution. The operation of varying the parameters of that distribution was found to have a number of implications for cross-sectional response analysis. The interested reader is referred to the above paper for a fuller description of the tests and their results. Here we wish to emphasize the flexibility of the simulation method for the theoretical analysis of choice models while recording its limitations in practical implementation which involve parameter estimation in choice models involving several alternatives.

Before drawing together the implications of the theoretical work outlined in the above sections, we turn to examine the application of alternative members of the logit family of models in context.

3.7 EXAMPLES OF THE FORMATION OF TRAVEL-RELATED DEMAND MODELS AND ECONOMIC EVALUATION MEASURES

3.7.1 Introduction

A number of applications which involve the construction of consistent micro-demand models and evaluation measures will now be discussed. These deal with models for multi-mode, and 'strategic' transport planning contexts. The examples will involve the structuring of demand functions appropriate to additive, separable utility functions for which the theory developed above may be appealed to directly.

The fundamental problem is one of devising a suitable form for the underlying utility function in any particular example, and developing a model structure appropriate to that function. Since the model represented by the fan structure in Figure 3.5(b) is a special case of that appropriate to Figure 3.5(a) in which all correlation effects represented by the intermediate branches in Figure 3.5(a) are zero, it is often desirable to adopt the nested structure and be guided by the results of the calibration process on the extent of the correlation. The appropriate common attributes or factors must be provided in the specification of the model. If the multinomial logit model is adopted there should be good reason to believe that the movements involved will be perceived as distinct with no appreciable commonality effects. In all the examples to follow it is important to recognize the fundamental inconsistency of assuming the existence of factors which give rise to dispersion, while neglecting the influence of these factors in the benefit calculation.

Although the main considerations of the section are with model structure, it is important to comment on the aggregation process. If we write the probability that an individual with characteristics **a** selects an alternative A_ρ as

$$P_\rho^{\mathbf{a}} = F_\rho^{\mathbf{a}}(\{\mathbf{Z}\}:\theta^{\mathbf{a}}) \tag{3.155}$$

then the total demand by individuals in class $\boldsymbol{a}$ selecting alternatives in class ψ

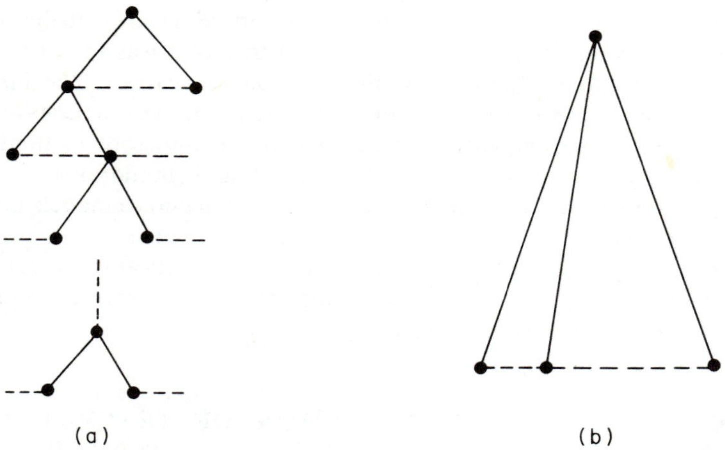

Figure 3.5 Sequential and simultaneous structures
(a) sequential
(b) 'joint choice'—simultaneous

may be written

$$T^{\alpha}_{\psi} = \underset{\mathbf{a} \in \alpha}{\mathrm{S}} \ \underset{\{\mathbf{Z}_{\rho}\} \in \psi}{\mathrm{S}} \ \phi(\mathbf{a}, \{\mathbf{Z}\}) F^{\mathbf{a}}_{\rho}(\{\mathbf{Z}\}:\boldsymbol{\theta}^{\mathbf{a}}) \tag{3.156}$$

in which $\phi(\mathbf{a}, \{\mathbf{Z}\})$ denotes the joint density function of the individual and travel-related attributes, and S denotes summation (addition over discrete variables, integration over continuous variables) over subclasses or submovements belonging to a particular aggregate class $\boldsymbol{\alpha}$ or ψ. If the classes ψ are defined with respect to an irregular zoning system, as for example when T refers to an interzonal trip matrix, $\{T_{ij}\}$ and the class ψ denoting all alternatives in pairs of zones $\{ij\}$, the integration over the attributes (times, costs, etc.) $\mathbf{Z}$, although straightforward in principle, will introduce some practical difficulties and requires the distribution of the attributes over the zoning system to be specified in the function $\phi(.)$.

Some approaches to the aggregation problem, varying from approximation theory to simulation, have been discussed in detail by Koppleman (1976), McFadden and Reid (1975), Bouthelier and Daganzo (1979), among others. In the remainder of this chapter we shall assume that an appropriate market segmentation has been defined, and that a sufficiently fine zoning system is adopted to render small any aggregation bias associated with the replacement of zonal attributes by mean values defined with respect to zonal centroids. We would emphasize, however, that if this practice is uncritically applied to an arbitrary zoning system, not only will aggregation bias be introduced by equating macro- and micro parameters, but also much of the benefit of developing micro models will be sacrificed.

3.7.2 Multi-mode travel-choice models

In any conurbation transport system the car competes with possibly several public transport modes, each of which consists of many services operating over different routes through a network. It is easily forgotten that the private and public modes in binary choice mode-split models are each composites or aggregates of several possible means of transportation. For example, in the Greater London Transportation Study (Havers, *et al.*, 1974) surveyed movements were initially grouped into eleven classes *by stage*: (1) car, van, or motor cycle driver; (2) goods vehicle driver; (3) car, van, or motorcycle passenger; (4) underground rail; (5) British Rail (surface train); (6) taxi or minicab; (7) scheduled bus; (8) coach; (9) walk or pedal cycle; (10) goods passenger; (11) other passenger.

For the distribution and assignment stages of the strategic passenger demand model, two *distribution* modes are defined: (a) travel by private transport including trips by car, van and motorcycle; and (b) travel by public transport, which may, for example, incorporate scheduled bus, underground, and surface rail. This initial two-fold classification recognizes the existence of common attributes, which, in this case, concerns the degree of personalization of the movement and the influence of comfort, convenience, and mode flexibility.

Traditionally the 'submodal split' calculation between public transport modes, for example bus and train, has involved a 'primary' split between the two composites in which the public transport disutility of the available submodes. Such a 'hierarchical' or 'nested' model may be underpinned by an additive separable utility function $u(k)$ defined over modal attributes in which 'public' modes contain common attributes $u_{\tilde{k}}$. Thus

$$u(k) = u_{\tilde{k}} + u_{k(\tilde{k})} \tag{3.157}$$

in which $u_{k(\tilde{k})}$ is that component of the utility function which varies over modes k in the 'public' movement composite $\tilde{k}$. Now it is clear that, for any individual in the choice-making population, the perception of specifically 'public transport' elements of the utility function would not be identical for bus and rail. What has been assumed in the specification of equation (3.157) is that the grouping criterion is satisfied for which the rank correlation associated with these elements is considerably higher for the public/private choice dichotomy than for any pair of public transport modes. In other words, if other factors such as journey times, money costs, etc. were considered equal, a group tendency to prefer car to bus *and* train, or vice versa, would be exhibited. The appropriate utility tree is shown in Figure 3.6(a) and the rejected structure is that of Figure 3.6(b). The theory developed in Section 3.6 may be appealed to directly to generate an appropriate model which may be expressed as follows:

$$M_{ij}^{nk(\tilde{k})} = \frac{\exp[-\lambda^n(\delta^{\tilde{k}} + \tilde{c}_{ij}^{n\tilde{k}})]}{\sum\limits_{\tilde{k}\in H(n)} \exp[-\lambda^n(\delta^{\tilde{k}} + \tilde{c}_{ij}^{n\tilde{k}})]} \cdot \frac{\exp[-\Delta^n c_{ij}^{k(\tilde{k})}]}{\sum\limits_{k\in Y(k)} \exp[-\Delta^n c_{ij}^{k(\tilde{k})}]} \tag{3.158}$$

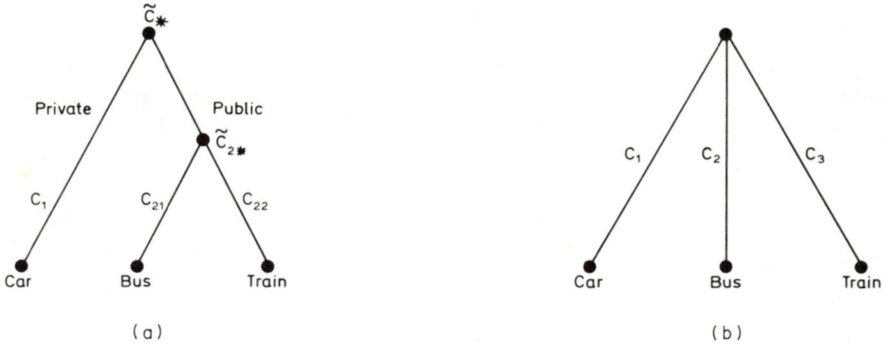

Figure 3.6 Composite costs in (a) sequential and (b) joint-choice structures

in which
$M_{ij}^{nk(\tilde{k})}$ is the share of travellers of type n between locations i and j who select mode k in the composite $\tilde{k}$. The combinations $1(\tilde{1})$, $1(\tilde{2})$, $2(\tilde{2})$ denote private car, bus, and rail transport respectively.
The following generalized and composite costs are defined:

$$c_{ij}^{k} = \sum_{\eta} \alpha_{\eta} Z_{ijk}^{n} \tag{3.159}$$

$$\tilde{c}_{ij}^{n\tilde{k}} = -\frac{1}{\Delta^{n}} \log \sum_{k \in Y(\tilde{k})} \exp[-\Delta^{n} c_{ij}^{k(\tilde{k})}] \tag{3.160}$$

Here $\mathbf{Z}_{ijk}$ denotes a set of variables, such as time, distance, and cost, which vary over location and modes. $\delta^{\tilde{k}}$ is the source of correlation between the submodes in the composite set $\tilde{k} = 2$. $\sum_{k \in Y(\tilde{k})}$ denotes summation over all modes in that composite set, while $\sum_{\tilde{k} \in H(n)}$ denotes summation over all composite modes available to persons of type n.

The quantities $\delta^{\tilde{k}}$ may be regarded as calibration parameters—their maximum-likelihood values ensuring that the predicted and observed modal-share at the composite level are equal over the whole system—or alternatively may be expressed in terms of proxy variables for comfort, convenience, etc. (Stopher, 1975). As $\delta^{\tilde{k}}$ tends to zero the model (3.158) collapses to one appropriate to a fan structure in which all modes are given equal 'weight'. There have now been many examples of the multinomial logit model applied to multi-modal choice contexts (see, for example, the studies of Rassam *et al.*, 1971; Richards, 1975; Lawson and Mullen, 1976; and the review by Speer, 1977). It can now be seen that these are special cases of the general framework presented above.

Other examples of the application of hierarchical logit models to multi-mode, mode-route and mixed-mode choice contexts may be found in Williams (1977) and Daly and Zachary (1978).

3.7.3. Transport planning models

We have indicated that the three most commonly applied model structures implemented in strategic studies have been of the forms G/MS/D/A, G/D/MS/A, and G/D–MS/A, in the notation described in section 3.5.2. From the considerations of Section 3.6 it is clear that such nested structures may be derived from choice contexts involving additive separable utility functions defined over choice combinations of trip frequency (F), location (D), mode (K), and route (R) dimensions. These models, together with their associated consumer surplus benefit forms, can now be consistently formulated within the theory developed above. We shall discuss the specification of the form G/D/MS/A and shall assume that Wardrop's conditions (see Chapter 2) govern the minimum route-cost assignment. We shall continue here to distinguish between positive and negative components of utility through the use of u and c, respectively. Because zero dispersion will be taken to characterize the route-choice stage, the interzonal travel (generalized) costs c_{ij}^k may be identified with the minimum cost on routes between i and j on mode k.

By a straightforward application of the formalism developed in section 3.5.3 we may derive from the additive utility function

$$U(F, D, K, R) = U^F + U^{DF} + U^{KDF} + U^{RKDF} \tag{3.161}$$

the (aggregate) choice model

$$T_{ij}^{nk} = G_i^n M_{ij}^n M_{ij}^{nk} \tag{3.162}$$

with

$$G_i^n = G_i^n (SE_i^n, \tilde{u}_{i*}^n) \tag{3.163}$$

$$M_{ij}^n = \frac{A_j \exp[\beta^n (u_{ij}^n - \tilde{c}_{ij}^n + \gamma_j)]}{\sum_j A_j \exp[\beta^n (u_{ij}^n - \tilde{c}_{ij}^n + \gamma_j)]} \tag{3.164}$$

$$M_{ij}^{nk} = \frac{\exp(-\lambda^n c_{ij}^k)}{\sum_{k \in H(n)} \exp(-\lambda^n c_{ij}^k)} \tag{3.165}$$

Here T_{ij}^{nk} denotes the number of trips between i and j by persons of type n on mode k, while G_i^n is the total number of trips from the spatial unit i by persons of type n. The latter is taken to be a function of the socio-economic attributes SE_i^n and the composite utility $\tilde{u}_{i*}$. The share functions M_{ij}^n and M_{ij}^{nk} correspond to location and mode choice models, respectively, underpinned by stochastic utility maximization with Weibull utility functions characterized by the parameters β and λ.

The appropriate composite cost functions $\tilde{c}_{ij}^n$ and $\tilde{u}_{i*}^n$ are given as follows:

$$\tilde{c}_{ij}^n = -\frac{1}{\lambda^n} \log \sum_{k \varepsilon H(n)} \exp(-\lambda^n c_{ij}^k) \tag{3.166}$$

$$\tilde{u}_{i*}^n = \frac{1}{\beta^n} \log \sum_j A_j \exp[\beta^n (u_{ij}^n - \tilde{c}_{ij}^n + \gamma_j)] \tag{3.167}$$

It is appropriate to comment on several aspects of the model and their significance within the theoretical framework outlined above. We discuss in turn: the dispersion parameters β and λ; the composite functions $\tilde{c}^n_{ij}$ and $\tilde{u}_{i*}$; consumer surplus user benefit measures accompanying system modifications; the location model; and programming formulations of the location model.

3.7.3.1 The dispersion parameters β and λ

In section 3.5.4 we derived conditions which must be satisfied by the parameters of a nested logit model if it is to be consistent with stochastic utility maximization. One condition in the present context was that β should not be greater than λ.

Senior and Williams (1977) have found that for those British Transportation Studies which have employed the nested logit model of the form G/D/MS/A this condition is invariably violated. For the work journey, β is typically 2–3 times the value of the mode choice parameter. Is this a setback for the theory or an indictment of practice? Whatever the criticisms of the above theoretical framework there are two (related) reasons why the violation of this inequality gives cause for concern. Firstly, it may be shown (Williams and Senior, 1978) that the parameter relation $\beta < \lambda$ is also (equivalently) derivable from the requirement that the cross-elasticities of the model be of the correct sign. A cross-elasticity of the wrong sign would imply that the demand for a given alternative may *increase* when the cost of a substitute *decreases*. Secondly, it is easily demonstrated (Williams and Senior, 1977) that the violation does in fact give rise to such anomalous demand-response results in rather common policy-testing contexts.

Among the possible reasons for this 'parameter size anomaly' are the treatment of short trips and poor market segmentation employed in conventional models (Williams and Senior, 1977; Hawkins, 1978). We shall return to this issue in our concluding remarks.

3.7.3.2 The composite functions $\tilde{c}^n_{ij}$ and $\tilde{u}_{i*}$

Another set of conditions which must be satisfied if the nested logit model is to be consistent with rational choice behaviour relates to the appropriate determination of composite utility (or cost) functions which link the partial share models M^n_{ij} and M^{nk}_{ij}. For multinomial logit share models these costs are defined by equation (3.166). Again it has been found (Williams, 1977; Senior and Williams, 1978) that the many forms for $\tilde{c}^n_{ij}$ used to link distribution and modal-split models in British studies have been inappropriate, and some in fact give rise to anomalous response characteristics.

It is clear from the above discussion that if a model of elastic trip generation is developed to correspond to the underlying utility structure (3.161) then an appropriate measure of net utility is defined by equation (3.167). This is seen to be proportional to the log transform of a modified Hansen measure of accessibility (Hansen, 1959) and is of similar structure to the index proposed by Koenig (1975). Indeed there may be a case for using $\tilde{u}_{i*}$ as a definition of the concept of accessibility (Williams, 1977; Williams and Senior, 1978; Ben–Akiva and Lerman, 1979).

3.7.3.3 Consumer surplus user benefit measures

If the composite utility functions are defined correctly an exact expression for the change in total perceived benefit accompanying a transport system change may be obtained from

$$\Delta S = -\sum_{ij}\sum_{n}\sum_{k\in H(n)}\int_0^1 d\hat{\sigma}\frac{dc_{ij}^k(\hat{\sigma})}{d\sigma}T_{ij}^{nk}(\hat{\sigma}) \tag{3.168}$$

After applying various transformations (Williams, 1977) this expression may be written

$$\Delta S = -\sum_{n}\sum_{i}\int_0^1 d\hat{\sigma}G_i^n(\tilde{u}_{i*}(\hat{\sigma}))\frac{d\tilde{u}_{i*}(\hat{\sigma})}{d\hat{\sigma}} + \sum_j A_j(\gamma_j^F - \gamma_j^I) \tag{3.169}$$

which for inelastic trip ends becomes

$$\Delta S = \sum_{n}\sum_{i} G_i^n(\tilde{u}_{i*}^{nF} - \tilde{u}_{i*}^{nI}) + \sum_j A_j(\gamma_j^F - \gamma_j^I) \tag{3.170}$$

This expression is none other than the difference in the composite utilities $\tilde{U}_{***}$ defined for the whole set of choices available when evaluated in the initial and final system states. We shall comment below on the interpretation of the cost-dependent term γ_j.

3.7.3.4 The location model

We have yet to define the terms u_{ij} and γ_j in equation (3.164) or the precise nature of the location choice process. Indeed, the former must be interpreted in the context of the latter. It is well known that the work-trip distribution segment in a land-use transport model may be interpreted either in terms of a residential or job choice process. Thus we can write for the work purpose, unembellished by market segmentation indices,

$$T_{ij} = H_i p_{ij} \qquad \text{(workplace choice model)} \tag{3.171}$$

$$T_{ij} = E_j p_{ji} \qquad \text{(residential choice model)} \tag{3.172}$$

in which T_{ij} is the number of work trips betwen i and j and H_i, E_j are the numbers of residents and workers (or stock units) respectively in the spatial units i and j. p_{ij} and p_{ji} may in turn be interpreted as the probability of a resident in i working in j and a worker in j living in i respectively. Senior (1977b) argues that in transport systems analysis the implicit interpretation of the distribution model is the former, while in land-use models (e.g. in the Lowry model) it is explicitly the latter.

If, then, the model is interpreted in a residential-choice context the location-specific component u_{ij}^n will be written u_i^n and will vary over residential opportunities. u_{ij}^n will in general involve both location and activity variable characteristics and should preferably have a substructure which allows such attributes as the tenure dimension to assert itself, with appropriate consequences for the model structure. Such disaggregate choice models of residential location have been considered by Lerman (1975), Quigley (1976), Gustaffson *et al.* (1977), McFadden (1978), Anas (1979), Los (1979), among others. Corresponding comments are appropriate if a workplace location interpretation is adopted (as discussed by Williams and Senior, 1978).

If supply-side opportunities are restricted, as would be the case in both residential and workplace location contexts, the allocation model will be underpinned by a particular mechanism of exchange, and in either competitive or centrally directed processes suitably defined shadow prices $\boldsymbol{\gamma}$ will be introduced in order to provide an appropriate interaction between consumers and suppliers (see Senior and Wilson, 1974b; Cochrane, 1975; and Champernowne *et al.*, 1976). We shall comment below on the programming derivation of the interaction (choice) model when preference dispersion exists.

It should be noted that the term $\exp(\beta u_{ij})$ in equation (3.164) may in either location context be treated as an attraction weight. The nature of this quantity should be carefully distinguished from the weight A_j which is a measure (or proxy) for the number of choice opportunities in each spatial unit.

3.7.3.5 A programming derivation of the location model

In Section 3.4 we noted that group surplus or composite utility expressions may be formed and used to derive through maximization the corresponding choice model. Here we shall comment briefly on the derivation of the allocation function T_{ij}, interpreted as a choice model, within a programming framework underpinned by an economic rationale. This generalization of the Herbert–Stevens–Harris model (Herbert and Stevens, 1960; Harris, 1962) to include preference dispersion within an economic framework is considered in greater detail by Williams and Senior (1978).

It may readily be affirmed that the residential location model

$$T_{ji}^{wh} = E_j^w p_{ji}^{wh} \tag{3.173}$$

$$= E_j^w \frac{H_i^h \exp[\beta^w(b_i^{wh} - \tilde{c}_{ij}^h - \gamma_i)]}{\sum_i \sum_h H_i^h \exp[\beta^w(b_i^{wh} - \tilde{c}_{ij}^h - \gamma_i)]} \tag{3.174}$$

with

$$\sum_{jw} T_{ji}^{wh} \leqslant H_i^h \tag{3.175}$$

may be generated from the program

$$\max_{\{\mathbf{T}\}} LS = LS_1 + LS_2 \tag{3.176}$$

$$= \sum_{w} \frac{1}{\beta^{w}} \sum_{ijh} T_{ji}^{wh} \log T_{ji}^{wh}$$

$$+ \sum_{jw} \sum_{ih} T_{ij}^{wh} (b_i^{wh} - \tilde{c}_{ij}^{h}) \qquad (3.177)$$

subject to

$$\sum_{ih} T_{ji}^{wh} = E_j^w \qquad (3.178)$$

$$\sum_{jw} T_{ji}^{wh} \leqslant H_i^h \qquad (3.179)$$

$$\mathbf{T} \geqslant 0$$

Here T_{ji}^{wh} denotes the number of workers of type w in zone j residing in a house of type h in zone i. E_j^w is the total number of locators of tye w in zone i, while H_i^h is the number of residential opportunities in zone i. b_i^{wh} is the bid (less composite travel cost which account for modal split in the journey to work) by w-type workers for (ih) residential opportunities.

The mathematical program (3.177)–(3.179) is formally similar to the entropy-maximizing model of Senior and Wilson (1974b) and reduces in the $\boldsymbol{\beta} \to \infty$ limit to the zero-dispersion linear programming model of Herbert–Stevens–Harris. In a competitive housing market the money transfers between locators (consumers) and landlords (producers) must be accounted for in the derivation of a total locational surplus function LS embracing consumer *and* producer contributions. The LS function may be interpreted within a random utility choice context (cf. equation 3.75) in terms of a mean bid-term LS_2 and a dispersion contribution LS_1 which includes a benefit contribution due to the distribution of utility values.

If $\boldsymbol{\alpha}$ and $\boldsymbol{\Upsilon}$ are the shadow prices associated with the constraints (3.178) and (3.179), respectively, it may readily be shown that

$$\alpha_j^w = \tilde{u}_{j*}^w \qquad (3.180)$$

with $\tilde{u}_{j*}^w$ the composite utility or consumer surplus accruing to residential locators working in zone j, while $\boldsymbol{\Upsilon}$ has the usual interpretation associated with the Herbert–Stevens model as the price actually paid for residential services when the market is cleared. The shadow prices $\boldsymbol{\alpha}$ and $\boldsymbol{\Upsilon}$ may be used to measure the spatial benefit accompanying land use and transport plans (Williams and Senior, 1978).

It should be emphasized that the restricted substitutability of choices for housing services defined on the dimension indices i and n may be relaxed through the use of nested logit functions as discussed above (see also Los, 1979).

3.8 CONCLUSION

In this chapter we have discussed the formation and application of probabilistic choice models within the framework provided by random utility theory. The 'nested' (and other) members of the logit family have been derived from stochastic utility maximization to give a (*post hoc*) rationalization of existing 'sequential' or 'recursive' model structures in terms of the optimal selection by individuals of alternatives from a choice set. The general extreme-value class extends this range of available models which do not suffer from the restrictive properties of cross-substitution exhibited by the widely used multinomial logit model.

We have argued that within this theoretical framework the structure of a model is intimately related by the structure of the variance–covariance matrix of the stochastic residuals in the utility functions, and indeed the structure is uniquely determined by the form of the utility function. This provides an alternative explanation for model structure to those prominent in the early 1970s. It is clear that what have been described as 'simultaneous' and 'sequential' structures are both consistent with optimal choice behaviour and are distinguished by the structure of similarity or correlation between alternatives. The development of alternative explanations of model structures within different theoretical frameworks remains an interesting research topic (Brand, 1973; McFadden, 1978; Williams and Ortuzar, 1979).

It has been shown that important restrictions exist on nested logit models if these are to be consistent with rational choice behaviour. The restrictions on the sizes of elasticity parameters, and on the composite utilities have been mentioned. The restrictions on the elasticity parameters derived in section 3.5.4 may be used to discriminate between alternative nested logit structures, e.g. G/D/MS/A and G/MS/D/A as discussed by Williams, 1977; Williams and Senior, 1977; and Ben–Akiva, 1977).

Within the theory of rational choice there is no flexibility in the selection of these composite utility functions (as suggested by practice). Indeed, the penalties for not including the correct form are considerable. Not only may counter-intuitive behaviour be generated in the choice process (Williams, 1977; Williams and Senior, 1977) but also, by forgoing the conditions of optimality in which the composite utilities play an essential role, the aggregation properties which are indispensable in the development of exact benefit measures are no longer attainable. The opportunity for developing consistent demand models and evaluation measures is then immediately sacrificed.

In the following chapter the above theory will be further developed in underpinning (modified) Lowry-type models with individual choice processes. We shall further exploit the fact that such random utility models may be derived from the maximization of a group surplus function as discussed in Sections 3.4 and 3.7.3. As stated in Chapter 2, appropriately specified mathematical programs may be underpinned by an economic theory of rational choice behaviour at the individual level.

4. *Programming approaches to activity location and land-use evaluation*

4.1 INTRODUCTION

Lowry's celebrated activity allocation model for Pittsburgh (Lowry, 1964) has, suitably embellished by Garin (1966), Goldner (1971), Wilson (1971b), and others, formed the basis for the development of operational urban and regional land-use models in many parts of the world. Its popularity has been attributed to the simplicity of its underpinning theory, the easily adaptable model structure, its relatively modest data requirements, and the possibilities for enlarging the framework. At the heart of the model two spatial-interaction models are used to generate a unique configuration of population and service employment, consistent with the assumptions of economic base theory, for an exogenous distribution of basic employment. Excellent reviews of the model, its variants and applications have been provided by Goldner (1971), Batty (1972a), Wilson (1974), and Putman (1974).

In this chapter we explore two main issues: the development of a framework within which the analysis, evaluation, and design of land use–transportation plans may be made; and secondly, the process of embedding within the economic base formalism a causal structure of location choice more realistic than that implicit in the integration of the spatial-interaction models at the heart of the Lowry framework. The work has been motivated firstly by the need to base spatial-interaction phenomena on sound behavioural principles of location choice; and secondly, by the weakness of the present land-use evaluation methodology.

Any proposed land-use alternative has a relative value to society arising partly by virtue of the transport system supporting it, and depends on the locational preferences of the behavioural units, or decision makers, in the system. When different configurations of basic employment and resultant activity distributions are assessed relative to one another, total travel cost (or time) or some exogenous measures of accessibility, have traditionally been used in a total benefit criterion. The work of Neuberger (1971) has shown that the former (i.e. travel cost) is an inappropriate component of a land-use plan evaluation function, and we shall show below that if accessibility measures are constructed exogenously to the model framework, they must be fully compatible with the decision criteria of consumers and suppliers of services who are themselves responsible for the resultant stock configuration. The approach taken in this work concerns the generation of spatial-interaction patterns and measures of comparative advantage which are internally consistent within a behavioural framework. These measures, having been identified, may then be exploited in the design process which is a natural outgrowth of the evaluation problem.

A mathematical programming model—the group surplus model (GSM)—which is based on a macro-behavioural economic approach and generates the spatial-interaction models that reproduce the variability existing at the micro level, through the maximization of a location surplus function for trip groups, is developed in this chapter. Since the GSM is explicitly founded on economic principles, it allows land-use transportation plans to be subject in a natural way to an examination of consumer efficiency.

We shall argue that the programming framework is endowed with several advantages over the Lowry framework and its derivatives in the following aspects: first, the interfacing of policies with the model is a natural part of the process since constraint handling is a trivial matter in this approach; second, it is fully consistent with the Hotelling economic measure of consumers' surplus for trip distribution (cf. Chapter 3); third, it produces an overall economic plan-evaluation indicator as a result of the optimization procedure; fourth, it is readily converted into the plan design mode; and fifth, it is easily extended to include both demand and supply of urban and regional commodities and therefore to approach the general problem of spatial equilibrium.

4.2 THE LOWRY MODEL: ITS ANALYTIC STRUCTURE

4.2.1 Theoretical foundations

The economic base assumption embedded in the Lowry model asserts that the level of urban and regional growth (or decline) is a function of the expansion (or contraction) in its exporting sector. The labour employed in this sector, in turn, gives rise through the well-known multiplier mechanism to the firms and outlets serving household consumption needs, and their location will be closely related to residential areas. In Chapter 7, we show how this can be integrated with a full input–output model representation. At an aggregate level a spatial configuration of employment and population is generated on the basis of the accessibility of workers to employment and of service centres to households. At the level of operation of the model the complex set of cause–effect processes resulting from a multitude of location decisions (taken essentially within a dynamic context) are condensed into the following closed set of simple equations, which express a total stock equilibrium and associated travel pattern at a given time:

$$\hat{T}_{ij} = \frac{E_j W_i^R \exp[-\beta^w c_{ij}^w]}{\sum_i W_i^R \exp[-\beta^w c_{ij}^w]} \tag{4.1}$$

$$P_i = \alpha \sum_j \hat{T}_{ij} \tag{4.2}$$

$$\hat{S}_{ij} = \frac{(\sigma P_i) W_j^s \exp[-\beta^s c_{ij}^s]}{\sum_j W_j^s \exp[-\beta^s c_{ij}^s]} \tag{4.3}$$

$$E_j^{NB} = \sum_i \hat{S}_{ij} \tag{4.4}$$

$$E_j = E_j^B + E_j^{NB} \tag{4.5}$$

in which the variables and parameters are defined as follows:

$\hat{T}_{ij}$—number of workers in i who are employed in zone j

$\hat{S}_{ij}$—number of service jobs in zone j created by service demand in zone i

W_i^R—weight attached to zone i measuring its attractiveness for the residential location

W_j^s—weight attached to zone j measuring its attractiveness for the location of service activities

E_j^B, E_j^{NB}, E_j—basic, non-basic, and total employment in zone j

α—the number of people supporting unit number of jobs (inverse activity rate)

σ—the number of service jobs created by unit population (service activity rate)

c_{ij}^w, c_{ij}^s—travel costs from zone i to j for work and service trips, respectively

β^w, β^s—parameters of the spatial-interaction submodels.

The residence/workplace submodel represented by (4.1) is such that it satisfies the condition

$$\sum_i \hat{T}_{ij} = E_j \tag{4.6}$$

which requires the total number of jobs in j to be equal to the number of jobs assigned from all residential zones to zone j. Likewise, the residence/services submodel defined by equation (4.3) satisfies the condition

$$\sum_j \hat{S}_{ij} = \sigma P_i \tag{4.7}$$

which expresses an equality between the total amount of service employment created by the population of zone i and the amount allocated from i to all service employment zones. The set of equations (4.1)–(4.7) forms a simple version of the Lowry model combining a number of features proposed by Wilson (1971), such as the explicit consideration of the allocation functions and the residential and services attractiveness weights. These equations may be solved by the following iterative procedure: first, the spatial distribution of employment is generated through (4.1), starting from the exogenously defined amount of basic employment; then, the location of population is determined from (4.2) and fed into (4.3) to generate the spatial distribution of service employment; the amount of service employment in each zone is calculated in (4.4) and added to the basic employment in (4.5) to produce the total employment allocation; finally, the total employment is fed back into (4.1) and the process repeated until convergence is obtained. This iterative procedure expresses the intuitive idea that the location of residences is a function of the employment pattern and creates the basic demand for community facilities and services. It is this 'straightforward and easily understood causal structure' that is one of the most outstanding and 'appealing characteristics' of the Lowry model, as Goldner (1971) has pointed out.

The employment variables $\hat{T}_{ij}$ and $\hat{S}_{ij}$ may be transformed into trip variables T_{ij} and S_{ij} through the simple relations

$$T_{ij} = \eta \hat{T}_{ij} \tag{4.8}$$

$$S_{ij} = \rho \hat{S}_{ij} \tag{4.9}$$

where the parameters η and ρ denote the number of work and service trips generated by unit of total and service employment, respectively. Where appropriate we shall switch between the variable sets $\{\hat{T}_{ij}, \hat{S}_{ij}\} \to \{T_{ij}, S_{ij}\}$ through the transformation defined above.

While the Lowry model, which predicts a stock equilibrium, is characteristically a cross-sectional conditional forecasting device, the iterative solution of equations (4.1)–(4.7) is often taken to represent a *causal* sequence which may be exploited in dynamic extensions (see, for example, Batty, 1972b). The *behavioural* interpretation of this sequence which involves the decision processes of workers and retailers is traditionally one in which a worker selects a residence on the basis of transport cost from his employment location and the residential attractiveness of the various zones. It is then assumed that the service centres locate to maximize access to purchasing power. The important point to be noted is that the equation, *if subject to this behavioural interpretation*, embodies a conditional decision structure at the micro level, in which the choice of shopping location is dependent on the location of residence, but the choice of residence itself is independent of the resultant spatial distribution of retail services. We present below an alternative model which avoids this restriction.

Finally, we note that eliminating the variables P_i and E_j^{NB}, equations (4.2) and (4.4)–(4.7) are equivalent to

$$\sum_j T_{ij} - \lambda_1 \sum_j S_{ij} = 0 \tag{4.10}$$

$$\sum_i T_{ij} - \lambda_2 \sum_i S_{ij} = \eta E_j^B \tag{4.11}$$

where λ_1 and λ_2 are parameters defined as $\lambda_1 = \eta/\rho\alpha\sigma$ and $\lambda_2 = \eta/\rho$. These equations represent the relation between work and service trips implicit in the equilibrium solution of the Lowry model and the economic base assumption, respectively, and form an important basis for mathematical programming representations of the model presented below.

4.2.2 The incorporation of planning constraints

The model as specified by equations (4.1)–(4.7) would predict the spatial repercussions of unimpeded urban growth (or decline) triggered by an initial distribution or changes in the distribution of basic employment. As a planning device such a model is of little use in assessing the implications of policy alternatives. Constraints which represent an interfacing device between the allocation mechanism and planning controls may, however, be included, as for example in the simple descriptor

$$P_i \leqslant \bar{P}_i \tag{4.12}$$

where P_i is the predicted population in zone i and $\bar{P}_i$ is an imposed upper bound on this quantity. The basic Lowry equations (4.1)–(4.7) must now be solved subject to this set of constraints.

Many heuristic procedures have been suggested for incorporating the effect of density constraints, and Batty *et al.* (1974) have noted the fundamental dilemma: 'inconsistent constraints procedures in such models are cheap and fast to operate in contrast to consistent procedures which tend to be expensive and slow'. Wilson (1971) has proposed a method for the consistent handling of inequality constraints of the type shown in equation (4.12) which is now widely applied in a British context. This involves the identification of two sets of zones $Z^{(1)}$ and $Z^{(2)}$ such that $Z^{(1)}$ contains all those zones in which the constraint is binding and $Z^{(2)}$ those zones in which a capacity surplus exists. The spatial-interaction variable is now taken to satisfy the following equations:

$$T_{ij} = A_i B_j E_j \bar{P}_i \exp[-\beta^W c_{ij}^W] \qquad i \in Z^{(1)} \tag{4.13}$$

$$T_{ij} = B_j E_j W_i^R \exp[-\beta^W c_{ij}^W] \qquad i \in Z^{(2)} \tag{4.14}$$

where

$$A_i = \left(\sum_j B_j E_j \exp[-\beta^W c_{ij}^W] \right) - 1 \tag{4.15}$$

$$B_j = \left(\sum_{i \in Z^{(1)}} A_i \bar{P}_i \exp[-\beta^W c_{ij}^W] + \sum_{i \in Z^{(2)}} W_i^R \exp[-\beta^W c_{ij}^w] \right)^{-1} \tag{4.16}$$

Equation (4.13) expresses the doubly constrained nature of the spatial-interaction model for trips to zones subject to the capacity limit, while the balancing factor B_j defined by (4.16) ensures that the employment total is consistent with the relation (4.6). This scheme involves an iterative solution of the Lowry equations in which zones are allocated between the sets $Z^{(1)}$ and $Z^{(2)}$ until a convergent solution is achieved. The details are provided by Wilson (1971) and the implementation of the process is described by Batty *et al.* (1974), who have reported accelerated methods for its operation.

The framework developed in Section 4.4 provides an alternative approach to handling planning constraints in activity allocation models through mathematical programming, and will be contrasted with the approach described above.

4.2.3 The spatial characteristics of the interaction variables

It has apparently not been recognized that the analytic structure of the Lowry model can be thought of as being characterized by two interlocking doubly constrained gravity submodels rather than the more evident singly constrained models. In order to show this, let us define balancing factors B_j^W and A_i^S, as follows:

$$B_j^W = \left[\sum_i W_i^R \exp[-\beta^W c_{ij}^W] \right]^{-1} \tag{4.17}$$

and

$$A_i^S = \sigma \left[\sum_j W_j^S \exp[-\beta^S c_{ij}^S] \right]^{-1} \tag{4.18}$$

Then, the spatial-interaction submodels (4.1) and (4.3) may be written

$$\hat{T}_{ij} = B_j^W E_j W_i^R \exp[-\beta^W c_{ij}^W] \tag{4.19}$$

and

$$\hat{S}_{ij} = A_i^S P_i W_j^S \exp[-\beta^S c_{ij}^S] \tag{4.20}$$

where the balancing factors B_j^W and A_i^S ensure that constraints (4.6) and (4.7) are satisfied, respectively. By inserting (4.19) an (4.20) into the equations (4.2) and (4.4) which define the allocated stock variables, it can be seen that

$$P_i = \alpha W_i^R \sum_j B_j^W E_j \exp[-\beta^W c_{ij}^W] \tag{4.21}$$

and

$$E_j^{NB} = W_j^S \sum_i A_i^S P_i \exp[-\beta^S c_{ij}^S] \tag{4.22}$$

Hence, defining balancing factors

$$A_i^W = \frac{1}{\alpha} \left[\sum_j B_j^W E_j \exp[-\beta^W c_{ij}^W] \right]^{-1} \tag{4.23}$$

and

$$B_j^S = \left[\sum_i A_i^S P_i \exp[-\beta^S c_{ij}^S] \right]^{-1} \tag{4.24}$$

we have

$$W_i^R = P_i A_i^W \tag{4.25}$$

$$W_j^S = E_j^{NB} B_j^S \tag{4.26}$$

where P_i and E_j^{NB} are of course the stock allocation values determined internally. This enables us to write (4.1) and (4.3) as two doubly constrained spatial interaction submodels,

$$\hat{T}_{ij} = A_i^W B_j^W P_i E_j \exp[-\beta^W c_{ij}^W] \tag{4.27}$$

and

$$\hat{S}_{ij} = A_i^S B_j^S P_i E_j^{NB} \exp[-\beta^S c_{ij}^S] \tag{4.28}$$

interlocked by the stock variables and the economic base equation. Converting the variable set $\{\hat{T}_{ij}, \hat{S}_{ij}\}$ to $\{T_{ij}, S_{ij}\}$ through the transformations (4.8) and (4.9), we have

$$T_{ij} = A_i^W B_j^W P_i E_j \exp[-\beta^W c_{ij}^W] \tag{4.29}$$

and

$$S_{ij} = A_i^S B_j^S P_i E_j^{NB} \exp[-\beta^S c_{ij}^S] \tag{4.30}$$

where the balancing factors are redefined as:

$$A_i^W = \frac{\eta}{\alpha}\left[\sum_j B_j^W E_j \exp[-\beta^W c_{ij}^W]\right]^{-1} \tag{4.31}$$

$$B_j^W = \eta\left[\sum_i A_i^W P_i \exp[-\beta^W c_{ij}^W]\right]^{-1} \tag{4.32}$$

$$A_i^S = \rho\sigma\left[\sum_j B_j^S E_j^{NB} \exp[-\beta^S c_{ij}^S]\right]^{-1} \tag{4.33}$$

$$B_j^S = \rho\left[\sum_i A_i^S P_i \exp[-\beta^S c_{ij}^S]\right]^{-1} \tag{4.34}$$

These equations will be used later for comparative purposes.

4.2.4 The interdependence of the stock variables

It is readily seen that the stock variables internally determined in the model are related as follows:

$$P_i = \alpha W_i^R \sum_j E_j \exp[-\beta^W c_{ij}^W] / \sum_i W_i^R \exp[-\beta^W c_{ij}^W] \tag{4.35}$$

$$E_j^{NB} = \sigma W_j^S \sum_i P_i \exp[-\beta^S c_{ij}^S] / \sum_j W_j^S \exp[-\beta^S c_{ij}^S] \tag{4.36}$$

$$E_j = E_j^B + E_j^{NB} \tag{4.37}$$

These equations display the interdependence between those variables and will also be used later when the structures of different models are compared.

4.2.5 Problems and limitations

The Lowry model (LM) has a number of very attractive features which include the simplicity of the iterative structure and the provision of the opportunity for many extensions. These have contributed to its wide application in the last decade. Nevertheless, there are several problems and limitations associated with the approach, the discussion of which helps to define the limits of its usefulness as a tool in urban and regional planning and suggest paths for potential developments.

Perhaps one of the most important limitations is the fact that the model is essentially a demand model and thereby it only attempts to simulate a 'half-portion' of the land-use system. Supply is implicitly assumed to meet the demand. The mechanisms involved in the supply/demand interaction are therefore excluded from the model framework. The assumption that supply follows the demand is only valid, however, in very limited circumstances. In reality, as Lee (1973, p. 100) stated: 'The location of activities is likely to be very sensitive to the supply side of the market, and the provision of stocks of accommodation is related to many factors in addition to demand.' Undoubtedly the demand pattern is shaped by continuous adjustments to supply, which heavily determines the final equilibrium. Thus, an integration of the supply/demand interaction mechanism is needed.

Since the residential and service allocation functions in the LM are gravity-type submodels, the limitations associated with these models apply also to the model's framework. In particular, we recall the highly aggregated nature of variables and relationships and the dependence of the parameters on inexplicit and therefore unmeasured factors. These also represent additional issues which call for investigation.

Another important limitation is related to the economic base mechanism itself. The economic base theory postulates that growth originates from the exogenously dependent basic employment sector. No feedback from the system to the basic sector is considered. It is widely accepted, however, that growth in an urban and regional system is also attracted from within the system itself by housing development, increases in the level of services, and local public investments that in addition are more easily under the control of planners. The LM is obviously of little use in assessing the impact of those strategies. Furthermore, the location of basic employment according to specified social welfare goals cannot also be tackled within the model framework properly.

A common criticism of the model is that it is static and generates an equilibrium allocation of activities at only one point in time. In Lowry's own words, it produces an 'instant metropolis'. The process of change is ignored in the original framework. However, this may not be considered an intrinsic limitation, as is shown by various dynamic versions that have meanwhile been formulated (see, for example Crecine, 1968; Batty, 1972b; Mackett, 1977; and Wilson, 1974).

An essential problem involves the conditional decision process implicit in the causal structure of the model which at a micro level embodies the behavioural assumption that the choice of residence is completely independent of the location of service facilities as discussed above.

In the ramainder of this chapter, and some subsequent ones, we will attempt to overcome some of these shortcomings. In particular, in the section that now follows a model framework based on behavioural concepts founded on random utility theory, and which improves the simulation of the decision process of residential location, is developed.

4.3 AN ACTIVITY LOCATION MODEL BASED ON PROBABILISTIC CHOICE THEORY

4.3.1 The probabilistic choice model (PCM)

The Lowry model (LM) is based on the concepts of spatial interaction and accessibility. Critics of the model argue that it is conceptually set within a non-behavioural (descriptive) framework. The first step in developing a framework within which demand for travel is internally consistent with land-use evaluation measures which are logically compatible with the choices which are made by the essential components of the systems, is to base the spatial-interaction structure on an improved theory of choice.

As was emphasized earlier, the implicit behaviour assumption underpinning the LM is that the decision about where to live is independent of the location of the service facilities. This particular modelling detail is upgraded in the PCM, to be developed below, by viewing the choice of residence location as the result of a decision process in which the net benefit derived from the shopping choices is transferred through an aggregate composite utility index to the mechanism of residential location choice. Thus, a joint process of residential and shopping (service) choice is embodied in the PCM; the distribution of shopping, and more generally service facilities surrounding the prospective residential location are taken into account in the process of residential choice (see Figure 4.1). It will be noted that the Lowry model is a particular case of the probabilistic choice model in which the residential and shopping location choices are made independently.

At the root of the PCM approach is the assumption of an additive separable random net utility function of the form:

$$U(i, j, k) = U^R(i, j) + U^S(k \mid i, j) \tag{4.38}$$

in which $U(i, j, k)$ is the net utility associated with living in zone i, working in zone j, and shopping at zone k, $U^R(i, j)$ is the net utility of living in zone i and working in zone j, and $U^S(k \mid i, j)$ is the utility associated with shopping at location k, conditional on residing in i and working in j. These 'net utilities', 'surplus' or 'bid values' are assumed to be random variables over the population of decision makers and the formalism of random utility described in Chapter 3 is applied for the representation of the process of choice.

The conditional probability that shopping location k is chosen by a person living in i and working in j is, of course,

$$\text{prob}\{U(i, j, k) > U(i, j, k'), \text{ for all } k'\} \tag{4.39}$$

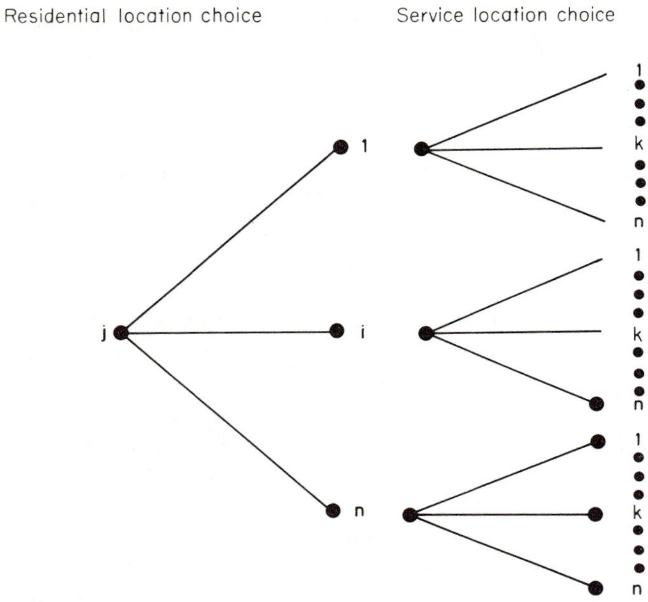

Figure 4.1 Residential location and service location: sequential choice structure

that is,

$$\text{prob}\{U^S(k|i,j) > U^S(k'|i,j), \text{ for all } k'\} \tag{4.40}$$

which means that the common net utility component $U^R(i,j)$ is irrelevant to the conditional choice probability of shopping location. For the purpose of comparing results with the LM version formulated in section 4.2.1, it will be assumed that $U^s(k|i, j)$ is independent of j, but this assumption can easily be relaxed. We shall write

$$U^S(k|i) = u_k^S - c_{ik}^S + \varepsilon_{ik}^S \tag{4.41}$$

where u_k^S is the 'fixed' trip-end utility associated with shopping in location k, c_{ik}^S is the travel cost from zone i to service activities in k, and ε_{ik}^S is a 'residual' random utility component. If ε_{ik}^S is Weibull distributed, then by invoking the formalism of random utility theory, the multinomial logit model

$$P(k|i) = \frac{\exp[\beta^S(u_k^S - c_{ik}^S)]}{\sum_k \exp[\beta^S(u_k^S - c_{ik}^S)]} \tag{4.42}$$

is generated, where $P(k|i)$ is the probability of shopping in k conditional on living in i, and β^S is a parameter related to the standard deviation of the residual component ε^S.

Let $\bar{u}_i^S$ denote the mean surplus (or, composite utility) associated with shopping from residence location i. We have

$$\bar{u}_i^S = \frac{1}{\beta^S} \ln \sum_k \exp[\beta^S(u_k^S - c_{ik}^S)] \tag{4.43}$$

which is a measure of the attractiveness of zone i for shopping (service) purposes. We now seek an aggregate index derived from the optimal shopping pattern which transfers to the residential choice model the comparative advantage of the zone for shopping. Since the ratio of work to service trips is λ_1, such an aggregate index will be in this case precisely

$$\tilde{u}_i^S = \frac{1}{\lambda_1} \bar{u}_i^S$$

(for a detailed discussion on the definition of composite utilities see Williams, 1977). Now by writing

$$U^R(i,j) = (u_i^R - c_{ij}^W) + \varepsilon_{ij}^W \tag{4.44}$$

where u_i^R is the net trip-end utility offered by zone i for residential location, c_{ij}^W is the travel cost from zone i to workplace in j, and ε_{ij}^W is a residual random utility component assumed to be Weibull distributed, the residential multinomial logit model is:

$$P(i,j) = \frac{\exp[\beta^W(u_i^R + \tilde{u}_i^S - c_{ij}^W)]}{\sum_i \exp[\beta^W(u_i^R + \tilde{u}_i^S - c_{ij}^W)]} \tag{4.45}$$

where $P(i,j)$ is the probability of residential location i being chosen by those having workplace in j, and β^W is, as before, a parameter related to the standard deviation of the sum of the random utility component ε_{ij}^W and ε_{i*}^S. The mean surplus attached to employment at location j is similarly given by

$$\bar{u}_j^R = \frac{1}{\beta^W} \ln \sum_i \exp[\beta^W(u_i^R + \tilde{u}_i^S - c_{ij}^W)] \tag{4.46}$$

which is a measure of comparative advantage of j resulting from the optimal joint residential and shopping location choice process.

On the assumption of rational choice behaviour and Weibull-distributed residual random utility component, the probabilistic choice model for joint residential and shopping location is as follows:

$$P(i,j,k) = P(i,j) \, . \, P(k \mid i,j) \tag{4.47}$$

$$= P(i,j) \, . \, P(k \mid i) \tag{4.48}$$

$$= \frac{\exp[\beta^W(u_i^R + \tilde{u}_i^S - c_{ij}^W)]}{\sum_i \exp[\beta^W(u_i^R + \tilde{u}_i^S - c_{ij}^W)]} \cdot \frac{\exp[\beta^S(u_k^S - c_{ij}^S)]}{\sum_k \exp[\beta^S(u_k^S - c_{ij}^S)]} \tag{4.49}$$

where $P(i,j,k)$ is the probability of residing in i, working in j, and shopping at k.

Defining the attractiveness weight W_i^R and W_k^S as follows:

$$W_i^R = \exp[\beta^W u_i^R] \tag{4.50}$$

and

$$W_k^S = \exp[\beta^S u_k^S] \tag{4.51}$$

the joint probability of residential and shopping location (4.49) becomes

$$P(i,j,k) = \frac{W_i^R \exp[\beta^W \bar{u}_i^S]\exp[-\beta^W c_{ij}^W]}{\sum_i W_i^R \exp[\beta^W u_i^S]\exp[-\beta^W c_{ij}^W]} \cdot \frac{W_k^S \exp[-\beta^S c_{ik}^S]}{\sum_k W_k^S \exp[-\beta^S c_{ik}^S]} \tag{4.52}$$

that is,

$$P(i,j,k) = \frac{\hat{W}_i^R \exp[-\beta^W c_{ij}^W]}{\sum_i \hat{W}_i^R \exp[-\beta^W c_{ij}^W]} \cdot \frac{W_k^S \exp[-\beta^S c_{ik}^S]}{\sum_k W_k^S \exp[-\beta^S c_{ik}^S]} \tag{4.53}$$

in which the attractiveness weight for residential location

$$\hat{W}_i^R = W_i^R \exp[\beta^W \bar{u}_i^S] = W_i^R \exp\left[\frac{\beta^W}{\lambda_1}\bar{u}_i^S\right] \tag{4.54}$$

includes the aggregate comparative advantage of the zone for shopping.

4.3.2 Its relationship with the Lowry model

Associated with the choice probabilities (4.42) and (4.45), we have in the notation of the Lowry model expressed in trip variables

$$T_{ij} = \eta E_j p_{ij} = \frac{\eta E_j \hat{W}_i^R \exp[-\beta^W c_{ij}^W]}{\sum_i \hat{W}_i{}^R \exp[-\beta^W c_{ij}^W]} \tag{4.55}$$

$$= B_j'^W E_j \hat{W}_i^R \exp[-\beta^W c_{ij}^W] \tag{4.56}$$

and

$$S_{ij} = \rho\sigma P_i p_{j|i} = \frac{\rho\sigma P_i W_j^S \exp[-\beta^S c_{ij}^S]}{\sum_j W_j^S \exp[-\beta^S c_{ij}^S]} \tag{4.57}$$

$$= A_i'^S P_i W_j^S \exp[-\beta^S c_{ij}^S] \tag{4.58}$$

where the balancing factors $B_j'^W$ and $A_i'^S$ are defined as follows:

$$B_j'^W = \eta\left[\sum_i \hat{W}_i^R \exp[-\beta^W c_{ij}^W]\right]^{-1} \tag{4.59}$$

$$A_i'^S = \rho\sigma \left[\sum_j W_j^S \exp[-\beta^S c_{ij}^S] \right]^{-1} \tag{4.60}$$

Therefore, embedding the residential and shopping location choice probabilities within the economic base framework, an aggregate model of similar structure to the Lowry model results, which is formed by equations (4.55), (4.51), and (4.3)–(4.9), and shall be called here the probabilistic choice model (PCM). By identifying the attractiveness weights W_i^R and W_k^S that figure in the Lowry structure with (4.50) and (4.51) respectively, the relation between the LM and the structurally equivalent PCM above is expressed through the utility transformation (4.54). The PCM is, of course, solved iteratively in the same way as the LM simply by replacing the attractiveness weights W_i^R by $\hat{W}_i^R$. The optimal shopping pattern generated from rational choice behaviour—which is a manifestation of the characteristics of the shopping (service) facilities and the transportation system—is transferred to the residential choice submodel through (4.54), and the relative advantage (or, disadvantage) of shopping from location i, represented by the mean surplus $\bar{u}_i^S$, is transformed into an additional attractiveness factor. Clearly, if the residential decisions are assumed to be taken independently of the net benefit resulting from the accessibility/attractiveness of the shopping outlets, then the probabilistic choice model reduces to the Lowry model.

4.3.3 A land-use evaluation index—the group surplus measure

The total group surplus associated with the joint decision process of residential and shopping location is given by

$$\begin{aligned} GS &= \eta \sum_j E_j \bar{u}_j^R \\ &= \frac{\eta}{\beta^W} \sum_j E_j \ln \sum_i \hat{W}_i^R \exp[-\beta^W c_{ij}^W] \end{aligned} \tag{4.61}$$

where the Euler constant has been dropped out of (4.61) as usual. Accompanying any change in the utility components from $\{u_i^{R(1)}, u_k^{S(1)}, c_{ij}^{W(1)}, c_{ik}^{S(1)}\}$ to $\{u_i^{R(2)}, u_k^{S(2)}, c_{ij}^{W(2)}, c_{ik}^{S(2)}\}$ corresponding to the two situations 1 and 2, is a change in total surplus determined as follows:

$$\Delta GS = \sum_j E_j (\bar{u}_j^{R(2)} - \bar{u}_j^{R(1)}) \tag{4.62}$$

$$= \frac{\eta}{\beta^W} \sum_j E_j \ln \frac{\sum_i \hat{W}_i^{R(2)} \exp[-\beta^W c_{ij}^{W(2)}]}{\sum_i \hat{W}_i^{R(1)} \exp[-\beta^W c_{ij}^{W(1)}]} \tag{4.63}$$

$$= -\frac{\eta}{\beta^W} \sum_j E_j \ln \frac{B_j'^{W(2)}}{B_j'^{W(1)}} \tag{4.64}$$

This represents the change in locational surplus accruing to the population as a result of the alteration of the probabilities of both residential and shopping location choices caused by a change in the valuation of alternatives. The optimal decision process underlying the probabilistic choice model reveals itself in the identification of a single measure $\bar{u}_j^R$ combining the net benefits resulting from residential and shopping location choices. The change in total surplus ΔGS may also be expressed in terms of the variables of the mathematical programs associated with the spatial-interaction models (4.55) and (4.57) as discussed in Section 3.4. It is clear that the model (4.55) can be generated by the program

$$\max_{\{T_{ij}\}} Z = -\frac{1}{\beta^W}\sum_{ij} T_{ij}\left(\ln\frac{T_{ij}}{\hat{W}_i^R} - 1\right) - \sum_{ij} T_{ij}c_{ij}^W \tag{4.65}$$

$$= -\frac{1}{\beta^W}\sum_{ij} T_{ij}\left(\ln\frac{T_{ij}}{W_i^R} - 1\right) + \sum_{ij} T_{ij}(\tilde{u}_j^S - c_{ij}^W) \tag{4.66}$$

subject to

$$\sum_i T_{ij} = \eta E_j \tag{4.67}$$

the dual of which is given by

$$\min_{\{\gamma_j'\}} U = \sum_{ij} W_i^R \exp\left[\beta^W(-\gamma_j' + \tilde{u}^S{}_i - c_{ij}^W)\right] + \eta\sum_j \gamma_j' E_j \tag{4.68}$$

where γ_j' is the dual variable associated with constraint (4.67). Recalling that the first summation of (4.68) is, at optimality, a constant equal to the total number of trips, and that from the optimality conditions of the above mathematical program

$$\gamma_j' = -\frac{1}{\beta^W}\ln(B_j'^W E_j)$$

we obtain

$$\Delta GS = \eta\sum_j E_j(\gamma_j'^{(2)} - \gamma_j'^{(1)}) \tag{4.69}$$

$$= \left[-\frac{1}{\beta^W}\sum_{ij} T_{ij}\left(\ln\frac{T_{ij}}{W_i^R} - 1\right) - \sum_{ij} T_{ij}c_{ij}^W\right]_{(1)}^{(2)} + \left[\sum_{ij} T_{ij}\tilde{u}_i^S\right]_{(1)}^{(2)} \tag{4.70}$$

On the other hand, we have, from (4.2), (4.7), (4.42), and (4.43):

$$\left[\sum_{ij} T_{ij}\tilde{u}_i^S\right]_{(1)}^{(2)} = \rho\sigma\left[\sum_i P_i\tilde{u}_i^S\right]_{(1)}^{(2)} \tag{4.71}$$

$$= \left[-\frac{1}{\beta^S}\sum_{ij} S_{ij}\left(\ln\frac{S_{ij}}{W_j^S} - 1\right) - \sum_{ij} S_{ij}c_{ij}^S\right]_{(1)}^{(2)} \tag{4.72}$$

in which the terms in the latter expression are evaluated for situations 1 and 2 and correspond to the terms in the following mathematical program:

$$\max_{\{S_{ij}\}} Z = -\frac{1}{\beta^S}\sum_{ij} S_{ij}\left(\ln\frac{S_{ij}}{W_j^S} - 1\right) - \sum_{ij} S_{ij}c_{ij}^S \tag{4.73}$$

subject to

$$\sum_j S_{ij} = \rho\sigma P_i \tag{4.74}$$

Thus, we have

$$\begin{aligned}\Delta GS &= \left[-\frac{1}{\beta^W}\sum_{ij} T_{ij}\left(\ln\frac{T_{ij}}{W_i^R} - 1\right) - \sum_{ij} T_{ij}c_{ij}^W\right]_{(1)}^{(2)} \\ &\quad + \left[-\frac{1}{\beta^S}\sum_{ij} S_{ij}\left(\ln\frac{S_{ij}}{W_j^S} - 1\right) - \sum_{ij} S_{ij}c_{ij}^S\right]_{(1)}^{(2)}\end{aligned} \tag{4.75}$$

where the bracket terms are associated with programs (4.65)–(4.67) and (4.73)–(4.74), respectively. A group surplus function may therefore be defined as follows:

$$\begin{aligned}GS = &-\frac{1}{\beta^W}\sum_{ij} T_{ij}\left(\ln\frac{T_{ij}}{W_i^R} - 1\right) - \sum_{ij} T_{ij}c_{ij}^W \\ &-\frac{1}{\beta^S}\sum_{ij} S_{ij}\left(\ln\frac{S_{ij}}{W_j^S} - 1\right) - \sum_{ij} S_{ij}c_{ij}^S\end{aligned} \tag{4.76}$$

Finally, we note that if other constraints exist, such as market clearing conditions, planning restraints, etc., then shadow prices have to be considered to accommodate the constraints. These shadow prices will emerge in the surplus equations (4.62)–(4.64) as additional cost components, whereas the surplus function (4.70) will be structurally invariant since the shadow prices are internally generated through the constraint set of the mathematical programs.

We wish to emphasize that in addition to the shopping trips, a number of other trip choices may readily be incorporated into the framework by including additional components in the separable random utility function governing the joint decision process. Thus, recreational choices, educational choices, etc., may also be represented through probabilistic choice submodels derived in the formalism of the random utility theory, in order to approximate more closely the intricate complexity of the 'bundle' of land-use location choices displayed in real-world systems.

4.4 THE GROUP SURPLUS MAXIMIZATION MODEL FOR ACTIVITY ALLOCATION

4.4.1 The construction of, and rationale for, the GSM

The group surplus maximization model—or in an abbreviated manner the group surplus model (GSM)—is a mathematical programming model which attempts to incorporate the multiple-choice features of the probabilistic choice framework. The key to its development is given by the measure of total surplus derived in conjunction with spatial-interaction models and submodels and embodies the Marshallian/Hotelling definition of consumers' surplus or, equivalently, the random utility concept of expected total surplus. At the macro level we can identify with the group of trips $\{T_{ij}\}$ a surplus function:

$$GS' = -\frac{1}{\beta^W}\sum_{ij} T_{ij}\left(\ln\frac{T_{ij}}{W_i^R} - 1\right) - \sum_{ij} T_{ij}c_{ij}^W \tag{4.77}$$

and with the group of trips $\{S_{ij}\}$ a function

$$GS'' = -\frac{1}{\beta^S}\sum_{ij} S_{ij}\left(\ln\frac{S_{ij}}{W_j^S} - 1\right) - \sum_{ij} S_{ij}c_{ij}^S \tag{4.78}$$

where the notation is the one introduced before. As we noted in Chapter 3, these are surplus functions in the sense that their maximization reproduces the variability in trip behaviour observed at the micro level. *It does not presuppose cooperative behaviour.* The group surplus model is now defined in terms of the maximization of

$$GS = GS' + GS'' \tag{4.79}$$

$$= -\frac{1}{\beta^W}\sum_{ij} T_{ij}\left(\ln\frac{T_{ij}}{W_i^R} - 1\right) - \sum_{ij} T_{ij}c_{ij}^W$$

$$-\frac{1}{\beta^S}\sum_{ij} S_{ij}\left(\ln\frac{S_{ij}}{W_{ij}^S} - 1\right) - \sum_{ij} S_{ij}c_{ij}^S \tag{4.80}$$

which is the counterpart of (4.75) for the probabilistic choice model, subject to the economic base equation

$$E_j = E_j^B + E_j^{NB} \tag{4.81}$$

and the LM trip-end relations

$$\sum_j T_{ij} = \frac{\eta}{\alpha} P_i \tag{4.82}$$

$$\sum_i T_{ij} = \eta E_j \tag{4.83}$$

$$\sum_j S_{ij} = \sigma\rho P_i \tag{4.84}$$

$$\sum_i S_{ij} = \rho E_j^{NB} \tag{4.85}$$

By eliminating the stock variables P_i, E_j^{NB}, and E_j from the trip-end relations and the economic base condition, the following mathematical program is obtained:

$$\max_{\{\mathbf{T},\mathbf{S}\}} GS = -\frac{1}{\beta^W}\sum_{ij} T_{ij}\left(\ln\frac{T_{ij}}{W_j^R} - 1\right) - \sum_{ij} T_{ij} c_{ij}^W$$
$$-\frac{1}{\beta^S}\sum_{ij} S_{ij}\left(\ln\frac{S_{ij}}{W_j^S} - 1\right) - \sum_{ij} S_{ij} c_{ij}^S \qquad (4.86)$$

subject to

$$\sum_j T_{ij} - \lambda_1 \sum_j S_{ij} = 0 \qquad (4.87)$$

and

$$\sum_i T_{ij} - \lambda_2 \sum_i S_{ij} = \eta\, E_j^B \qquad (4.88)$$

where the parameters λ_1 and λ_2 have the definition given in section 4.2.1. The constraint (4.87) is a consistency or accounting equation requiring that the population estimated from the trip matrix $\{T_{ij}\}$ is consistent with that estimated from $\{S_{ij}\}$, while (4.88) is equivalent to the economic base condition.

The GSM and the PCM have a common behavioural background provided by the rational choice framework underpinning the surplus measures (4.86), but these two approaches give rise to different equilibrium solutions (and are also distinct in the numerical techniques applied in determining these equilibrium states). The source of the difference may be traced to an aggregation issue. In the PCM the actions of individuals are considered at the micro level and the total surplus associated with groups are obtained by integration. The GSM objective function is composed of two parts which *separately* comprise a consistent aggregation of micro-relations associated with the journey to work and the journey to shop. The maximization of the *sum* of these *aggregate* expressions, subject to the accounting and economic base constraints, will not in general result in the same formal expressions as the former aggregation process. This implies that different results will be obtained with these two models although both generate spatial-interaction patterns of the type defined by (4.25)–(4.30), as shown in the next section.

An important feature of the GSM approach results from the fact that the objective function GS is a strictly concave and separable function (see Section 2.6). Therefore, the GSM is a linearly constrained separable and concave mathematical optimization model which may be solved computationally by several available methods, such as those provided by Rosen (1960), Wolfe (1965), Faure and Huard (1965), Zangwill (1967), and McCormick (1970a and 1970b).

The addition of planning constraints into the GSM is, by virtue of its optimization structure, a natural and straightforward step. On the contrary, it is well known that the introduction of planning constraints is a cumbersome business in the traditional LM framework and this is also the case for the PCM introduced in the previous section.

The GSM is a residential and service activity allocation model—as are the LM and PCM. The extension of the GSM to embrace the design of basic employment configurations by exploiting its economic evaluation properties is, however, easily achieved, as will be shown later in Chapter 8.

Several important properties of the group-surplus approach will result from the discussion of the Kuhn–Tucker conditions and the mathematical dual program associated with the GSM. In particular, the following properties will be derived:

(a) The residence–work and residence–service spatial-interaction variable $\{T_{ij}\}$ and $\{S_{ij}\}$ respectively, generated solving the GSM, will satisfy two interlocked doubly constrained gravity submodels like those observed in the LM (and PCM); this shows that GSM and LM (PCM), while generating different stock variables, create the same types of spatial-interaction patterns;
(b) the dual of the GSM is an unconstrained minimization problem which determines the solution of the GSM uniquely and thereby provides an efficient procedure for computational purposes;
(c) the surplus function GS is consistent with the Marshallian/Hotelling measure of consumers' surplus derived from the joint travel demand–spatial-interaction submodels $\{T_{ij}\}$ and $\{S_{ij}\}$.

4.4.2 Kuhn–Tucker conditions

The Kuhn–Tucker (K–T) optimality conditions will now be explored in detail to give a further insight into the nature of the GSM approach. For the program (4.86)–(4.88) these conditions are obtained from the Lagrangian

$$L = GS + \sum_i v_i(\lambda_1 \sum_j S_{ij} - \sum_j T_{ij}) + \sum_j \gamma_j(\eta E_j^B - \sum_i T_{ij} + \lambda_2 \sum_i S_{ij}) \qquad (4.89)$$

Additional components will be associated with planning constraints. The K–T conditions are (cf. Chapter 2):

$$\frac{\partial L}{\partial T_{ij}} \leqslant 0, \quad \frac{\partial L}{\partial S_{ij}} \leqslant 0 \qquad (4.90)$$

together with the complementary slackness relations

$$T_{ij}\frac{\partial L}{\partial T_{ij}} = 0, \quad S_{ij}\frac{\partial L}{\partial S_{ij}} = 0 \qquad (4.91)$$

and

$$\frac{\partial L}{\partial v_i} = 0, \quad \frac{\partial L}{\partial \gamma_j} = 0 \qquad (4.92)$$

which give rise to constraints (4.87) and (4.88). For finite β^W and β^S, $\{T_{ij}\}$ and $\{S_{ij}\}$ are always positive, and therefore (4.91) implies that the equality sign will prevail in (4.90). Hence

$$\frac{\partial L}{\partial T_{ij}} = 0 = -\frac{1}{\beta^W}\ln\frac{T_{ij}}{W_i^R} - v_i - \gamma_j - c_{ij}^W \qquad (4.93)$$

$$\frac{\partial L}{\partial S_{ij}} = 0 = -\frac{1}{\beta^S}\ln\frac{S_{ij}}{W_j^S} + \lambda_1 v_i + \lambda_2 \gamma_j - c_{ij}^S \tag{4.94}$$

that is,

$$T_{ij} = W_i^R \exp[-\beta^W(v_i + \gamma_j + c_{ij}^W)] \tag{4.95}$$
$$S_{ij} = W_j^S \exp[\beta^S(\lambda_1 v_i + \lambda_2 \gamma_j - c_{ij}^S)] \tag{4.96}$$

It may now readily be shown that $\{T_{ij}\}$ and $\{S_{ij}\}$ will satisfy interlocked doubly constrained gravity models as the LM and the PCM (cf. equations (4.27) and 4.28) in Section 4.2.3 above). Let P_i, E_j^{NB}, and E_j represent the population and employment variables defined *internally*, that is,

$$P_i = \frac{\alpha}{\eta}\sum_j T_{ij} = \frac{1}{\sigma\rho}\sum_j S_{ij} \tag{4.97}$$

$$E_j^{NB} = \frac{1}{\eta}\sum_i T_{ij} - E_j^B = \frac{1}{\rho}\sum_i S_{ij} \tag{4.98}$$

$$E_j = \frac{1}{\eta}\sum_i T_{ij} = \frac{1}{\rho}\sum_i S_{ij} + E_j^B \tag{4.99}$$

Now, define 'balancing factors' A_i^W, B_j^W, A_i^S, and B_j^S as

$$A_i^W = W_i^R \exp[-\beta^W v_i]/P_i \tag{4.100}$$
$$A_i^S = \exp[\beta^S \lambda_1 v_i]/P_i \tag{4.101}$$
$$B_j^W = \exp[-\beta^W \gamma_j]/E_j \tag{4.102}$$
$$B_j^S = W_j^S \exp[\beta^S \lambda_2 \gamma_j]/E_j^{NB} \tag{4.103}$$

The spatial interaction models (4.95) and (4.96) will then take the more conventional form

$$T_{ij} = A_i^W B_j^W P_i E_j \exp[-\beta^W c_{ij}^W] \tag{4.104}$$
$$S_{ij} = A_i^S B_j^S P_i E_j^{NB} \exp[-\beta^S c_{ij}^S] \tag{4.105}$$

Finally, inserting (4.104) and (4.105) into (4.97)–(4.98), it is readily seen that

$$A_i^W = \frac{\eta}{\alpha}\left[\sum_j B_j^W E_j \exp[-\beta^S c_{ij}^S]\right] \tag{4.106}$$

$$A_i^S = \rho\sigma\left[\sum_j B_j^S E_j^{NB} \exp[-\beta^S c_{ij}^S]\right] \tag{4.107}$$

$$B_j^W = \eta\left[\sum_i A_i^W P_i \exp[-\beta^W c_{ij}^W]\right] \tag{4.108}$$

$$B_j^S = \rho\left[\sum_i A_i^S P_i \exp[-\beta^S c_{ij}^S]\right] \tag{4.109}$$

By examining the identical results obtained in section 4.2.3, it is clear that the LM, PCM, and GSM create spatial gravity patterns of the same *type*. Because the population and employment variables are implicitly determined within the models by different mechanisms, it does not mean, however, that the LM, PCM, and GSM are equivalent.

4.4.3 Relationship to the Lowry model

An analytical relationship between LM and GSM may be established by comparing the stock variables in both models. A few simple manipulations of (4.100)–(4.103) show that

$$P_i = \alpha W_i^R \exp[-\beta^W v_i] \sum_j E_j \exp[-\beta^W c_{ij}^W] / \sum_i W_i^R \exp[-\beta^W v_i] \exp[-\beta^W c_{ij}^W] \tag{4.110}$$

$$E_j^{NB} = \sigma W_j^S \exp[\beta^S \lambda_2 \gamma_j] \sum_i P_i \exp[-\beta^S c_{ij}^S] / \sum_j W_j^S \exp[\beta^S \lambda_2 \gamma_j] \exp[-\beta^S c_{ij}^S] \tag{4.111}$$

Thus, it is clear by comparing these expressions with (4.35) and (4.36) that replacing the LM weight W_i^R and W_j^S by

$$\tilde{W}_i^R = W_i^R \exp[-\beta^W v_i] \tag{4.112}$$

and

$$\tilde{W}_j^S = W_j^S \exp[\beta^S \lambda_2 \gamma_j] \tag{4.113}$$

respectively, then both models will generate the same activity allocation. This fact has been confirmed empirically by running the LM with the weights (4.112) and (4.113); a solution identical to the GSM has then been obtained.

4.4.4 Relationship to the probabilistic choice model

The weights $\tilde{W}_i^R$ and $\tilde{W}_j^S$ which are introduced in the LM to generate the GSM solution have been derived above. Similarly, it has been shown in section 4.3.2 that the LM weights which produce the PCM outcome are

$$\hat{W}_i^R = W_i^R \exp\left[-\beta^W \frac{\bar{u}_i^S}{\lambda_1}\right] \tag{4.114}$$

$$\hat{W}_j^S = W_j^S \tag{4.115}$$

where $\bar{u}_i^S$ is the composite utility of shopping from location i. From (4.112)–(4.115), we obtain:

$$\frac{\tilde{W}_i^R}{\hat{W}_i^R} = \exp\left[-\beta^W\left(v_i - \frac{\bar{u}_i^S}{\lambda_1}\right)\right] \tag{4.116}$$

and

$$\frac{\tilde{W}_i^S}{\hat{W}_i^R} = \exp[\beta^S \lambda_2 \gamma_j] \tag{4.117}$$

Hence, we have

$$v_i = \frac{\bar{u}_i^S}{\lambda_1} - \frac{1}{\beta^W} \ln \frac{\hat{W}_i^R}{\hat{W}_i^R} \tag{4.118}$$

$$\gamma_j = \frac{1}{\beta^S \lambda_1} \ln \frac{\hat{W}_j^S}{\hat{W}_j^S} \tag{4.119}$$

These equations express the relationship between both sets of weights, the Lagrangian multipliers of GSM, and the composite utility arising from the shopping from zone i.

Equation (4.118) suggests that the meaning of the Lagrangian multiplies $\mathbf{v}$ is related to the mean surplus associated with travel to service activities from zone i.

4.5 DUALITY AND THE GSM

4.5.1 The dual program

It will be shown below that the dual of the GSM program is an unconstrained convex minimization problem. This feature has been exploited in numerical approaches to the solution of the model (cf. Section 2).

Wolfe's (1961) formulation for the dual of a mathematical program is followed here. Invoking the duality relationships described in section 2.2.4 it may readily be shown that the dual of the GSM program is

$$\min_{\{\mathbf{T},\mathbf{S},\mathbf{v},\boldsymbol{\gamma}\}} U(\mathbf{T},\mathbf{S},\mathbf{v},\boldsymbol{\gamma}) = \frac{1}{\beta^W} \sum_{ij} T_{ij} + \frac{1}{\beta^S} \sum_{ij} S_{ij} + \eta \sum_j \gamma_j E_j^B \tag{4.120}$$

subject to the K–T optimality conditions

$$v_i + \gamma_j = -c_{ij}^W - \frac{1}{\beta^W} \ln \frac{T_{ij}}{W_i^R} \tag{4.121}$$

$$\lambda_1 v_i + \lambda_2 \gamma_j = c_{ij}^S + \frac{1}{\beta^S} \ln \frac{S_{ij}}{W_j^S} \tag{4.122}$$

that is

$$T_{ij} = W_i^R \exp[-\beta^W (v_i + \gamma_j + c_{ij}^W)] \tag{4.123}$$

$$S_{ij} = W_j^S \exp[\beta^S (\lambda_1 v_i + \lambda_2 \gamma_j - c_{ij}^S)] \tag{4.124}$$

where v_i and γ_j are the dual variables (or Lagrangian multipliers). In the formation of the dual constraints (4.121) and (4.122) we have exploited the fact that the variables $\mathbf{T}$ and $\mathbf{S}$ are non-zero for finite β^S and β^W. By inserting (4.123) and (4.124) into (4.120) the GSM dual program will become an unconstrained minimization problem in the variables $\mathbf{v}$ and $\boldsymbol{\gamma}$, as follows:

$$\operatorname*{Min}_{\{\mathbf{v},\boldsymbol{\gamma}\}} U(\mathbf{v},\boldsymbol{\gamma}) = \frac{1}{\beta^W} \sum_{ij} W_i^R \exp[-\beta^W (v_i + \gamma_j + c_{ij}^W)]$$

$$+\frac{1}{\beta^S}\sum_{ij} W_j^S \exp[\beta^S(\lambda_1 v_i + \lambda_2 \gamma_j - c_{ij}^S)]$$

$$+\eta\sum_j \gamma_j E_j^B \tag{4.125}$$

At this stage, a first comparison of the computational difficulty of solving the primal and dual GSM programs is necessary: if n is the number of origin and destination zones, then the GSM primal program (4.86)–(4.88) has $2n^2$ variables and $2n$ constraints, while the dual program (4.125) is *unconstrained* and contains only $2n$ variables. Numerical work is therefore naturally performed in the dual program representation. Suitable computer programs based on unconstrained minimization methods such as the Newton–Raphson or the Fletcher–Reeves methods are generally available.

4.5.2 Duality and the allocation process

In the GSM approach as in the Lowry model, an estimate is made of stock allocated as a result of the basic employment configuration $\{E_j^B\}$. Determining the stock allocation variables P_i, E_j^{NB}, and E_j from the dual GSM solution is rather simple in computational terms. In fact, the dual variables v_i and γ_j determine, through (4.123) and (4.124), the spatial distribution of trips $\{T_{ij}\}$ and $\{S_{ij}\}$, respectively. Then, the allocation variables are readily obtained from the trip-end conditions (4.97)–(4.99).

Conversely, we note that the dual solution is determined by the primal outcome: if the spatial-interaction variables $\{T_{ij}\}$ and $\{S_{ij}\}$ are known, then the system of linear equations (4.121) and (4.122) may be solved to give v_i and γ_j. This system may be solved either directly, or through an iterative scheme analogous to the Furness method.

4.5.3 Duality and evaluation

The evaluation of land-use transport plans generated through the GSM may be carried out by exploring the meaning of the objective function GS and its dual as measures of the total expected surplus. Thus, the change in total surplus resulting from a change between two situations 1 and 2 is, as follows:

$$\Delta GS = GS^{*(2)} - GS^{*(1)} \tag{4.126}$$

$$= \left[-\frac{1}{\beta^W}\sum_{ij} T_{ij}\left(\ln\frac{T_{ij}}{W_i^R} - 1\right) - \sum_{ij} T_{ij}c_{ij}^W - \frac{1}{\beta^S}\sum_{ij} S_{ij}\left(\ln\frac{S_{ij}}{W_j^R} - 1\right) - \sum_{ij} S_{ij}c_{ij}^S \right]_{*(1)}^{*(2)} \tag{4.127}$$

$$= U^{*(2)} - U^{*(1)} \tag{4.128}$$

$$= \eta\sum_j E_j^B(\gamma_j^{*(2)} - \gamma_j^{*(1)}) \tag{4.129}$$

where the superscripts 1 and 2 refer to the initial and final states, and the asterisk indicates that the quantities are determined at optimality. It may readily be shown (Coelho and Williams, 1978; Coelho, 1977) that the concept of locational surplus applied to the demand functions implicitly generated in the GSM approach also produces the evaluation measures (4.126)–(4.129). The method followed in those calculations was described in section 4.3.3.

In summary, the GSM provides the following measures for land-use transport plans evaluation: one, the primal objective function expresses the total surplus† in terms of the trip distributions $\{T_{ij}\}$ and $\{S_{ij}\}$; two, the dual is also a measure of the total surplus expressed in terms of the configuration of basic employment; and three, the dual variables γ_j, associated with the economic base equation, multiplied by η give the incremental surplus arising from trips generated by an additional unit of basic employment at location j. This measure has a considerable significance in the design of basic employment configurations.

4.6 A VARIANT: THE GROUP ENTROPY MAXIMIZING MODEL (GEM)

Corresponding to the GSM is a group entropy maximizing model (GEM) which is obtained by replacing the objective function GS by

$$EM = -\sum_{ij} T_{ij}\left(\ln\frac{T_{ij}}{W_i^R} - 1\right) - \sum_{ij} S_{ij}\left(\ln\frac{S_{ij}}{W_j^S} - 1\right) \tag{4.130}$$

and adding the two constraints to the set (4.87) and (4.88)

$$\sum_{ij} c_{ij}^W T_{ij} = c^W \tag{4.131}$$

$$\sum_{ij} c_{ij}^S S_{ij} = c^S \tag{4.132}$$

where c^W and c^S are the total travel costs in the residence–work and residence–services transportation submodels respectively. The objective function EM is composed of two 'entropy functions' which generate the spatial-interaction submodels $\{T_{ij}\}$ and $\{S_{ij}\}$ respectively. If we now denote by β^W and β^S the dual variables associated with constraints (4.131) and (4.132) respectively, and assume them to be known, then it is easily shown that, for constant β^W and β^S, the GEM is equivalent to the model obtained by maximizing

$$\begin{aligned} EM^1 = &\left[-\sum_{ij} T_{ij}\left(\ln\frac{T_{ij}}{W_i^R} - 1\right) - \beta^W \sum_{ij} c_{ij}^W T_{ij}\right] \\ &+ \left[-\sum_{ij} S_{ij}\left(\text{n}\frac{S_{ij}}{W_j^S} - 1\right) - \beta^S \sum_{ij} c_{ij}^S S_{ij}\right] \end{aligned} \tag{4.133}$$

† Since the surplus is a function of location variables the expression 'total locational surplus' is also appropriate.

subject to the consistency and economic base constraints alone. Although the GSM and GEM are founded upon two theoretically distinct approaches, the only 'structural' difference between them is clearly on the weighting of bracket terms of (4.133): if those terms in brackets were multiplied by $1/\beta^W$ and $1/\beta^S$ respectively, then it would become equal to the objective function *GS*. However, it may be shown (Coelho, 1977) that by virtue of the apparently minor difference the exact evaluation measures derived for the GSM are no longer obtained in the GEM context. Nevertheless, if the parameters β^W and β^S are of similar value, then a good approximation for the change in consumers' surplus is provided by†

$$CS = \frac{\eta}{\tilde{\beta}}\sum_j E_j^B(\gamma_j^{*1} - \gamma_j^{*2}) \tag{4.134}$$

where $\tilde{\beta}$ is a weighted average of the parameters β^W and β^S:

$$\frac{1}{\tilde{\beta}} = \frac{1}{\beta^W} + \frac{1}{\beta^S} \tag{4.135}$$

The absence of the exact evaluation measures established for the GSM in the GEM framework emphasizes the role played by the factors $1/\beta^W$ and $1/\beta^S$ in the conversion of the utility associated with travel demand models into the same (money) units. It should be noted, however, that within the theoretical frameworks which involve the 'entropy maximizing' and 'random utility' interpretations of dispersion, the interpretation of the parameters themselves are fundamentally different.

4.7 DISAGGREGATION OF THE MATHEMATICAL PROGRAMMING ACTIVITY ALLOCATION MODEL

In conceptual terms, there is virtually no conceivable limit to model disaggregation although it would only be reasonable to consider disaggregations that are behaviourally significant. The limitations are concerned instead with data availability and implementation capacity. The level of disaggregation is often determined by a balance between the natural desire for further disaggregation and accuracy and the availability of data, time, and computational resources.

As formulated above, the group surplus model is essentially a two-sector model. In the above model no person-type, house-type, or mode of transportation were provided. But, of course, there is in principle no obstacle to defining a fully disaggregated GSM. Let us consider, for example, the situation where N trip submodels $\{X_{ij}^k\}$ $(k = 1, \ldots, N)$ are defined. The surplus function associated with those 'groups' of trips are then

$$GS = -\sum_k \frac{1}{\beta^k}\sum_{ij}\left(\ln\frac{X_{ij}^k}{W_{ij}^k} - 1\right) - \sum_{ij} c_{ij}^k X_{ij}^k \tag{4.136}$$

† See Williams (1976) for a similar example with an entropy maximizing trip-distribution model disaggregated by type of person.

where β^k is the parameter of the trip distribution submodel $\{X_{ij}^k\}$, c_{ij}^k are travel costs, and W_{ij}^k are attraction weights for the group of trips k and the trip end zones i and j. The program representing the maximization of GS subject to the consistency, economic base, and any other planning constraints will be a multisector version of the GSM. Disaggregation by type of person, for example, will be achieved by defining variables X_{ij}^{kn} to represent the number of trips type k, from location i to location j, by persons of type n. The aggregate group surplus is then

$$GS = -\sum_{kn} \frac{1}{\beta^{kn}} \sum_{ij} X_{ij}^{kn} \left(\ln \frac{X_{ij}^{kn}}{W_{ij}^{kn}} - 1 \right) - \sum_{ijkn} c_{ij}^{kn} X_{ij}^{kn} \tag{4.137}$$

where β^{kn} are the parameters of interaction submodels, W_{ij}^{kn} are attractiveness weights, and c_{ij}^{kn} travel costs for the group of trip k, type of person n, and trip end zones i and j. The consistency and economic base equations will take the general form:

Consistency: $$\sum_k \lambda_k^n \sum_j X_{ij}^{kn} = 0 \qquad \text{for all } i \text{ and } n \tag{4.138}$$

Economic base: $$\sum_k \delta_k^n \sum_i X_{ij}^{kn} = E_j^{Bn} \qquad \text{for all } j \text{ and } n \tag{4.139}$$

where δ_k^n and λ_k^n are coefficients and E_j^{Bn} is the amount of basic employment in zone j providing occupation for persons type n. The dual is formed as above. It can be written as follows:

$$\min_{\{v_i^n, \gamma_j^n\}} U(v_i^n, \gamma_j^n) = \sum_{ijkn} W_{ij}^{kn} \exp[-(\lambda_k^n v_i^n + \delta_k^n \gamma_j^n + \beta^{kn} c_{ij}^{kn})] + \sum_{nj} \gamma_j^n E_j^{Bn}$$
$$+ \text{terms from planning constraints} \tag{4.140}$$

where v_i^n and γ_j^n are the dual variables associated with constraints (4.138) and (4.139) respectively. As before, duals would be used for computation and evaluation.

Clearly, no matter what disaggregation is defined for the group surplus model, it can always be easily adapted to the GEM context in view of the *structural* similarity between both approaches.

4.8 THE CALIBRATION PROCESS

The method of calibrating the GSM is based on the same process that is followed to calibrate the LM. Hyman (1969) and Evans (1971) have shown that the maximum-likelihood estimators for the parameters of exponential spatial-interaction submodels, like those integrated in the LM, PCM, and GSM, are determined by solving the cost equation. Thus, the parameters β^W and β^S are obtained by repeatedly solving the GSM and adjusting the parameter values to ensure that

$$\left[\sum_{ij} T_{ij} c_{ij}^W - \sum_{ij} T_{ij}^{\text{obs}} c_{ij}^W \right] = 0 \tag{4.141}$$

and

$$\left[\sum_{ij} S_{ij}c_{ij}^S - \sum_{ij} S_{ij}^{obs} c_{ij}^S\right] = 0 \tag{4.142}$$

where $\{T_{ij}^{obs}\}$ and $\{S_{ij}^{obs}\}$ are the observed residence–work and residence–services trip-distribution matrices. Batty (1970) has exhaustively analysed this and other methods of calibrating the LM model and has concluded through empirical observation that travel cost equations (4.141) and (4.142) are the most sensitive statistics for the calibration of the LM. Although further research will be necessary to extend Batty's empirical evidence to the GSM approach, it seems plausible that this conclusion will remain valid.

A solution of the non-linear equations (4.141) and (4.142)—as functions of β^W and β^S—may be obtained by the Newton–Raphson method. The speed of convergence will be dependent, of course, on the closeness of the initial starting values. An indirect method to generate close estimates of the parameters β^W and β^S is provided by the GEM. Since the cost equations are incorporated explicitly as constraints in the GEM, the solution of its dual program:

$$\begin{aligned}\min_{\{\mathbf{v}, \boldsymbol{\gamma}, \beta^W, \beta^S\}} \quad U(\mathbf{v}, \boldsymbol{\gamma}, \beta^W, \beta^S) = &\sum_{ij} W_i^R \exp[-(v_i + \gamma_j + \beta^W c_{ij}^W)] \\ &+ \sum_{ij} W_j^S \exp[\lambda_1 v_i + \lambda_2 \gamma_j - \beta^S c_{ij}^S] \\ &+ \eta \sum_j \gamma_j E_j^B + \beta^W c^W + \beta^S c^S\end{aligned} \tag{4.143}$$

produces automatically the maximum-likelihood estimates of the parameters β^W and β^S for this latter model. These values will provide estimates that may be used to initialize the iterative process for the calibration of the GSM.

5. *Entropy and economic accounts: some new models*

5.1 INTRODUCTION

Various forms of economic account-based analysis (input–output analysis) are familiar cornerstones of regional and locational analysis, particularly the spatially disaggregated forms. In this chapter we turn to a particular class of economic account-based analysis (specifically, the analysis of rectangular input–output systems) and a particular class of mathematical programming methods (entropy-maximizing methods) and develop an interesting new class of entropy-maximizing account-based models for rectangular input–output systems. We are thus exploiting optimization in its technical, mathematical, sense, using methods of non-linear programming of a specific kind to extend a range of models whose conventional solution in terms of exact matrix algebra entails some rather limiting assumptions. Of particular importance in aiming to relax these assumptions is to establish an input–output multiplier mechanism within a mathematical programming framework. Since this multiplier operates primarily at the sectoral rather than the spatial level, the spatial component will be incorporated at a relatively late stage in the chapter (in Section 5.6).

In concentrating specifically on rectangular input–output models, we are concerned only with those input–output models which allow the economic sectors that are identified to produce any number of products; in particular, we do not consider those which lump each sector's output into one homogeneous product (even though these are perhaps more familiar) except as a brief starting point for the developments that follow. Thus the type of input–output models with which we are mainly concerned are those favoured, for example, by Stone *et al.* (1963), Gigantes (1970), Rosenbluth (1968), Victor (1972), Armstrong (1975) and Schinnar (1978). Existing models in this class will be reviewed briefly in the following section as a starting point for the developments below.

The combination of entropy-maximizing methods with this brand of input–output analysis reveals both a new area of application for these methods, and perhaps more importantly brings to light a far more flexible approach to modelling than could be possible with the rigid matrix algebra methods previously used. The constraint/objective function format of mathematical programming models, of course, immediately invites model modification in terms of alternative system objectives, and the incorporation of additional available information in further constraint equations.

From a different perspective, the mathematical programming rectangular input–output models to be presented may be seen to form a bridge between two existing modelling fields: on the one hand, the existing families of exact algebraic rectangular input–output models already referred to; and on the other, the range of mathematical programming activity-analysis models of Takayama and Judge (1972) and others. Thus over the range of economic processes of interest, the sector–product distinction identified in rectangular input–output models can be mapped exactly onto an appropriate activity–commodity classification in activity-analysis terms (sectors corresponding to activities and products to commodities).

5.2 A REVIEW OF THE EXISTING MODELS

We begin by considering the familiar input–output model

$$\begin{aligned} X^1 &= a^{11}X^1 + a^{12}X^2 + \ldots a^{1n}X^n + Y^1 \\ &\vdots \\ X^n &= a^{n1}X^1 + \ldots\ldots a^{nn}X^n + Y^n \end{aligned} \tag{5.1}$$

The a^{ij}'s are technical coefficients defining the input requirement of sector i's product per unit output of sector j's product. X^j is the current level of product output of sector j, and Y^j is the current final or exogenous demand for sector j's product. This exogenous demand effectively drives the system since the essential concern of input–output multiplier analysis in its most basic form is to determine necessary future levels of production of the endogenous sectors (X^i), given known or hypothesized future levels of final demand (Y^j), and assuming constant technical coefficients (a^{ij}). This is done by forming the familiar Leontief inverse, achieved by manipulating equations (5.1) to give:

$$\mathbf{X} = (\mathbf{I} - \mathbf{A})^{-1}\mathbf{Y} \tag{5.2}$$

where $\mathbf{X}$ is a vector of the endogenous sectors' output, $\mathbf{Y}$ is a vector of exogenous demand, $\mathbf{A}$ is the matrix of technical coefficients, and $\mathbf{I}$ the unit matrix.

A simple but important conceptual distinction can be made immediately. The distinction is between the assembly of available information that depicts the prevailing pattern of intersectoral flows in some base period and from which, incidentally, the technical coefficients a^{ij} are derived, and the manipulation and use of these coefficients in equations such as (5.2) in conjunction with known or hypothesized future final demands (or other exogenous terms) to determine future levels of outputs or transactions. Any set of terms depicting base-year intersectoral flows will be referred to below as base-period transactions, and any modelling required to determine such flows (typically to force data consistency, or fill gaps in data) will be termed base-period input–output modelling. Projection-period input–output modelling, on the other hand, will refer to the determination of future levels of outputs and transactions, typically involving known or hypothesized exogenous terms in conjunction with assumed constant technical coefficients which pick up a multiplier mechanism. This, of course, is usually performed using the classic matrix inverse form (as in (5.2)).

The developments below involve maximum-entropy representations of these two stages, with an emphasis on the second stage, because it is alternative representations of the multiplier mechanism that are of particular interest to us here.

Only the most rudimentary version of existing input–output methodology was given above, as this highlighted well enough the intended distinction between base- and projection-period models. This conceptual distinction can be applied equally in cases where refinements to the above basic form have been made, and it is well known that many such refinements exist in the literature, though they are typically based very closely on the fundamental form noted above. The particularly limiting feature of that form for present purposes is that equation (5.2) assumes a square matrix of technical coefficients **A** and with it the implicit assumption that each sector produces just one homogeneous product. This assumption ensures above (amongst other things) that matrix inversion and exact algebraic solution to the equations is possible.

This one-to-one sector–product assumption central to the above model is not a natural one to make, and quite apart from the fundamental intention throughout the framework of analysis below to abandon this assumption completely, the basic collection and tabulation of published input–output data itself acknowledges a fundamental sector–product distinction, filling the non-shaded partitions of the table in Figure 5.1 (a similar table is presented in United Nations (1968) and Gigantes (1970), for example). The only necessarily square partitions (namely the product × product and sector × sector partitions) in this table have no entries at all; thus any attempt to use a model as simple as (or of a similar form

	Products	Sectors	Final demands	Totals
Products		**U**	**e**	**q**
Sectors	**V**			**g**
Primary inputs		**y'**		
Totals	**q'**	**g'**		

U is an absorption matrix; a typical element u_{ij} represents the amount of product i used up in production by sector j
V is a make matrix; a typical element v_{ij} represents the amount of project j produced by sector i
q is a vector of total product outputs (**q**′ its transpose)
g is a vector of total sector outputs (**g**′ its transpose)
e is a vector of final demands for products
y′ is a vector of primary inputs.

Figure 5.1 Basic input–output accounts

to) equation (5.2) will need either to make matrix **U** square (by using the same labels for sectors as for products), or to manipulate the data from matrices **U** and **V** to provide entries for the square product × product and sector × sector matrices. In the former case, coefficients a^{ij} in matrix **A** could be given by u^{ij}/g^{j}, the classification of sectors and of products would match exactly, and since under these assumptions sectors can produce only one homogeneous product, matrix **V** is redundant. In the latter case, sectors can produce any number of products, matrices **U** and **V** need not be square, and in order to relate endogenous sector output to exogenous sector final demand or to relate endogenous level of product output to exogenous level of product demand in what has here been identified as the 'second stage', assumptions regarding the internal structures of the square product × product or sector × sector matrices are required. It may be appropriate to note that the more realistic sector–product assumptions it is possible to build into these manipulations can go some way in answering earlier criticisms over the naivity of technical coefficients derived directly from square input–output tables; possible derivations of the more intricate matrix inverse multipliers, in place of $(\mathbf{I}-\mathbf{A})^{-1}$ in equation (5.2) are discussed in Stone, Bates and Bacharach (1963), given in Gigantes (1970) based on extensive use of rectangular input–output tables in Canada, and also in United Nations (1968), Central Statistical Office (1973), and Armstrong (1975). They are summarized briefly in the following paragraphs.

The models all use linear rate coefficients derived from the base-period absorption and make matrices, and these coefficients are represented in the following arrays:

B with elements b^{kn}, where $b^{kn} = U^{kn}/g^{n}$, giving the input of commodity k per unit output of sector n;

C with elements c^{mk}, where $c^{mk} = V^{mk}/g^{m}$, fixing the proportions in which a particular sector produces various commodities;

D with elements d^{mk}, where $d^{mk} = V^{mk}/q^{k}$, fixing a given sector's share in the total production of a particular commodity.

Used in various combinations, these three matrices of coefficients can give rise to a number of different input–output models. These are presented in detail elsewhere (United Nations, 1968; Gigantes, 1970; Central Statistical Office, 1973; Armstrong, 1975) and therefore need only a brief summary at this stage. Their characteristic aim is to find projected levels of sector and commodity outputs (given by vectors **g** and **q**) corresponding to hypothesized or known future levels of commodity final demands, **e**, given particular combinations of the assumptions represented in matrices **B**, **C**, and **D**.

The so-called 'industry technology' model uses coefficients from matrices **B** and **D** and assumes that commodity inputs to sectors are proportional to their total outputs, weighted for each commodity produced according to the market shares of each sector. This implies the following relationships:

$$\mathbf{q} = \mathbf{B}\mathbf{g} + \mathbf{e} \tag{5.3}$$

and

$$\mathbf{g} = \mathbf{Dq} \tag{5.4}$$

These produce the following forms for $\mathbf{q}$ and $\mathbf{g}$:

$$\mathbf{q} = (\mathbf{I} - \mathbf{BD})^{-1}\mathbf{e} \tag{5.5}$$

$$\mathbf{g} = (\mathbf{I} - \mathbf{DB})^{-1}\mathbf{De} \tag{5.6}$$

This model assumes that the input structure of given commodities will be different according to the sectors in which they are produced.

It is interesting to note that Schinnar (1978) has recently devised a new method for calculating the inverse matrix $(\mathbf{I} - \mathbf{DB})^{-1}$ used in the model given by equation (5.6) here. This exploits the matrix product $\mathbf{DB}$ and calculates the inverse matrix $(\mathbf{I} - \mathbf{DB})^{-1}$ in a finite number of iterations; a technological-change interpretation is offered for the successive terms in the series thus produced.

The so-called 'commodity technology' model uses coefficients from matrix $\mathbf{C}$ rather than $\mathbf{D}$ on the assumption that input structures of commodities are stable and unique; thus it uses the relationship:

$$\mathbf{q} = \mathbf{Cg} \tag{5.7}$$

rather than (5.4), giving the models:

$$\mathbf{q} = (\mathbf{I} - \mathbf{BC}^{-1})^{-1}\mathbf{e} \tag{5.8}$$

$$\mathbf{g} = (\mathbf{I} - \mathbf{C}^{-1}\mathbf{B})^{-1}\mathbf{C}^{-1}\mathbf{e} \tag{5.9}$$

This model therefore assumes that a given commodity has the same input structure in whichever sector it is produced.

Apart from these two pure-technology models there are a number of standard 'mixed-technology' models which, as their name suggests, use a combination of the above assumptions in order that various elements of production can be treated on different assumptions. In order to formulate these models it is necessary to divide the make matrix $\mathbf{V}$ into two partitions $\mathbf{V}_1$ and $\mathbf{V}_2$ such that

$$\mathbf{V}_1 + \mathbf{V}_2 = \mathbf{V} \tag{5.10}$$

where $\mathbf{V}_1$ consists of elements for which the hypothesis of constant market share is to be used (thus using coefficients $\mathbf{D}_1$ given by $d_1^{mk} = V_1^{mk}/q_1^k$) and $\mathbf{V}_2$ consists of elements for which the hypothesis that sectors produce commodities in fixed proportions is to be used (thus using coefficients $\mathbf{C}_2$ given by $c_2^{mk} = V_2^{mk}/g_2^m$ and carrying over a brand of commodity technology assumption). The exact form of model to result then depends on the way in which particular elements of subsidiary production are to be treated (see, for example, Armstrong, 1975, for more details on this). If, for instance, the production of by-products follows fixed market shares, the model

$$\mathbf{q} = \{\mathbf{I} - \mathbf{B}[\mathbf{C}_1{}^{-1}(\mathbf{I} - \mathbf{D}_2'\mathbf{i}) + \mathbf{D}_2]\}^{-1}\mathbf{e} \tag{5.11}$$

is obtained. On the other hand, if the outputs of by-products in $\mathbf{V}$ is linked to the outputs of the producing sectors, then in place of (5.11) the model

$$\mathbf{q} = \{\mathbf{I} - [\mathbf{B}\mathbf{C}_1{}^{-1}(\mathbf{I} - \mathbf{C}_2\mathbf{H}) + \mathbf{B}\mathbf{D}_2]\}\mathbf{e} \tag{5.12}$$

with

$$\mathbf{H} = (\mathbf{I} - \mathbf{C}_2'\mathbf{i} + \mathbf{D}_1\mathbf{C}_2)^{-1}\mathbf{D}_1 \tag{5.13}$$

is obtained (in fact $\mathbf{H}$ relates $\mathbf{g}$ to $\mathbf{q}$, namely $\mathbf{g} = \mathbf{H}\mathbf{q}$). The reader is referred to Gigantes (1970) or Armstrong (1975) for the intermediate details of these models. Further alternative models are derivable but will not be repeated explicitly here, and it may be noted that in any case the difference in results obtained between models (5.11) and (5.12) or other alternative 'mixed' models are generally accepted as being insignificant (Armstrong, 1975).

Before turning to the mathematical programming multiplier models which are to be developed below as alternatives to the above models, a few additional observations in connection with the matrix inverse multiplier models summarized above will be made. Note firstly that models (5.8), (5.9), (5.11), and (5.12) all require inversion of the matrix $\mathbf{C}$ and therefore require this matrix to be square. This means that for all these models (in fact for all existing models using wholly or partly a commodity technology assumption) the number of commodities must exactly match the number of sectors or activities. This is considered to be a weakness of these models (though Schinnar (1978) has shown that the use of generalized inverses can avoid this problem), and no such requirement will be needed for the alternative models to be developed in the following sections. Apart from this specific point, it may be noted more generally that choice between the above matrix inverse models in any particular context will depend on the plausibility of the technology assumptions they embody, the relative ease of handling and solution of each, and the likely stability of the different coefficient combinations they involve; commodity technology models assume the stability of the unique input structure of each commodity, industry technology models assume effectively the stability of each of the coefficient matrices $\mathbf{B}$ and $\mathbf{D}$ in their own right (Armstrong, 1975).

As we discuss in more detail later, the merit of offering possible alternatives to these existing methods, apart from being of independent theoretical and practical interest, are that they arise within a framework that is automatically more flexible in its ability to incorporate a greater variety of assumptions (including the possibility of having fewer) than can be used with existing methods. This is because with a suitable choice of system representation and a mathematical programming approach, the linear rate coefficients derived from basic input–output accounts appear in the context of constraint equations defining a feasible region of possible solutions for endogenous terms rather than in the context of exactly determined equations that give rise to a unique solution for the endogenous terms. Thus the flexibility of the new method lies in the immediate scope offered to add to or replace the constraint set defined initially by the chosen linear rate coefficients, and also in the range of alternative possibilities of objective functions needed to complete the mathematical programming formulation.

5.3 THE DEVELOPMENT OF AN ALTERNATIVE APPROACH

The system representation to be used involves the array of variables $\{u^{mnk}\}$, which define the flow product (or commodity) k from sector (or activity) m to sector (or activity) n. This is an aggregate of an array first introduced in Cripps, Macgill, and Wilson (1974). It can be related directly to the entries in the absorption and make matrices in the table in Figure 5.1 by noting that summation over the entire m index immediately defines elements in an absorption matrix and summation over the entire n index, after allowing for relevant final demand flows, defines elements in a make matrix. The use of a single three-dimensional array simply recognizes that the functioning of an economy takes place in terms of specific interactions between producers and consumers, and that therefore the most immediate representation of this functioning is in an array of the form $\{u^{mnk}\}$, where both producers (m's) and receivers (n's) are explicitly defined, as well as the specific entity (some commodity k) in terms of which any particular transaction may be identified. The familiar make matrix effectively defines the sources (producers) of these interactions and the familiar absorption matrix similarly defines the sinks (receivers), but the natural focus of interest is in the complete interaction array $\{u^{mnk}\}$.

In relation to the terms in the previous section, in particular Figure 5.1, the following identities will be satisfied (where an asterisk denotes summation over the relevant index set):

$$U^{kn} = u^{*nk} \tag{5.14}$$

$$V^{mk} = u^{m*k} + \tilde{e}^{mk} \tag{5.15}$$

$$q^{k} = u^{**k} + \tilde{e}^{*k} \tag{5.16}$$

$$g^{m} = u^{m**} + \tilde{e}^{m*} \tag{5.17}$$

The new term $\tilde{e}^{mk}$ in these identities represents the final demand for commodity k from sector or activity m, thus elements in the vector $\mathbf{e}$ in Figure 5.1 are now written

$$e^{k} = \tilde{e}^{*k} \tag{5.18}$$

The next stage is to write down relationships that exist between the assumed constant rate coefficient from the make and absorption matrices and the system state variables (corresponding to equations (5.3), (5.4), and (5.7)). In previous input–output models these latter variables were the vectors $\mathbf{q}$ and $\mathbf{g}$ giving total commodity and sector outputs; in the present approach they are the terms in the array $\{u^{mnk}\}$. An important difference between the present and previous approaches may be observed here, namely that it is now possible to derive two kinds of relationships from each of the matrices **B**, **C**, and **D**. These will be considered in turn.

Firstly, it is possible to derive relationships directly from the definitions of the terms in the matrices **B**, **C**, and **D**, or in other words to define the coefficients of **B**, **C**, and **D** directly in terms of the new state variables $\{u^{mnk}\}$. This gives

$$b^{kn} = \frac{u^{*nk}}{u^{n**} + \tilde{e}^{n*}} \tag{5.19}$$

$$c^{mk} = \frac{u^{m*k} + \tilde{e}^{mk}}{u^{m**} + \tilde{e}^{m*}} \tag{5.20}$$

$$d^{mk} = \frac{u^{m*k} + \tilde{e}^{mk}}{u^{**k} + \tilde{e}^{*k}} \tag{5.21}$$

and the model relationships they imply are therefore

$$u^{*nk} = b^{kn}(u^{n**} + \tilde{e}^{n*}) \tag{5.22}$$

$$u^{m*k} + \tilde{e}^{mk} = c^{mk}(u^{m**} + \tilde{e}^{m*}) \tag{5.23}$$

$$u^{m*k} + \tilde{e}^{mk} = d^{mk}(u^{**k} + \tilde{e}^{*k}) \tag{5.24}$$

On the reasonable assumption that $\tilde{e}^{mk} = c^{mk}\tilde{e}^{m*}$ in (5.23), and $\tilde{e}^{mk} = d^{mk}\tilde{e}^{*k}$ in (5.24), the latter two equations may be rewritten

$$u^{m*k} = c^{mk}u^{m**} \tag{5.25}$$

$$u^{m*k} = d^{mk}u^{**k} \tag{5,26}$$

Secondly, it is possible to derive a further set of relationships by performing simple summations on each of the equations (5.22), (5.25) and (5.26). The most useful of such summations are considered to be the following:
sum (5.22) over n, giving

$$u^{**k} = \sum_{n} b^{kn}(u^{n**} + \tilde{e}^{n*}) \tag{5.27}$$

sum (5.25) over m, giving

$$u^{**k} = \sum_{m} c^{mk}u^{m**} \tag{5.28}$$

sum (5.26) over k, giving

$$u^{m**} = \sum_{k} d^{mk}u^{**k} \tag{5.29}$$

Thus (5.27), (5.28), and (5.29) provide a second set of relationships between the assumed constant make and absorption matrix rate coefficients and the system state variables.

A close comparison of the two kinds of input–output relationships defined in terms of a $\{u^{mnk}\}$ array, ((5.22), (5.25), (5.26) and (5.27), (5.28), (5.29)) with those used in existing models ((5.3), (5.4), and (5.7)) reveals that it is in fact (5.27), (5.28), and (5.29) that are directly equivalent to the earlier relationships in existing models, even though (5.22), (5.25), and (5.26) seem in many ways to have a closer correspondence with the coefficients in matrices **B**, **C**, and **D**. This aspect merits some deliberation when it comes to choosing the most appropriate model relationships for particular applications as it would clearly be unreasonable to use (5.22) as well as (5.27), (5.23) as well as (5.28), or (5.24) as well as (5.29). Before giving it explicit consideration in the present chapter, however, it is necessary to complete a suitable model framework within which the above type of relationship may be used. Combinations of the above $\{u^{mnk}\}$ input–output relationships do not yet form a complete model framework since most sensible combinations of them at present produce fewer equations than unknowns (equations (5.17) in combination with (5.24), for instance, give rise to $L \times K + L$ equations in L^2K endogenous unknowns). In other words, we now require an objective function to use in connection with the above as constraint equations.

There will generally be great scope in the choice of objective function, but at the same time there are a number of criteria it should be expected to satisfy. Prominent here is the need for a suitable underlying theory for the optimization, and this will be given due consideration below. Other matters for consideration are that the function should be non-linear rather than linear in order that the number of non-zero terms at the optimum is not unduly restricted, and preferably convex (for a minimization, or alternatively concave for a maximization) in order to guarantee absolute uniqueness of solution. Apart from this a final useful quality of the programming formulation is that it should restrict all terms in the $\{u^{mnk}\}$ array to be non-negative.

With the aim of focusing firmly on all relevant ideas, a single model satisfying all these criteria will be presented and discussed. This model is notable for its sparing use of technology assumptions and as such provides a particularly suitable starting point for the full range of $\{u^{mnk}\}$ rectangular input–output multiplier models to be developed later.

The model is the following:

$$\text{choose } \{u^{mnk}\} \text{ to maximize } S = -\sum_{mnk} u^{mnk} \ln \frac{u^{mnk}}{\bar{u}^{mnk}} \tag{5.30}$$

$$\text{subject to} \quad \sum_{m} u^{mnk} = b^{kn}\left(\sum_{m'k'} u^{nm'k'} + e^{n*}\right) \tag{5.31}$$

with $u^{mnk} \geqslant 0$ and $u^{mnk} = 0$ whenever $\bar{u}^{mnk} = 0$. The terms in the array $\{\bar{u}^{mnk}\}$ which appear in the objective function (5.30) give the amount of k produced by m and used by n in the base period, and it will be assumed that these terms are known (or, can be found; see below). The constraint equations (5.31) are simply a repetition, for convenience, of input–output relations (5.22).

The purpose of this model is to find terms $\{u^{mnk}\}$ representing the amount of each k produced by each m and used by each n in the projection period, given the base period (prior) values for these terms, $\{\bar{u}^{mnk}\}$, and the absorption matrix input–output relationships (5.31) driven in the usual way by elements of final demand (though given in this case by sector rather than by commodity). It has all the essential properties of an input–output multiplier model in that it produces a unique estimate of sector and commodity outputs in the projection period, with a multiplier effect pulled through due to linear rate coefficients from the base-period absorption table. It also has qualities that are not familiar to existing input–output multiplier models, and some of these will be given further elaboration, in particular, more discussion on the underlying theory of the above model, the explicit position and magnitude of the base-year terms $\{\bar{u}^{mnk}\}$, and suitable numerical solution methods for the model.

The objective function, expression (5.30), is an entropy function (strictly concave, and defined only over non-negative values for $\{u^{mnk}\}$), and its position in the model will ensure that the values of $\{u^{mnk}\}$ produced will be the most probable of all possible solutions that satisfy the constraint equations (5.31), weighted by the prior values $\{\bar{u}^{mnk}\}$. Thus the derivation of the above maximum-entropy estimate will be achieved by assuming that for each unit of demand generated within the multiplier chain, there will be a proportional contribution from all possible suppliers, weighted according to the prior values $\{\bar{u}^{mnk}\}$. In principle this mode of assignment can be traced throughout the multiplier chain defined by constraint equation (5.31).

The weighting by the prior values will ensure that all flows will be weighted according to their magnitude in the base period. This is desirable for a number of reasons. It will hold at zero all flows that were zero in $\{\bar{u}^{mnk}\}$, typically representing the fact that particular sectors do not use all commodities (either in the base or the future period) and similarly that particular sectors typically produce only a relatively small subset of all the commodities that may be identified within any rectangular input–output framework. Thus both the array $\{\bar{u}^{mnk}\}$ and also the array $\{u^{mnk}\}$ will typically have several zero entries. A further aspect of the weighting of individual terms in $\{u^{mnk}\}$ according to their previous magnitude in $\{\bar{u}^{mnk}\}$ is that it will ensure that the intersectoral commodity flows that were of a relatively high magnitude in $\{\bar{u}^{mnk}\}$ will again tend to be of a relatively high magnitude in $\{u^{mnk}\}$. This is desirable if it is expected that, for instance, sectors supplying a large proportion of a particular commodity will continue to supply a large proportion of that demand in the future (with the opposite situation holding for small-scale suppliers). Rather than amplifying this point any further at this stage, it is considered more appropriate to complete an initial summary of the present model, and turn to the question of the magnitudes of the base-period terms $\{\bar{u}^{mnk}\}$, which it is assumed are known.

The data in the base-year rectangular input–output accounts give the amount of each commodity produced by each sector or activity (in the make matrix **V**) and the amount of each commodity used by each sector or activity (in the absorption matrix **U**), and what is needed for present purposes is the corresponding amount of each commodity produced by a particular sector or activity and used by each other sector or activity.

The most obvious maximum-entropy approach to the generation of such base-period terms $\{\bar{u}^{mnk}\}$ would match directly the terms in an absorption matrix **U** to their most likely counterpart in the make matrix **V**, having taken due account in the make matrix of outputs destined for final consumers. This gives the following model:

$$\bar{u}^{mnk} = \frac{U^{kn}(V^{mk} - \hat{e}^{mk})}{Z^k} \tag{5.32}$$

where

$$Z^k = \sum_n U^{kn} = \sum_m (V^{mk} - \hat{e}^{mk}) \tag{5.33}$$

This model again uses for endogenous components of the system the simple assumption that for each unit of demand for a particular commodity by a particular activity, represented by the absorption matrix terms **U**, there will be a proportional allocation to (or contribution from) all possible suppliers, represented by the make-matrix terms **V** having taken account of the degree of closure of the system by subtracting relevant final-demand terms $\breve{e}^{mk}$.

The most important remaining aspect of the above model (given by equations (5.30)–(5.33)) to consider concerns appropriate numerical solution methods. The model as it stands is a concave non-linear maximization in up to L^2K variables with linear constraints. As with models in previous chapters, it is most efficient to derive numerical solutions via the dual formulations. This is given by the following:

$$\text{choose } \gamma^{kn} \text{ to minimize } D = \sum_{mnk} \exp\left[-1 + \gamma^{kn} - \sum_{k'} \gamma^{k'm} b^{k'm}\right] \bar{u}^{mnk} - \sum_{nk} \gamma^{kn} b^{kn} \breve{e}^{n*} \tag{5.34}$$

where γ^{kn} are dual variables (one for each of the constraints), and in terms of these variables, the elements in the required $\{u^{mnk}\}$ array are given by

$$u^{mnk} = \exp\left[-1 + \gamma^{kn} - \sum_{k'} \gamma^{k'm} b^{k'm}\right] \bar{u}^{mnk} \tag{5.35}$$

Thus substitution of the optimal value for γ^{kn} from (5.34) into (5.35) will give the required values for $\{u^{mnk}\}$. While it must be admitted that the numerical solution of the simple non-linear programming rectangular input–output multiplier model is more difficult than the straightforward matrix algebra required for existing models such as (5.5), (5.6) or (5.8), (5.9), its reduction to an unconstrained minimization in just $L \times K$ variables makes its solution even in relatively large-scale applications a viable proposition.

5.4 MORE ALTERNATIVE RECTANGULAR INPUT–OUTPUT MODELS

The overall principles of the new rectangular input–output models have now been outlined and in the present section we formulate a whole family of rectangular input–output models using these general principles. Different members of this family will be generated by using alternative forms of constraint equations, depending on the different make and absorption matrix coefficient assumptions required, and alternative forms of objective function. The constraint equations will always be taken from the basic relationships (5.22), (5.25), (5.26), (5.27), (5.28), and (5.29), and though (5.30) is considered to be a particularly suitable type of objective function, the following alternatives will also be used:

$$\text{choose } \{u^{mnk}\} \text{ to maximize } S' = -\sum_{mnk} u^{mnk} \ln u^{mnk} \tag{5.36}$$

or choose $\{u^{mnk}\}$ to minimize $L = \sum_{mnk} \frac{(u^{mnk} - \bar{u}^{mnk})^2}{\bar{u}^{mnk}}$ (5.37)

Expression (5.36) is a further form of entropy function where in contrast to (5.30) no prior values are present. It will again have the property of producing maximally unbiased estimates subject to whatever information is used, but in this case all information to be used will need to be in the form of constraint equations. This will enable, if desired, sector × commodity outputs to be controlled by make matrix coefficient assumptions such as (5.28) and (5.29), rather than by a base-year weighting as in the model of the previous section. Expression (5.37), a least-squares form, is included as a further candidate objective function, firstly because it is an obvious contender, given the reasoning that the projection-period values $\{u^{mnk}\}$ should be close, in some sense, to the base-period values $\{\bar{u}^{mnk}\}$, and also because it may be considered to be both conceptually and computationally easier to handle than the entropy functions. The above is, of course, by no means the first context in which least-squares and entropy-maximizing formulations have arisen as alternatives to each other. In row and column balancing of a contingency matrix, for example, least-squares and entropy-maximizing formulations have both been widely used. Thus the least-squares form above may be compared with the Friedlander (1961) approach and the entropy-maximizing form may correspondingly be compared with Bacharach's (1970) RAS form (the latter, in turn, being derivable from entropy-maximizing principles).

Note in connection with the former aspect raised in the previous paragraph that the entropy function (5.30) embodies its own criterion of nearness between $\{u^{mnk}\}$ and $\{\bar{u}^{mnk}\}$ both at a theoretical level as mentioned and also because:

$$\ln \frac{u^{mnk}}{\bar{u}^{mnk}} \simeq \frac{(u^{mnk} - \bar{u}^{mnk})^2}{\bar{u}^{mnk}} \quad (5.38)$$

for small $(u^{mnk} - \bar{u}^{mnk})$.

There is a final aspect to consider before presenting the family of alternative rectangular input–output models, and this arises out of the intended nature of these models as non-linear programming formulations. It is simply that any constraint set used should define a feasible region of possible solutions rather than determine exactly a unique solution for the state variables. This is in direct contrast with existing matrix inverse multiplier models, summarized above, where most of the modelling effort may be attributed essentially to generating exactly determined model equations under particular assumptions. Thus in the present case it is deemed unsuitable to attempt to use simple combinations of equations (5.27) with only (5.28) and/or (5.29) since these will tend to produce exactly determined estimates for u^{m**} and u^{*nk} (and in fact reproduce the earlier matrix inverse solutions), or equations (5.25) and (5.26) since these again produce exactly determined estimates, this time for u^{m*k} without any embodiment at all of a multiplier effect. With this in mind the following alternative models are proposed; the objective functions and constraint equations used in these models are summarized for convenience in Figure 5.2.

Objective function (5.30)	$\text{Max } S = -\sum_{mnk} u^{mnk} \ln \frac{u^{mnk}}{\bar{u}^{mnk}}$	Maximally unbiased weighted entropy
Objective function (5.36)	$\text{Max } S' = -\sum_{mnk} u^{mnk} \ln u^{mnk}$	Maximally unbiased unweighted entropy
Objective function (5.37)	$\text{Min } L = \sum_{mnk} \frac{(u^{mnk} - \bar{u}^{mnk})^2}{\bar{u}^{mnk}}$	Weighted least squares

Constraint equation (5.22)	$u^{*nk} = b^{kn}(u^{n**} + \tilde{e}^{n*})$	Commodity inputs proportonal to sector outputs, for each sector
Constraint equation (5.27)	$u^{**k} = \sum_n b^{kn}(u^{n**} + \tilde{e}^{n*})$	Commodity inputs proportional to sector outputs, aggregated over all sectors
Constraint equation (5.28)	$u^{**k} = \sum_m c^{mk} u^{m**}$	Sectors produce commodities in fixed proportions, aggregated over all sectors
Constraint equation (5.29)	$u^{m**} = \sum_k d^{mk} u^{**k}$	Sectors maintain fixed market shares in the production of commodities

Figure 5.2 Objective functions and constraints used in the eight alternative rectangular input–output models

Model 1: objective function (5.30)
constraint equations (5.22)

This is simply the model developed in the previous section. It will produce maximally unbiased estimates for the terms $\{u^{mnk}\}$, weighted by their base-period values and with the multiplier effect determined by equations (5.22). Equations (5.22) serve also to determine the input of commodities to sectors or activities while the weights $\{\bar{u}^{mnk}\}$ on the entropy function serve to control the outputs of commodities from sectors or activities.

Model 2: objective function (5.37)
constraint equations (5.22)

Here we control sector × commodity inputs and outputs in a similar way to model 1, but rather than the theoretical underpinnings of a weighted entropy model, we now form a simple weighted least-squares estimate.

Model 3: objective function (5.36)
constraint equations (5.22)

The constraint equations in this model are the same as those in models 1 and 2, but it takes as its objective function an unweighted entropy function. This amounts in one respect to assigning all the prior values $\{\bar{u}^{mnk}\}$ equal to 1.0 in equation (5.30), and the model has been computed to confirm beyond doubt that such prior values contribute non-redundant information to model 1. The obvious drawback of this model for a practical application is that it has no control at all over commodity outputs from activities, either by weighting or by exact specification.

Model 4: objective function (5.30)
constraint equations (5.27)

This is again a maximally unbiased weighted maximum-entropy model but uses absorption coefficients in a more aggregate form than model 1, in fact copying the absorption-rate equation from existing matrix inverse models (compare equations (5.27) with equation (5.3)). As before, the absorption-rate coefficients will control the input of cmmodities from sectors or activities.

Model 5: objective function (5.37)
constraint equations (5.27)

Here we control sector × commodity inputs and outputs as in model 4, again with absorption-rate coefficients used as in equation (5.3), but use a weighted least-squares rather than a maximum-entropy estimate.

Model 6: objective function (5.36)
constraint equations (5.22)
(5.28)

This is a maximum-entropy model which uses absorption-rate coefficients to embody the multiplier effect and control sector × commodity inputs in the same way as model 1, but rather than controlling sector × commodity outputs by weighting them according to their base-period values, it uses an unweighted entropy function and the fixed proportional-output assumption via coefficients $\{c^{mk}\}$ in constraint (5.28) for this purpose. In this respect it is similar to the existing matrix inverse model given by equations (5.8) and (5.9) above.

Model 7: objective function (5.36)
constraint equations (5.22)
(5.29)

This model simply replaces coefficients c^{mk} in constraint (5.28) and the fixed sector × commodity proportions assumption, by coefficients d^{mk} in constraint (5.29) and the fixed market shares assumption, and is therefore an obvious alternative to model 6. It may be compared directly with the matrix inverse model in equations (5.5) and (5.6).

Model 8: objective function (5.36)
constraint equations (5.22)
(5.28) $m, k \in \mathbf{V}_2$
(5.29) $m, k \in \mathbf{V}_1$

This is the last of the alternative input–output models to be presented at this stage and, as in the case of existing matrix inverse multiplier models, recognizes that it may not be appropriate to apply either matrix **D** or matrix **C** to all elements of production, but rather to use a mixture of the assumptions they embody. Thus model 8 is a maximally unbiased maximum-entropy model that controls the inputs of commodities to each sector or activity and incorporates the familiar multiplier effect via coefficients b^{kn} in equation (5.22), taken as a constraint, and controls the output of commodities from sectors or activities via coefficients c^{mk} and equation (5.28) if they are to be treated on the fixed commodity × sector proportional output assumption, and via coefficients d^{mk} and equation (5.29) if they are to be treated on the fixed market share assumption.

Due to its more intricate nature it is appropriate to show more explicitly the exact form of model 8. In order to achieve this, the special characteristic of this model, namely the partitioning of the elements of production into sets $\mathbf{V}_1$ and $\mathbf{V}_2$, needs to be carried over explicitly to the state variables $\{u^{mnk}\}$. Thus the array $\{u^{mnk}\}$ is now to be made up of two new arrays $\{u_1^{mnk}\}, \{u_2^{mnk}\}$, where $\{u_1^{mnk}\}$ are the outputs of commodities k from sectors or activities m and used by sectors or activities n which are to be treated on a fixed market shares assumption (via coefficients $\mathbf{D}_1$), and $\{u_2^{mnk}\}$ are corresponding terms to be treated on a fixed commodity composition assumption (via coefficients $\mathbf{C}_2$). The full model is then given by

$$\underset{\{u_1^{mnk},\, u_2^{mnk}\}}{\text{maximize}}\ S''' = -\sum_{mnk} u_1^{mnk} \ln u_1^{mnk} - \sum_{mnk} u_2^{mnk} \ln u_2^{mnk} \tag{5.39}$$

subject to

$$u_1^{*nk} + u_2^{*nk} = b^{kn}(u_1^{n**} + u_2^{n**} + \tilde{e}^{n*}) \tag{5.40}$$

$$u_2^{**k} = \sum_m c_2^{mk} u_2^{m**} \tag{5.41}$$

$$u_1^{m**} = \sum_k d_1^{mk} u_1^{**k} \tag{5.42}$$

No further alternative models will be given explicitly at this stage as the above eight are considered to be the most useful of such models and retain a relatively simple structure. However, a rough indication of how further models could arise will be given. The unweighted entropy function in models 6, 7, and 8 could be replaced by a weighted entropy or least-squares function, and this could be done either for the whole of these models, or for certain subsets of components within these models. Further models may be generated by using different types of weights, a more sophisticated way of mixing the proportional sector output and fixed market share assumptions, or even by mixing equations (5.22) and (5.27) in some non-redundant way. In all cases it is necessary to use equations (5.22) or (5.27) as long as a multiplier effect is required in any model, because it is these equations that pull through additional sector or activity inputs once an increased demand for their outputs has been generated.

Numerical solution to any of the above models can be achieved via the corresponding non-linear duals, which will always turn out to be unconstrained convex minimizations in as many variables as there are constraints in the primals.

5.5 DISCUSSION AND FURTHER DEVELOPMENTS

A useful point to note at the outset is that entropy maximization as a modelling tool is not a complete stranger to general input–output methods. The RAS updating or regionalisation technique is familiar and widely used (Bacharach, 1970; Hewings, 1971; Morrison and Smith, 1977) and though not originally presented as such, turns out to be the maximum-entropy estimate of the projected matrix of transactions given the information that is used (some known base matrix of transactions and the specified row and column totals for the projected matrix).

None of the models presented in the previous subsection will give results identical to those of the existing rectangular input–output models summarized in Section 5.2, though they are all based on the same kind of information as the existing models. As mentioned earlier, the only way to reproduce exactly the existing matrix-inverse models would be to use direct combinations of equations (5.27) with (5.28) and/or (5.29) as constraints, but these would then determine uniquely a solution for u^{m**} and u^{**k} and thus make a mathematical programming approach redundant.

Apart from the different theoretical bases in the maximum-entropy and weighted least-squares models, choice between the range of models given above must depend to a large extent on whether commodity outputs from sectors should be controlled simply by a base-period weighting, or whether by explicit make-matrix coefficient assumptions through the use of elements from **D** and **C** in equations (5.28) and (5.29), taken as constraints. In this respect, some comments made by Armstrong (1975) in his summary of corresponding existing matrix-inverse models are relevant. That author points out the need within such existing models to be clear on specifying the roles of matrices **C** and **D**: 'There seems a danger of them being applied in such a way that they act as output structure assumptions when it is intended that they should be acting as weights on input structures.' In the present context it would appear that such a danger can be resolved by using either the weighted entropy or the weighted least-squares objective function rather than constraints (5.28) or (5.29) when weights on input structures are required, and using (5.28) or (5.29) and some unweighted objective function when explicit output structure assumptions are required.

Some of the most interesting aspects of the new models concern their scope for modification and improvement about the central core that has now been developed. These modifications and improvements may take the form of replacing the objective functions given above, or by augmenting the functions already given. They may alternatively take the form of augmenting the constraint sets given in the above models. These aspects will be illustrated immediately.

We have noted above that the models in this chapter can be viewed as forming a bridge between the existing matrix-inverse rectangular input–output models and some of the existing mathematical programming activity-analysis models. In searching for alternative objective functions for the current models it is therefore of interest to review the type used by Takayama and Judge (1972) in their range of activity-analysis models. The linear minimization of total production costs is the most obvious possibility. In the present modelling context, this could produce a model of the form:

$$\text{minimize } P = \sum_{mnk} \rho^{mk} u^{mnk} \tag{5.43}$$

subject to (5.22), say, where ρ^{mk} is some average cost for producing commodity k by sector m. However, the linear programming characteristics of this model reduce its appeal, and it may therefore be more desirable to add a dispersion term to (5.43), giving the following:

$$\text{minimize } P' = \sum_{mnk} \rho^{mk} u^{mnk} + \sum_{mnk} u^{mnk} \ln \frac{u^{mnk}}{\bar{u}^{mnk}} \tag{5.44}$$

subject to (5.22), say. Further alternatives can be generated in a similar way but these will not be pursued here.

The first constraint-set augmentation introduces an alternative way of handling the exogenously specified driving terms behind the familiar multiplier effect. It is a particularly appropriate alternative in cases where there is a degree of closure on the accounts in Figure 5.1. The key driving terms in all the above models are the final demand elements $\tilde{e}^{n*}$ in constraint equations (5.22) and (5.27) and so far it has been necessary to specify an exact value for them in the projection period in order to get a working model. However, if the previously exogenous terms $\tilde{e}^{nk}$ are instead absorbed into an enlarged (m, n, k) index set on the state variables $\{u^{mnk}\}$, then rather than appearing within terms $\tilde{e}^{nk}$, the household demand for commodity k' from sector m', for instance, or the total household demand for commodity k', will now be given respectively by $u^{m'n'k'}$ or $u^{*n'k'}$, where n' represents the household sector and k' a commodity it uses. Thus if it is expected that household demands for various commodities will rise above some specified level ($P^{n'k'}$, say) in the projection period, or alternatively will fall below some specified level ($Q^{n'k'}$, say), expressions of the form

$$u^{*n'k'} > P^{n'k'} \quad \text{or} \quad u^{*n'k'} < Q^{n'k'} \tag{5.45}$$

or whatever, should be added to the existing constraint set of the relevant programming model. This notion can then be applied in a number of different ways as an alternative method of driving the input–output multiplier chain.

As well as the use of additional constraint equations for the purpose of driving the models it is also possible to use additional equations to constrain various productive capacities of sectors or activities. This may be done in either relative or absolute terms. Additional constraints of the form

$$u^{m''n''k''} < \tilde{u}^{m''n''k''} \quad \text{or} \quad u^{m''*k''} < \tilde{u}^{m''*k''} \tag{5.46}$$

where $\tilde{u}^{m''n''k''}$ and $\tilde{u}^{m''*k''}$ are known absolute upper limits for $u^{m''n''k''}$ and $u^{m''*k''}$, will effectively constrain the production of particular commodities by particular sectors or activities by the given absolute upper limits (and lower limits can clearly be handled in an analogous way). Additional constraints of the form

$$u^{m''*k''} < g^{m''} u^{m''*k''} \tag{5.47}$$

would similarly constrain the production of particular commodities from particular sectors or activities, relative to the production of other commodities from those sections or activities. Apart from their use as additional constraints, equations (5.47) could also potentially be used in place of output structure constraints such as (5.28) and (5.29), or weights in objective functions.

A further alternative modelling possibility emerges if it is attempted to combine the flexibility of the mathematical programming formulations of Section 5.4 with some of the computational efficiency of the original matrix-inverse models of Section 5.2. Consider the following model:

$$\text{choose } g^n \text{ and } q^k \text{ to maximize } S' = -\sum_n g^n \ln g^n - \sum_k q^k \ln q^k \tag{5.48}$$

subject to

$$q^k = \sum_n b^{kn} g^n + e^k \tag{5.49}$$

$$g^n, q^k > 0 \tag{5.50}$$

The key multiplier equations (5.49) are simply a repetition of equations (5.3), but rather than combining those underdetermined equations with further relationships in order to derive an exact algebraic solution for g^n and q^k, in the model above they are used instead as constraints within an entropy-maximizing formulation. The entropy function (5.48) is again favoured for its maximally unbiased properties, though as in the case of the earlier rectangular programming models, base-year values $\bar{g}^n$ and $\bar{q}^k$ to weight the endogenous g^n and q^k could be used. Thus (5.48) may be replaced by

$$S^n = -\sum_n g^n \ln \frac{g^n}{\bar{g}^n} - \sum_k q^k \ln \frac{q^k}{\bar{q}^k} \tag{5.51}$$

or some further alternative.

The obvious computational efficiency of formulation (5.48)–(5.50) over that given by (5.30) with (5.22), say, must be weighed against the ability of the latter, (5.30) with (5.22), to incorporate a richer variety of information (not least, the full base-year flows $\bar{u}^{mnk}$), and indeed it would probably be desirable to add to (5.48)–(5.50) some information on sector × commodity outputs, but it is now possible to use coefficients d^{mk} (cf. equation (5.4) above) for particular selected subsets of terms. The most recent formulation, (5.48)–(5.50), retains the flexibility for including further constraints (e.g. capacity constraints) and objectives, with greater computational ease.

A particularly striking feature to emerge from the above demonstrations is the immense number of mathematical programming alternatives that can be devised. The input–output constraints presented in Section 5.3 always provide the core of these models, and as such ensure that the numerical results of alternative models will not be unduly different from each other. However, the various modifications to this core that have been suggested each introduce their own distinctive bias. It would be a laborious task, and one with doubtful rewards, to attempt to document all the possible alternative models that could be derived from the above foundations. We therefore offer the models given so far as the main core of a new family of rectangular input–output approaches.

Finally in this seection it is relevant to add a qualification to some of the new models that have been introduced, specifically to models 1, 3, 4, 6, 7, and 8 of Section 5.4. This pertains to their identification as strict 'maximum entropy' forms. The problem is that it is not clear how to determine the probabilities p^{mnk}, say, associated with the terms u^{mnk}. It is appropriate to note explicitly in this connection that since the above $\{u^{mnk}\}$ 'maximum entropy' formulations are cast in terms of magnitudes of flows rather than probabilities of flows, it is strictly necessary to ensure that the model results obtained will be directly equivalent to those obtained from the alternative probability representations. As long as this is ensured then the choice between magnitude and probability representation can be left largely to personal taste, and it can generally be ensured as long as a constraint such as

$$\sum_{mnk} u^{mnk} = X \tag{5.52}$$

where X is a known constant, appears within the constraints of a model given in terms of magnitudes, since we then have $p^{mnk} = u^{mnk}/X$. This raises a problem, because in fact, X is *a priori* unknown in the context of the models of the present chapter. The matter will be left unresolved here, but it is important to note that different values of X in (5.52) will produce different numerical results for the terms $\{u^{mnk}\}$. The easiest approach to adopt is that already taken in the model presentations above, merely to omit an explicit counterpart to constraint (5.52) and implicitly accept the value of X implied by the rest of the model (see Macgill, 1977a, Appendix 4, for further discussion of this and of a similar problem).

5.6 SPATIAL MODELS

In this section we investigate the formulation of a spatially disaggregated model based on the developments of the previous sections. There would appear to be two obvious ways of achieving suitable models, the first being to use the type of modelling approaches of the previous sections at the activity and commodity level and then to disaggregate the results obtained over the spatial component, and the second being to introduce the spatial component directly into the type of model equations given in the previous sections.

Relatively little will be said about the former method, though there will clearly be several alternative ways of zonally scaling given projection-period activity and commodity inputs and outputs (one of which is given below), and the neglect of a spatial component when deriving the core model results will enormously reduce the dimensions of the arrays involved—an important consideration in an application of any reasonable size.

The second method will be approached by summarizing four additions to the formulations of previous sections in order to accommodate sufficiently the spatial components on the core model variables. These are:

(1) Generation of base-year estimates for a $\{\bar{u}_{ij}^{mnk}\}$ array (defining the amount of commodity k produced by activity m in zone i and used by activity n in zone j) for use in place of the non-spatially disaggregated terms $\{\bar{u}^{mnk}\}$ as prior values in the entropy function. An entropy-maximizing model based derivation of these terms could take a variety of forms (see, for example, Macgill, 1977a, Chapter 4). We could, for example, follow the classic principles behind the entropy-maximizing production-attraction constrained spatial-interaction models (Wilson, 1970, 1971), and generate $\{\bar{u}_{ij}^{mnk}\}$ in the following way:

choose $\{\bar{u}_{ij}^{mnk}\}$, $\{\bar{u}_{io}^{mnk}\}$, $\{\bar{u}_{oj}^{*nk}\}$to maximize

$$S = -\sum_{ijmnk} \bar{u}_{ij}^{mnk} \ln \bar{u}_{ij}^{mnk} - \sum_{imk} \bar{u}_{io}^{m*k} - \sum_{jnk} \bar{u}_{oj}^{*nk} \tag{5.53}$$

subject to;

(a) production constraint, given known terms X_i^{mk},

$$\sum_{jn} \bar{u}_{ij}^{mnk} + \bar{u}_{io}^{m*k} = X_i^{mk} \tag{5.54}$$

(b) an attraction constraint, given known terms Y_j^{nk},

$$\sum_{im} \bar{u}_{ij}^{mnk} + \bar{u}_{oj}^{*nk} = Y_j^{nk} \tag{5.55}$$

(c) a spatial-interaction cost constraint, given known average cost terms c_{ij}^k, c_{io}^k, c_{oj}^k,

$$\sum_{ijmn} c_{ij}^k \bar{u}_{ij}^{mnk} + \sum_{im} c_{io}^k \bar{u}_{io}^{m*k} + \sum_{jn} c_{oj}^k \bar{u}_{oj}^{*nk} = C^k \tag{5.56}$$

A classical Lagrangian solution produces the estimates:

$$u_{ij}^{mnk} = A_i^{mk} B_j^{nk} \exp[-\beta^k c_{ij}^k] \tag{5.57}$$

$$u_{io}^{m*k} = A_i^{mk} \exp[-\beta^k c_{io}^k] \tag{5.58}$$

$$u_{oj}^{*nk} = B_j^{nk} \exp[-\beta^k c_{oj}^k] \tag{5.59}$$

with

$$A_i^{mk} = \frac{X_i^{mk}}{\sum_{jn} B_j^{nk} \exp[-\beta^k c_{ij}^k] + \exp[-\beta^k c_{io}^k]} \tag{5.60}$$

$$B_j^{nk} = \frac{Y_j^{nk}}{\sum_{im} A_i^{mk} \exp[-\beta^k c_{ij}^k] + \exp[-\beta^k c_{oj}^k]} \tag{5.61}$$

In the numerical solution to this model it is fortunately possible to solve equations (5.60) and (5.61) (typically using the usual type of balancing factor iterative routine) separately for each k.

(2) Use, where possible, of zonal specific input–output coefficients in the key constraining equations, namely a zonal disaggregation of matrices **B**, **C**, and **D**.

(3) Use, where desired, of a spatial-interaction cost constraint in order that the familiar distance deterrence effect may operate in the projection as well as the base period. Thus although the presence of spatially disaggregated prior values $\{\bar{u}_{ij}^{mnk}\}$ in the entropy function will already have the effect of introducing spatial weighting, an equation such as (5.56) (given now in terms of $\{u_{ij}^{mnk}\}$) could also be included in the constraint set so that the multiplier process within the projection-

period model is quite explicitly influenced by distance cost effects.

(4) Adequate regard for system export and import terms, with particular deliberation over whether they are to be treated endogenously or exogenously.

By way of a more explicit presentation of these notions, a spatially disaggregated model based on model 1 of Section 5.4 could take the following form:

$$\text{choose } \{u_{ij}^{mnk}\} \text{ to maximize } -\sum_{ijmnk} u_{ij}^{mnk} \ln \frac{u_{ij}^{mnk}}{\bar{u}_{ij}^{mnk}}$$

$$-\sum_{imk} u_{io}^{m*k} \ln \frac{u_{io}^{m*k}}{\bar{u}_{io}^{m*k}} - \sum_{jnk} u_{oj}^{*nk} \ln \frac{u_{oj}^{*nk}}{\bar{u}_{oj}^{*nk}} \tag{5.62}$$

subject to:

$$\sum_{im} u_{ij}^{mnk} + u_{oj}^{*nk} = b_j^{kn}\left(\sum_{i'm'k'} u_{ji'}^{mn'k'} + \sum_{k'} u_{jo}^{n*k'}\right) \tag{5.63}$$

$$\sum_{ijmn} u_{ij}^{mnk} c_{ij}^k + \sum_{im} u_{io}^{m*k} c_{io}^k + \sum_{jn} u_{oj}^{*nk} c_{oj}^k = C^k \tag{5.64}$$

$$\sum_{ijmnk} u_{ij}^{mnk} + \sum_{jk} u_{oj}^{*nk} = Y^n \qquad \text{for } n = \text{households} \tag{5.65}$$

for $u_{ij}^{mnk} > 0$ with $u_{ij}^{mnk} = 0$ whenever $\bar{u}_{ij}^{mnk} = 0$.

The objective function (5.62) is self-explanatory; constraint (5.63) is the key input–output multiplier equation, with the absorption rate coefficients from an array **B** now disaggregated, if possible, over the spatial component; constraint (5.62) is an appropriate spatial-interaction cost constraint; constraint (5.65) specifies the exogenous demand that is the driving force for the multiplier reactions to be picked up by this model, and its appearance as a separate constraint, rather than including terms Y^n within the parentheses on the right-hand side of equation (5.63), indicates the above model is assuming closure at the (m, n, k) levels.

It is straighforward to see that it will be equally feasible to derive spatially disaggregated models based on any other of the eight alternative models from Section 5.4, and also to add to and refine the resulting models along the lines indicated in Section 5.5 or otherewise.

As a final point, and returning to the first method of spatial disaggregation indicated in this section, the model given by equations (5.62)–(5.65) above prompts the suggestion of an interesting way of scaling projection-period values $\{u^{mnk}\}$ over a spatial component. We may derive projection-period values $\{u^{mnk}\}$ using the type of methods discussed in Sections 5.3 and 5.4, or some existing rectangular multiplier methods (Gigantes, 1970), and other sources given earlier in Section 5.2, and then use them within a model such as the following:

choose $\{u_{ij}^{mnk}\}$, $\{u_{io}^{m*k}\}$, $\{u_{oj}^{*nk}\}$. to maximize

$$-\sum_{ijmnk} u_{ij}^{mnk} \ln \frac{u_{ij}^{mnk}}{u^{mnk}} - \sum_{imk} u_{jo}^{m*k} \ln \frac{u_o^{m*k}}{u^{m*k}} - \sum_{jnk} u_{oj}^{*nk} \ln \frac{u_{oj}^{*nk}}{u^{*nk}} \tag{5.66}$$

subject to

$$\sum_{jn} u_{ij}^{mnk} + u_{io}^{m*k} = X_i^{mk} \tag{5.67}$$

$$\sum_{im} u_{ij}^{mnk} + u_{oj}^{*nk} = Y_j^{nk} \tag{5.68}$$

$$\sum_{ijmn} u_{ij}^{mnk} c_{ij}^k + \sum_{im} u_{io}^{m*k} c_{io}^k + \sum_{jn} u_{oj}^{*nk} c_{oj}^k = C^k \tag{5.69}$$

In this model, constraints (5.67) and (5.68) are, respectively, production and attraction constraints in the projection period, and (5.69) is, of course, a spatial-interaction cost constraint. The spatially aggregated projection period values $\{u^{mnk}\}$ appear as weights on the entropy function (5.66), and the purpose of this model is clearly to scale (disaggregate) these over the spatial component; note that X_*^{mk} and Y_*^{nk} from (5.67) and (5.68) must be consistent with these $\{u^{mnk}\}$ values. The algebraic form of the solution to this model will show that it amounts, in effect, to a multiplicative scaling of the $\{u^{mnk}\}$ values: the classical Lagrangian manipulations for this model produce the expressions:

$$u_{ij}^{mnk} = A_i^{mk} X_i^{mk} B_j^{nk} Y_j^{nk} \exp[-\beta^k c_{ij}^k] u^{mnk} \tag{5.70}$$

$$u_{io}^{m*n} = A_i^{mk} X_i^{mk} \exp[-\beta^k c_{io}^k] u^{m*k} \tag{5.71}$$

$$u_{oj}^{*nk} = B_j^{nk} Y_j^{nk} \exp[-\beta^k c_{oj}^k] u^{*nk} \tag{5.72}$$

where A_i^{mk}, B_j^{nk}, and β^k ensure, respectively, that constraints (5.67), (5.68), and (5.69) are satisfied. Thus it is seen that the above model consists, in effect, of a multiplicative zonal disaggregation of projection-period $\{u^{mnk}\}$ values. Although it may appear to be less elegant than, for instance, the model given by (5.62)–(5.65), the more recent model has a dimensional advantage due to the fact that having calculated the projection-period $\{u^{mnk}\}$ values, the spatially disaggregated $\{u_{ij}^{mnk}\}$ projection-period values may be calculated separately for each k in turn. Iterative solution routines for the A_i^{mk} and B_j^{nk} balancing factors may be derived in the usual way by substituting (5.70)–(5.72) into equations (5.67) and (5.68), and solving for each k in turn. These will behave like the modified biproportional schemes in Macgill (1979). Choice between the models will, as ever, depend on individual circumstances, since their theoretical, computational, and data properties all need to be taken into consideration.

5.7 CONCLUDING REMARKS

Our chosen focus of interest above has been to develop alternative ways of embedding a rectangular input–output multiplier mechanism within a mathematical programming framework. In claiming to cover new ground here, it is important to note a particular difference between the type of mathematical programming models offered above and existing mathematical programming input–output models in the literature. The latter seem to have been of three broad types: firstly, models which use inequality rather than equality constraints in order to ensure sufficient degrees of freedom in the constraint set (Moses, 1955; Mathur, 1972); secondly, models in which the multiplier mechanism is calculated *a priori*, leaving the mathematical programming component only to distribute the result between producing and receiving sectors; and finally, those which, although using input–output coefficients to represent technologies of production, do not attempt to pick up a multiplier mechanism.

6. An extended Lowry model as an input–output model

6.1 INTRODUCTION

In the two preceding chapters we have focused respectively on various behavioural developments of the Lowry model (in Chapter 4) and on a specific type of input–output model framework (the rectangular input–output models in Chapter 5). Here, we focus on particular aspects of both these modelling approaches by formally presenting the Lowry model as an input–output model, and then working towards a mathematical programming formulation of this representation.

The input–output representation of the Lowry model has apparently not until recently been used by other authors (the recent exception being Williams, 1979), though the foundations for such an approach have been evident in a number of rather earlier papers (Broadbent, 1973; Romanoff, 1974). It was considered worth while in this chapter to describe the input–output representation in full (in Section 6.2) as it is of interest in its own right and raises a number of aspects worth some discussion. It is a particularly useful representation through its ability to offer a basis for various extensions to be made to the Lowry model within a comprehensive, internally consistent, but conceptually simple framework. The most immediate of these extensions (to be considered in Section 6.3) is the incorporation of full inter-activity relations, rather than being restricted to three all-embracing activity types—basic, retail, and household—with only partial interaction between them. This in turn invites several comparisons to be made between what have become two much-favoured though hitherto essentially independently utilized model frameworks, the input–output model and the Lowry model.

Alongside the representation of full inter-activity relations in an input–output based Lowry model in Section 6.3, a mathematical programming version of the model will be given and this will be compared with the existing algebraic approach. The discussion here will have some parallels with that given in Chapter 4, where some comparison between mathematical programming and other formulations of the Lowry model was made. Furthermore, in Chapter 10 we return to this line of argument yet again, in the context of a fuller discussion of similarities and equivalences between different methodological approaches for given models.

6.2 THE LOWRY MODEL AS AN INPUT–OUTPUT MODEL

6.2.1 Summary of the Lowry model

Although a summary of the Lowry model has already been given at the beginning of Chapter 4, it was given in the context of the developments sought within that chapter. It is appropriate to begin here with a further brief summary, as the developments to be made are now of a rather different nature. We follow the exposition of Wilson (1974).

A distribution of basic employment in various zones (E_i^B) is the initial driving force in the model, significant not for its potential to increase industrial output or for any direct or indirect consequence of such an increase, but rather for its immediate demand for residence. This may be typically represented in an equation of the form:

$$T_{ij} = A_i E_i f_H(c_{ij}) \tag{6.1}$$

(with $E_i = E_i^B$), which may be called a reverse journey-to-work model, where zones i are employment zones, j are residential zones, $f_H(c_{ij})$ is some spatial-interaction deterrence function, and A_i a balancing factor to ensure that the given employment totals E_i are satisfied. A_i is given explicitly by

$$A_i = \frac{1}{\sum_j f_H(c_{ij})} \tag{6.2}$$

The population generated in each zone follows immediately:

$$P_j = g \sum_i A_i E_i f_H(c_{ij}) \tag{6.3}$$

given some known inverse activity rate g.

The retail requirements of the resident population and also possible job-based retail utilization induces retail employment, distributed in relation to the likely shopping patterns of the resident population and employees. A number of separate stages may be recognized here: total retail employment (E^{R_k}) is proportional to total population (P):

$$E^{R_k} = a^{R_k} P \tag{6.4}$$

where a^{R_k} is a suitable set of constants: the shopping patterns of the resident population may be given by a typical singly constrained shopping model:

$$S_{ij}^{R_k} = D_i e_i^{R_k} P_i f_{R_k}(c_{ij}) \tag{6.5}$$

where $S_{ij}^{R_k}$ is the flow of cash from residents in zones i to retail sectors R_k in zones j, $e_i^{R_k}$ is expenditure per head in i on the output of retail sector k, and D_i is a balancing factor given by:

$$D_i = \frac{1}{\sum_j f_{R_k}(c_{ij})} \tag{6.6}$$

Thus the zonal distribution of retail employment may be given by:

$$E_j^{R_k} = b^k \left[c^k \sum_i D_i e_i^{R_k} P_i f_{R_k}(c_{ij}) + d^k E_j \right] \tag{6.7}$$

where c^k and d^k are home-based and job-based retail utilization rates and b^k a balancing factor to ensure that

$$E^{R_k} = \sum_j E_j^{R_k} \tag{6.8}$$

The resulting retail employment creates a further demand for residence, thus a reverse journey-to-work model of the form of equation (6.1) (with E_i replaced by $E_i^B + E_i^{R_k}$) is again brought into use and a fresh round of reactions through the above equations. Successive rounds continue until changes in employment and population totals are negligible, and each round of cycles may be subject to land-use density constraints as related in, for example, Wilson (1974).

In the developments that follow, two simplifications to the outline of the model as given above will be made, in that there will be no attempt to incorporate job-based retail utilization or any land-use density constraints until a later section of the chapter. Furthermore, the developments will concentrate initially on the sectoral component (basic–retail–household interactions), introducing the spatial component at a later stage. This will both ease the overall presentation and allow for a preliminary discussion in Section 6.4 of several points arising from a spatially aggregated input–output representation of the Lowry model.

6.2.2 Spatially aggregated input–output representation of the Lowry model mechanism

In order to achieve an input–output representation of the Lowry model we first identify the interactions from the Lowry model summary given above as particular cells in the usual form of input–output accounting framework. The particular classification of sectors used in the Lowry model gives rise to row and column labels for this framework as shown in Figure 6.1 (and it may be seen that these reflect a different emphasis than that usually associated with input–output accounts).

Given the convention that cells in the table in Figure 6.1 record a sale by the relevant row and a purchase by the appropriate column, the entries may be derived directly from the Lowry model summary as follows.

The household row records sales of labour to the retail sectors and to the basic sector, namely

$$w^{R_k} T_{**}^{R_k} \text{ for retail labour } (= w^{R_k} E_*^{R_k}) \tag{6.9}$$

$$w^B T_{**}^B \text{ for basic labour } (= w^B E_*^B) \tag{6.10}$$

	Hshds	Retail sectors	Basic sectors	Row totals
Hshds	0	$w^{R_1}T^{R_1}_{**}\, w^{R_2}T^{R_2}_{**} \ldots w^{R_N}T^{R_N}_{**}$	$w^B T^B_{**}$	$\bar{w}E$
	$\bar{S}^{R_1}_{**}$	0.	0	$E^{R_1}R_1$
	$\bar{S}^{R_2}_{**}$	. . . .	.	$E^{R_2}R_2$
		. . . .	.	
	.	. . . .	.	.
	.	. . . .	.	.
Retail	.	. . . .	.	.
sectors	.	. . . .	.	.
	$\bar{S}^{R_N}_{**}$	0	0	$E^{R_N}R_N$
		endogenous	exgenous	totals

Figure 6.1 Lowry model flows in an input–output accounting framework

where w^{R_k} and w^B are retail- and basic-sector wage rates. These expressions have been entered in the household row of Figure 6.1, and an alternative (more explicit) form for them can be used by substituting for $T^{R_k}_{**}$ and T^B_{**} from equation (6.1).

The retail-sector rows record sales of retail goods to households; since equation (6.5) represents cash flow from households to retail sectors, the equation:

$$\bar{S}^{R_k}_{ij} = D_j e^{R_k}_j P_j f_{R_k}(c_{ij}) \tag{6.11}$$

with

$$D_j = \frac{1}{\sum_i f_{R_k}(c_{ij})} \tag{6.12}$$

(the 'reverse' of equations (6.5) and (6.6)) will give the value of goods sold by retail sectors in i to households in j. This gives entries $\bar{S}^{R_k}_{**}$ in the retail–household cells of Figure 6.1, and as before, an alternative (more explicit) form for $\bar{S}^{R_k}_{**}$ may be found directly, this time from equation (6.11).

The row totals corresponding to the entries described immediately above will be needed later for deriving coefficients, and it is convenient to give them here. Thus for the household row, the sum is:

$$\sum_k w^{R_k}T^{R_k}_{**} + w^B T^B_{**} = \text{total wage bill} = \bar{w}E \tag{6.13}$$

where $\bar{w}$ is the average total wage rate and E the total number of employees. For the retail-sector row, the sums are simply $S^{R_k}_{**}$, which for present purposes, it is more convenient to write as:

$$S^{R_k}_{**} = e^{R_k}_* P \tag{6.14}$$

The entries described above represent all the transactions evident in the Lowry model mechanism (at a spatially aggregated level)—employees (output from households) are used by basic and retail sectors, retail goods (output from retail sectors) are used by households.

Since input–output models rest on coefficients derived from the endogenous part of the input–output accounting framework, the second stage in an input–output modelling representation of the Lowry model is to decide how equations (6.1)–(6.8) may be derived from the flows given in Figure 6.1. In particular, it is necessary to establish that in the Lowry model (and more specifically in each iteration round the Lowry model equations) the magnitude of inputs to each sector is linearly dependent on the output of that sector. It is straightforward to show that this is in fact true.

Consider the input of employees to retail sectors; in the Lowry model these depend linearly on the output of retail sectors (via coefficients a^{R_k} in equation (6.4)), thus it is valid to use corresponding coefficients for an equivalent input–output representation. Similarly, the input of retail goods to the household sector depends in the Lowry model on the size of the resident population, which is in turn directly related via (linear) coefficients g in equation (6.3) to the output of employees; thus again, the input of goods to the household sector is linearly dependent on the output of that sector, that is, on the employees produced.

The coefficients discussed here are shown schematically in Figure 6.2 and more explicit expressions in place of the a^{R_kH}'s and a^{HR_k}'s may be derived from Figure 6.1 (via the identities (6.9) and (6.10)) simply by dividing each cell by the receiving sector's row sum, viz:

$$a^{R_kH} = \frac{\bar{S}^{R_k}_{**}}{\bar{w}E} = \frac{e^{R_k}_{*}P}{\bar{w}E} \tag{6.15}$$

$$a^{HR_k} = \frac{W^{R_k}T^{R_k}_{**}}{E^{R_k}_{*}h^{R_k}} = \frac{W^{R_k}E^{R_k}_{*}}{E^{R_k}_{*}h^{R_k}} = \frac{W^{R_k}}{h^{R_k}} \tag{6.16}$$

At this point it may be noted that the 'suitable set of constants' a^{R_k} for the employment inducement in equation (6.4) in the initial Lowry model summary are themselves a combination of the terms e^{*R_k} (giving the per capita demand for

	Hshlds	Retail sectors	Basic sectors	Totals
Hshlds	0	$a^{HR_1}a^{HR_2}\ldots a^{HR_N}$	$w^BT^B_{**}$	$\bar{w}E$
Retail sectors	a^{R_1H}	0.	0	$E^{R_1}R_1$
	a^{R_2H}	. . . .	0	$E^{R_2}R_2$
	.	. . . .	.	.
	a^{R_NH}	0	0	$E^{R_N}R_N$

Figure 6.2 The pattern of input–output coefficients for the Lowry model

retail goods R_k) and h^{R_k} (giving the output per employee of retail goods R_k), namely:

$$a^{R_k} = e_*^{R_k}/h^{R_k} \tag{6.17}$$

and it has been found convenient for the input–output representations to break down the constants a^{R_k} into their two components e^{*R_k} and h^{R_k}.

Coefficients such as (6.15) and (6.16) are the key to an input–output model and in the present case they subsume an identical set of assumptions to those present at the sector level in the Lowry model summary given in equations (6.1)–(6.8). The fundamental mechanism is that changes in a given sector's input requirements are pulled through by changes in that sector's outputs; the former changes are themselves changes in other sectors' outputs, and these in turn pull through changes in input requirements, thus generating a whole chain of successive multiplier reactions through all linked sectors in the Lowry as in the input–output model. In order to emphasize this mechanism further, this chain of multiplier reactions for the Lowry model will be related more explicitly to the summary of that model given earlier. The initial exogenous demand for basic employment (E_i^B in equation (6.1)), creates an increase in output from the households (via equation (6.1), giving $w^B T_{**}^B$ in Figure 6.2)). This increase causes an increase in household input requirements, that is, of goods from the retail sector (manifested in the shopping model in equation (6.5) and by coefficients $a^{R_k H}$ in Figure 6.2). The ensuing change in retail output gives rise to new non-basic employment requirements (as given by equation (6.7) in the Lowry model and picked up via coefficients a^{HR_k} in Figure 6.2), that is, a further change in output from households. Thus a fresh round of reactions through equations (6.1), (6.5), and (6.7) (alternatively, coefficients $a^{R_k H}$ and a^{HR_k}) is generated, and successive rounds will continue until the changes pulled through the system are of negligible magnitude.

By way of completing the input–output representation, we may note that the identities that may be formed from the coefficients in Figure 6.2 along with the initially given exogenous distribution of basic employment immediately give rise to the usual form of input–output inverse multipliers.

From Figure 6.2 (and taking X^H, X^{R_k} as the total outputs of the household and retail sectors),

$$\begin{aligned} X^H &= a^{HR_1} X^{R_1} + a^{HR_2} X^{R_2} + \ldots + a^{HR_N} X^{R_N} + w^B T_{**}^B \\ X^{R_1} &= a^{R_k H} X^H \\ &\vdots \\ X^{R_N} &= a^{R_N} X^H \end{aligned} \tag{6.18}$$

and therefore

$$\mathbf{X} = (\mathbf{I} - \mathbf{A})^{-1} \mathbf{Y} \tag{6.19}$$

where in this case $\mathbf{A}$ is the matrix of coefficients displayed in the endogenous partition of Figure 6.2, viz:

$$\mathbf{A} = \begin{bmatrix} 0 & a^{HR_1} \ldots & a^{HR_N} \\ a^{R_1H} & 0 & \ldots \\ & \vdots & \vdots \\ a^{R_NH} & & \ldots \end{bmatrix} \tag{6.20}$$

Y is the exogenous vector

$$\mathbf{Y} = \begin{bmatrix} w^B E_*^B \\ 0 \\ \vdots \\ 0 \end{bmatrix} \tag{6.21}$$

and **X** the endogenous vector of sector outputs

$$\mathbf{X} = \begin{bmatrix} X^H \\ X^{R_1} \\ \vdots \\ X^{R_N} \end{bmatrix} \tag{6.22}$$

to be predicted by the model.

Thus, rather than working round equations (6.1)–(6.8) in order to produce a Lowry model solution, equation (6.19), with (6.20), (6.21), and (6.22) defining the terms involved, may be used instead, at a spatially aggregated level.

6.2.3 Discussion of the input–output representation

The immediate conclusion of the previous section is that the successive round of sectoral interactions evident in the Lowry model may be interpreted as a conventional (Leontief-type, demand-driven) input–output multiplier model, hinging on coefficients which say that sector or activity inputs are linearly dependent on the output of these sectors, but taking an initial distribution of basic employment as the initial driving force. This section discusses a number of points arising from the above representations.

It is considered useful to clear up a possible objection to the above presentation, arising from the observation that, at a spatially aggregated level, total sector outputs (the far right-hand column of Figures 6.1 and 6.2) are known *a priori*. A conventional input–output model would state its main aim as the estimation of these terms, but for a spatially aggregated Lowry model, they are known from the start, thus apparently making the above representation redundant. In reply we would simply point out that the purpose of the above section was mainly to throw insights on the underlying model mechanism (rather than to suggest a new numerical solution method), namely to recognize this mechanism as an input–output model. Two further comments may be noted, however. Firstly, the final sector outputs are not necessarily the only elements of interest in the model; the progress through the successive rounds of multiplier cycles may also be of interest to follow, and here the above representation will give an exact parallel to familiar input–output multiplier rounds. Secondly, the redundancy of the new representation, caused by an ability to fix total sector outputs *a priori*, will not occur in a spatially disaggregated version, because it is then not possible to prefix the corresponding totals (by zone); indeed, it is, of course, the estimation of the spatial distribution of employment and population totals (which follow directly given total sector outputs) that is the overall aim of the Lowry model.

Turning to a different aspect, it has been noted that the Lowry model input–output representation in equations (6.19)–(6.22) above is of the conventional Leontief-type demand-driven kind (with the key coefficients derived by dividing input–output transactions by the receiving sectors' total outputs). There is, however, an alternative though less popular form of input–output model (favoured, for example, by Augustinovics, 1970; Ghosh, 1958; and Giarratani, 1976), where the key coefficients are derived by dividing input–output transactions by the selling sectors' total outputs. The former and more popular type of input–output model is often referred to as being demand-driven (or an 'input' approach in the terminology of Augustinovics, 1970), and the latter, less familiar type is similarly referred to as being supply-driven (or an 'output' approach in the terminology of Augustinovics, 1970).

The use of exogenously given employment totals to drive an input–output model is in fact characteristic of supply-driven schemes, and not of demand-driven schemes. Thus, although it is exogenous employment that triggers the chain of reactions in the Lowry model above, it appears in the context of a demand-driven scheme, because as noted on several occasions above, in the Lowry model sector inputs are pulled through by a change in output of the receiving sector and not by a change in output of the producing sector. Thus, it is directly comparable to the original Leontief formulation, as suggested above. Whether it would seem suitable to restructure the whole model in line with Ghosh's (1958) input–output representation is a separate matter for investigation.

This aspect leads conveniently into a further point, namely that it is strange to find two models (the Lowry model and Leontief's input–output model) which have each found considerable popularity in their own right, and which have now been shown to have comparable internal structures, to be driven by two quite different forces. The Lowry model chooses a given requirement of basic employment, while the input–output model chooses a prespecified level of final demand—two quite different entities. The two cannot be reconciled by arguing that basic employment (in the Lowry model) may be export employment, and exports in turn often traditionally appear in a (Leontief-type) final demand vector. This line of reconciliation is unsound because the pattern of non-zero entries in the Lowry model exogenous vector does not cover manufactured goods at all (for export or any other possible component of 'final demand') because the only non-zero cell in this vector is the household row cell. The paper of Romanoff (1974) can be used to amplify any discussion on this point.

The simplicity of the inter-activity interactions present in the Lowry model is clearly illustrated by the pattern of entries in Figures 6.1 and 6.2 (non-zero entries in just two diagonally opposite partitions). Once the initial distinction between basic and retail has been made, these sectors operate quite independently of each other and moreover require only one input, labour, to sustain their production. Any increase in basic output is apparently of no endogenous interest to the system, and fundamental technologies of production are further ignored in allowing no retail–basic or basic–retail interaction at all. Significant modification to the overall model mechanism rather than any amount of redefinition of basic and retail sectors is required in order to overcome these criticisms, an exercise which in principle is quite a simple matter if attempted within the context of an input–output framework. An input–output representation of the Lowry model can immediately accommodate a full set of interactions both within and between basic and retail sectors, merely through the appearance of all sectors in the endogenous partition of the accounting table and appropriate non-zero coefficients for all sectors it is desired to link. A suitable table of coefficients is given in Figure 6.3. Whether the use of such a table within an urban model structure will retain recognizable Lowry model characteristics or will blur into conventional input–output analysis will depend to a large extent on the pattern of entries prespecified by the user in the 'exogenous sectors' column. In principle the choice is a wide one. Existing Lowry model characteristics could be retained by filling just the upper cell (as basic employment) and leaving the rest blank; this would

	Hshlds	Retail sectors	Basic sectors	Exog sectors	Total outputs
Employees (hshlds)		$a^{HR_1} \,.\,.\, a^{HR_N}$	$a^{HB_1} \,.\,.\, a^{HB_M}$	√	X^H
Retail goods (retail sectors)	a^{R_1H} · · · $a^{R_NH_N}$	$a^{R_1R_1}$ . . . . $a^{R_NR_N}$	$a^{R_1B_1}$ . . . . $a^{R_NB_M}$	(√)	X^{R_1} · · · X^{R_N}
Basic goods (basic sectors)	$a^{B_1H_1}$ · · · a^{R_MH}	$a^{B_1R_1}$ . . . . $a^{B_MR_N}$	$a^{B_1B_1}$ . . . . $a^{B_MB_M}$	(√)	X^{B_1} · · · X^{B_M}

Figure 6.3 Schematic representation of full sectoral interactions coefficients for an extended input–output Lowry Mcdel

have the effect of driving the model with basic employment, as before, though including the improvement of acknowledging basic–retail interactions in the endogenous part of the table (assuming relevant data were available). A more traditional input–output approach may prefer to leave the upper cell blank and fill the rest, thus driving the model with something approaching the usual final demand. Such a traditional approach may also prefer to remove the first row and column from the endogenous part of the table. Such considerations go back to Artle (1961) and beyond, and need no further comment at this stage.

In the next section we return to the familiar (spatially disaggregated) Lowry model and complete an input–output representation of it; the position reached in the previous paragraph, a discussion of a full range of intersectoral relations, is picked up again in Section 6.3.

6.2.4 Spatially disaggregated input–output Lowry model representation

Figure 6.4 is a spatially disaggregated version of Figure 6.1 where, in order to sharpen the discussion, two retail sectors and three zones have been identified. The entries will be discussed in the order that a Lowry-type formulation would require them.

Of initial interest is the exogenously prespecified basic employment. This appears in the table according to the assumed placed of residence of the basic employees (that is, they appear as outputs of households, by zones), and adjusted via their wage rates to bring them into money-based units. Thus, from a given distribution of basic employment E_j^B and a model as given in equations (6.1) and (6.2), the value of labour supplied by each zone of residence may be found. This is given by

$$w^B T_{i*}^B = w^B \sum_j \frac{E_j^B f_H(c_{ij})}{\sum_i f_H(c_{ij})} \tag{6.23}$$

Expression (6.23) has been entered in the basic-employment column of Figure 6.4 for each household row; all other cells in this column are zero.

The increase in output from the household sector, causing an increase in the income in the household sector, will induce a flow of goods from the retail sectors to the households (as given in equations (6.11) and (6.12)); thus the next partition of interest is the lower left-hand one. The flows of retail goods to households are given by the $\bar{S}_{ij}^{R_k}$ in equation (6.11), and these therefore form the entries of the latter partition.

The change in production within the retail sectors that will be pulled through via these latter flows will cause a change in the employment requirements of the retail sectors; thus flows in the upper right-hand cells of the endogenous partitions of the table are needed. The conventional journey-to-work model (the reverse of equation (6.1)) in conjunction with the retail-sector wage rates, w^{R_k}, give the value of the input from households by zone to the retail sectors by zones,

		Hshlds			Retail 1			Retail 2			Basic totals	Row totals
		zone 1	zone 2	zone 3	zone 1	zone 2	zone 3	zone 1	zone 2	zone 3		
Hshlds	zone 1	0	0	...	$w^{R_1}T_{11}^{R_1}$	$w^{R_1}T_{12}^{R_1}$	...	$w^{R_2}T_{11}^{R_2}$	$w^{R_2}T_{12}^{R_2}$	...	$w^B T_{1*}^B$	X_1^H
	zone 2	0			$w^{R_1}T_{21}^{R_1}$			$w^{R_2}T_{21}^{R_2}$			$w^B T_{2*}^B$	X_2^H
	zone 3	⋮			⋮			⋮			$w^B T_{3*}^B$	X_3^H
Retail 1	zone 1	$\bar{S}_{11}^{R_1}$	$\bar{S}_{12}^{R_1}$	...	0	0	...				0	$X_1^{R_1}$
	zone 2	$\bar{S}_{21}^{R_1}$			0						0	$X_2^{R_1}$
	zone 3	⋮			⋮						⋮	$X_3^{R_1}$
Retail 2	zone 1	$\bar{S}_{11}^{R_2}$	...	...	0							$X_1^{R_2}$
	zone 2	⋮			⋮					⋮	⋮	$X_2^{R_2}$
	zone 3	⋮			⋮			...	...	0	0	$X_3^{R_2}$

Figure 6.4 Input–output transactions for the Lowry model

namely:

$$w^{R_k} T_{ij}^{R_k} \left(= w^{R_k} \frac{E_j^{R_k} f_H(c_{ij})}{\sum_i f_H(c_{ij})} \right) \tag{6.24}$$

All other cells in Figure 6.4 are zero.

In order to complete the presentation of the spatially disaggregated Lowry model as an input–output model, it is necessary to translate the assumptions made in the model mechanism into input–output coefficients, paralleling the corresponding stage in section 6.2.2. In doing so it is important to ensure that the resulting coefficients for the spatially disaggregated version are consistent with those given earlier in connection with Figure 6.2.

Demand-driven input–output coefficients may be derived by dividing each element in Figure 6,4 by the receiving sectors' total outputs. The pattern of entries that will result is displayed in Figure 6.5, and these are described more fully in the following paragraphs.

To produce Figure 6.5, the magnitudes given in the lower left-hand column of Figure 6.4 need to be divided by an expression giving the total value of output from the households in zone j with population P_j. This expression may be derived as follows: division of P_j by g (the inverse activity rate) gives the total number of employees by zone of residence, and therefore multiplication by the average wage rate $\bar{w}$ gives the total value of output of the households in zone j. From this it may be deduced that:

$$a_{ij}^{R_k H} = \frac{g \bar{S}_{ij}^{R_k}}{P_j \bar{w}} = \frac{g e_j^{R_k} f_{R_k}(c_{ij})}{\bar{w} \sum_i f_{R_k}(c_{ij})} \tag{6.25}$$

and these coefficients embody the Lowry model assumption of equations (6.5) and (6.6) that any increase in output from the household sector (by zone) that may occur (that is, any change in employment) will induce a corresponding change in demand for retail goods (by zone).

The magnitudes given in the upper right-hand rows of the endogenous partition of Figure 6.4 need to be divided by an expression giving the total output of the retail sectors (by zone) in order to get corresponding entries for Figure 6.5. Assuming constants h^{R_k} giving the output per employee, as before, the total output of the retail sectors (by zone) is given by $e_*^{R_k} P$ and therefore the coefficients $a_{ij}^{HR_k}$ in Figure 6.5 by

$$a_{ij}^{HR_k} = \frac{w^{R_k} T_{ij}^{R_k}}{h^{R_k} E_j^{R_k}} = \frac{w^{R_k} f_{R_k}(c_{ij})}{h^{R_k} \sum_i f_{R_k}(c_{ij})} \tag{6.26}$$

These coefficients embody the Lowry model assumption of equation (6.7), and also equation (6.1), that any change in production (output) within the retail sectors will cause a change in employment requirements (inputs from households) of the retail sectors.

Since the coefficients described above embody all the key assumptions present

		Hshlds			Retail 1			Retail 2			Basic totals	Row totals
		zone 1	zone 2	zone 3	zone 1	zone 2	zone 3	zone 1	zone 2	zone 3		
Hshlds	zone 1	0	0	. . .	$a_{11}^{HR_1}$	$a_{12}^{HR_1}$	. . .	$a_{11}^{HR_2}$	. . .	⋮	$w^B T_{1*}^B$	X_1^H
	zone 2	0			$a_{21}^{HR_1}$					⋮	$w^B T_{2*}^B$	X_2^H
	zone 3	⋮							. . .	$a_{33}^{HR_2}$	$w^B T_{3*}^B$	X_3^H
Retail 1	zone 1	$a_{11}^{R_1H}$	$a_{12}^{R_1H}$	. . .	0	0	. . .			.	0	$X_1^{R_1}$
	zone 2	$a_{21}^{R_1H}$	. . .		0					⋮	0	$X_2^{R_1}$
	zone 3	$a_{31}^{R_1H}$								⋮	⋮	$X_3^{R_1}$
Retail 2	zone 1	$a_{11}^{R_2H}$								⋮	⋮	$X_1^{R_2}$
	zone 2	$a_{21}^{R_2H}$								0	⋮	$X_2^{R_2}$
	zone 3	⋮		$a_{33}^{R_2H}$		. . .	. . .	. . .	0	0	0	$X_3^{R_2}$

Figure 6.5 Input–output coefficients for the Lowry model

in the Lowry model summary of equations (6.1)–(6.8), completion of the input–output presentation of the Lowry model is straightforward. As in section 6.2.2, it is left only to derive the familiar matrix inverse multiplier.

Identities equivalent to equation (6.18) are

$$
\begin{aligned}
X_1^H &= a_{11}^{HR_1} X_1^{R_1} + a_{12}^{HR_1} X_2^{R_1} + \dots + w^B T_{1*}^T \\
&\vdots \\
X_I^H &= a_{I1}^{HR_1} X_1^{R_1} + a_{I2}^{HR_1} X_2^{R_1} + \dots + w^B T_{I*}^B \\
X_1^{R_1} &= a_{11}^{R_1 H} X_1^H \quad \dots \quad a_{1I}^{R_1 H} X_I^H \\
&\vdots \\
X_I^{R_N} &= a_{I1}^{R_N H} X_1^H \quad \dots \quad a_{II}^{R_N H} X_I^H
\end{aligned}
\tag{6.27}
$$

from which the familiar inverse multiplier

$$\mathbf{X} = (\mathbf{I} - \mathbf{A})^{-1}\mathbf{Y} \tag{6.28}$$

may be derived where, in this case, **A** is the matrix of coefficients displayed in the endogenous partition of Figure 6.5, viz:

$$
\mathbf{A} = \left[\begin{array}{c|c}
0 & a_{11}^{HR_1}\ a_{12}^{HR_1}\ \dots \\
 & \vdots \\
\hline
a_{11}^{R_1 H}\ a_{12}^{R_1 H}\ \dots & 0\ \dots \\
a_{21}^{R_1 H} & \vdots \qquad\qquad \vdots \\
\vdots & \qquad \dots
\end{array}\right]
\tag{6.29}
$$

and these internal coefficients are given more explicitly in expressions (6.24) and (6.25), **Y** is the endogenous vector

$$
\mathbf{Y} = \begin{bmatrix}
w^B T_{1*}^B \\
w^B T_{2*}^B \\
\vdots \\
w^B T_{I*}^B \\
0 \\
\vdots \\
0
\end{bmatrix}
\tag{6.30}
$$

and **X** the vector

$$
\mathbf{X} = \begin{bmatrix}
X_1^H \\
X_2^H \\
\vdots \\
X_1^{R_1} \\
X_2^{R_1} \\
\vdots \\
X_I^{R_N}
\end{bmatrix}
\tag{6.31}
$$

to be endogenously determined by the model (that is, by equation 6.28).

Equation (6.28), with the help of (6.29), (6.30), and (6.31) for defining various terms, is a representation of the Lowry model in the form of a traditional (demand-driven, Leontief-type) input–output multiplier model. In principle it could be used for numerical calculation of successive multiplier rounds or of the final solution for a Lowry model. This new formulation of the Lowry model will be used as a basis for the extensions to that model that are to be made in later sections. Before such extensions are demonstrated, however, a particularly obvious exercise is to relate the form given in equations (6.12)–(6.14) to Garin's (1966) matrix formulations of the Lowry model, since this also apparently consists of an inverse multiplier in conjunction with a vector of exogenously prespecified basic employment.

In view of a comment made in Section 6.2.2, over the *a priori* knowledge of total sector outputs in the spatially aggregated input–output Lowry model representation (and hence a degree of redundancy of equations (6.19)–(6.22)), it is appropriate to end this section by remarking that in the spatially disaggregated version of the present section the corresponding totals (the elements in expression (6.31) and (equivalently) the far right-hand columns of Figures 6.4 and 6.5) are not known *a priori*. Therefore, as asserted above, any redundancy no longer holds in the spatially disaggregated form. Note further that the total sector outputs that were used to derive the coefficients in expressions (6.25) and (6.26) ($P_j\bar{w}$ and $e^{R_k}E_j^{R_k}$, respectively) are endogenous (as yet unknown) quantities, but they nevertheless give rise to known constant input–output coefficients, since the unknown P_j and $E_j^{R_k}$ in the denominators cancel with identical P_j and $E_j^{R_k}$ in the numerators of expressions (6.25) and (6.26).

6.2.5 Relation to Garin's matrix representation

The purpose of this rather brief comparison is to illuminate aspects of the internal structure of the Lowry model that the input–output representation above can identify; aspects that are easily hidden in existing representations. One of the main points of interest here is to verify that a genuine addition to existing representations has been achieved in the analysis above. In doing so it should be noted that any apparent differences that may be evident are in presentation rather than content; the new representations above introduced no new assumptions over those already present in the Lowry model, other than in the use of wage rates to value employment, and two simplifying assumptions (neglecting work-based retail trips and land-use densities) that the final section will remedy.

Garin's (1966) representation may be written

$$\mathbf{E} = (\mathbf{I} - \mathbf{BA})^{-1}\mathbf{E}^B \tag{6.32}$$

where $\mathbf{E}$, $\mathbf{E}^B$ are vectors of total and basic employment (respectively) by zone, $\mathbf{B}$ is a matrix with elements

$$\frac{e^{R_k} f_{R_k}(c_{ij})}{\sum_j f_{R_k}(c_{ij})}$$

and **A** a matrix with elements

$$\frac{g f_{R_k^R}(c_{ij})}{\sum_i f_H(c_{ij})}$$

where i is residence and j workplace zones for both **A** and **B**.

The intended comparison between Garin's representation in equation (6.32) and the input–output representation given in equations (6.28)–(6.31) above is now straightforward: the full input–output representation produces a vector of total sector outputs for all sectors identified, and requires a full matrix of technical coefficients to pick up methodically all individual intersectoral interactions. The matrix representation of Garin, on the other hand, deals with employment totals only (equivalent just to the X_i^H partition of the vector given by expression (6.31), i ranging over all zones) and an inverse multiplier of considerably smaller dimension, the internal structure of which amalgamates (via the product **BA** in equation (6.32)) individual intersectoral interactions.

Computational requirements for the Garin representation are clearly less, but the new representation, being a genuine input–output model, is directly suitable for extension to incorporate the fullest possible intersectoral interactions.

6.3 EXTENSION OF THE LOWRY MODEL TO INCORPORATE FULL INTER-ACTIVITY RELATIONS

6.3.1 The matrix or algebraic representation

The analysis of previous sections has laid an immediate foundation for the specification of a full set of intersectoral relations within a Lowry model framework. It has been shown that the output–input coefficients are the key to the representations of previous sections, and the developments from here onwards will work exclusively in terms of such coefficients. Indeed, the aim of the present section will be achieved as soon as it can be recognized that full intersectoral interactions will be picked up in the output–input representation of the Lowry model, as long as a suitable pattern of terms in the input–output coefficient matrix is used. Figure 6.3 gave an indication of the appearance of a relevant interaction coefficient matrix for a spatially aggregated model; Figure 6.6 gives a spatially disaggregated version. A distinction between basic and retail sectors has been made in this figure in order to retain recognizable Lowry model characteristics.

The matter of determining the values of the coefficients in this figure will be addressed later, but first we will summarize the model mechanism associated with the pattern of entries given in Figure 6.6.

The main factor is that it has the structure of an input–output model, and will thus generate successive rounds of reactions (in terms of changed levels of output) through all interlinked sectors, following some initial exogenously prespecified change. For a Lowry model, this initial exogenous change will be an increase or decrease in the employment requirement of certain industrial sectors. The

			Hshlds	Retail sectors	Basic sectors	Exog sectors	Total outputs
				R_1 R_2	B_1 B_2		
			zone 1 zone 2 . . .	zone 1 zone 2 . . . zone 1 zone 2	zone 1 zone 2 . . . zone 1 zone 2		
Employees (hshlds)		zone 1	0	$a_{11}^{HR_1}$	$a_{11}^{HB_1}$	Y_1^H	X_1^H
		zone 2	⋮	⋮		⋮	X_2^H
		⋮		⋮			⋮
Retail goods (retail sectors)	R_1	zone 1	$a_{11}^{R_1H}$	$a_{11}^{R_1R_1}$	$a_{11}^{R_1B_1}$	$Y_1^{R_1}$	$X_1^{R_1}$
		zone 2	$a_{21}^{R_1H}$	⋮		⋮	$X_2^{R_1}$
		⋮	⋮	⋮		⋮	⋮
	R_2	zone 1	⋮	⋮		$Y_1^{R_2}$	$X_1^{R_2}$
		zone 2				⋮	$X_2^{R_2}$
		⋮					
		⋮					
Basic goods (basic sectors)	B_1	zone 1	$a_{11}^{B_1H}$	$a_{11}^{B_1R_1}$	$a_{11}^{B_1B_1}$	$Y_1^{B_1}$	$X_1^{B_1}$
		zone 2	$a_{21}^{B_1H_1}$	⋮	⋮	⋮	$X_2^{B_1}$
		⋮				⋮	⋮
	B_2	zone 1				$Y_1^{B_2}$	$X_1^{B_2}$
		zone 2				⋮	$X_2^{B_2}$
		⋮				⋮	⋮
		⋮					

Figure 6.6 Input–output coefficient (spatially disaggregated) accommodating all intersectoral flows

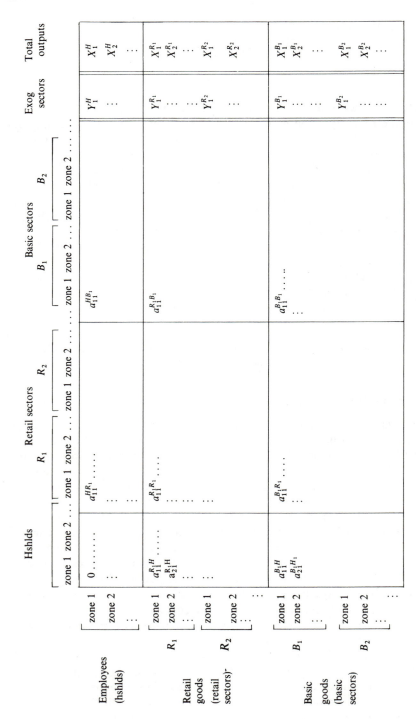

immediate response to this change is from the household sector, picked up through coefficients in the uppermost rows of Figure 6.6; these coefficients adjust household input requirements according to the outputs produced (employees) and this pulls through (via appropriate coefficients) a second round of reactions from all sectors (by zone) with which they are linked (that is, all sectors with non-zero coefficients in the far left, households, columns); successive rounds continue until they produce changes of negligible magnitude, thus completing the familiar effect. The appearance of any coefficients linking retail and basic sectors will pick up changes in interaction between these sectors which are missing from the original Lowry model.

The chain of multiplier reactions described above was assumed to be initiated by an exogenously prespecified change in basic employment, as in the original Lowry model. However, having the structure of an input–output framework, there is an immediate invitation in the representation above to consider alternatives to basic employment (an obvious such alternative being the use of 'final demand' as favoured in Leontief's original input–output suggestions) as mentioned in a related context in section 6.2.3. With this in mind, the familiar input–output inverse multiplier for this expressing the unknown (endogenous) sector outputs by zone in terms of known (exogenous) demands and interactions coefficients may be written as

$$\mathbf{X} = (\mathbf{I} - \mathbf{A})^{-1}\mathbf{Y} \tag{6.33}$$

where **A** is the matrix of input–output coefficients given in Figure 6.6, **Y** is the exogenously prespecified demand which drives the model, given in the penultimate column of Figure 6.6 (and may be basic employment, 'final demand' in the conventional input–output sense, or something quite different) and **X** the unknown sector outputs:

$$\mathbf{X} = \begin{bmatrix} X_1^H \\ X_2^H \\ \\ X_1^{R_1} \\ X_2^{R_1} \\ \\ X_1^{B_1} \\ X_2^{B_1} \\ \\ X^{B_L} \end{bmatrix} \tag{6.34}$$

to be determined by the model. As mentioned before, the distinction between basic and retail sectors has been made in order to relate this representation directly to the original Lowry model, but it should be recognized that the distinction does not effectively influence the algebraic form of the model and any classification of sectors can in principle be used with this representation.

To end this section we will indicate how more explicit entries for the cells in matrix **A** in equation (6.33) (and therefore for corresponding cells in Figure 6.6) may be derived by following the general pattern that can be identified from the discussion of section 6.2.4. Explicit expressions for the cells in Figure 6.5 (see equations (6.25) and (6.26)) may be seen to consist of the relevant input–output 'technical coefficients' (in aggregate matching those given earlier in Figure 6.2) in combination with an attraction-constrained spatial-interaction model. Thus if matrix **A** in equation (6.33) (and therefore the coefficients in Figure 6.6) is intended to contain transactions coefficients for an extended model of the same kind, it would be reasonable to assume that these should also consist of the assumed known technical coefficients for any intersectoral interactions it is intended to pick up (z_j^{mn}, say, where m and n are now general sector labels in place of the more specific H, R_k and B_l, and the zonal subscript j for the receiving sector is optional, depending on data availability) in combination with an attraction-constrained spatial-interaction model. This will give entries of the form

$$a_{ij}^{mn} = \frac{z_j^{mn} f_m(c_{ij})}{\sum_i f_m(c_{ij})} \tag{6.35}$$

where

$$\frac{f_m(c_{ij})}{\sum_i f_m(c_{ij})}$$

represents an attraction-constrained spatial-interaction model for the i–j transfer of the output of sector m. It would be possible to derive more explicit spatial-interaction submodels for all these cells corresponding to equation (6.1) for the household–retail interaction, but they will not be given explicitly here. Similarly, for the interactions in this representation that also appear in the original Lowry model, the coefficients z_j^{mn} will be expressible directly in terms of constants from the earlier model (compare expressions (6.25) and (6.26)), but other z_j^{mn}'s will require additional data.

6.3.2 A maximum-entropy (mathematical programming) representation

The recognition and formulation of the Lowry model as an input–output model, and its later extension to embody a full set of intersectoral relations, encourage further exploration of models with similar underlying assumptions. In particular, we will consider an entropy-maximizing model utilizing the same general assumptions that produced the model of the previous section and using methods similar to those of Chapter 5. The developments that follow can be considered to give a fresh interpretation of the utility of this earlier model.

In order to produce a suitable set of constraint equations, it is useful to take further explicit note of the two stages that were involved in producing the model of the previous subsection: the first (treated separately in section 6.2.2) was to concentrate on the technical relations of production and consumption of the endogenous sectors, expressible in terms of coefficients of an input–output type: the second was to introduce a spatial component, in effect, to spatially disaggregate the latter coefficients, and this came in the familiar form of cost functions. A model presented much earlier (Wilson, 1970, chapter 3) estimating terms X_{ij}^m defining the output from sector m in zone i used in zone j, has a suitable structure for present purposes in that it uses just this form of constraint equations, and is therefore summarized and re-interpreted here.

Input–output coefficients z_j^{mn} defining the requirements of sector m's output per unit output of sector n in zone j (previously used in expression (6.35)) give rise to the equations:

$$\sum_i X_{ij}^m = \sum_{in} z_j^{mn} X_{ji} + G_j^m \tag{6.36}$$

where G_j^m is the exogenously prespecified demand for m in zone j.

In the same way as above, a model driven from the familiar Lowry model starting point would choose G_j^m to be an initial distribution of basic employment (by zone of workplace), but in a more general input–output context, it is by no means essential to follow this lead (the opposite extreme would be to choose G_j^m as the traditional final demand for m in zone j). Coefficients z_j^{mn} have been given a zonal subscript to allow for the possibility of these being available at the zonal level. This will not necessarily be true, in which case the same coefficients must be used for all zones, but in the original input–output representation of the Lowry model, the availability of terms such as $e_i^{R_k}$ (expenditure on k per head in zone i) would make possible the use of zonal-specific coefficients z_j^{mn}.

The second set of constraints have the role of introducing spatial-interaction cost functions, and the usual way to do this in an entropy-maximizing model is via the familiar form of cost constraint:

$$\sum_{ij} c_{ij}^m X_{ij}^m = C^m \tag{6.37}$$

where c_{ij}^m is the average cost of movement of m between i and j and C^m is the total expenditure on such movements for m.

Since equations (6.36) and (6.37) embody what have been found to be the key assumptions of an input–output representation of the Lowry model, an entropy-maximizing model corresponding to this, consists of maximizing

$$S = -\sum_{ijm} X_{ij}^m \ln X_{ij}^m \tag{6.38}$$

(for $X_{ij}^m \geqslant 0$ with $X_{ij}^m \ln X_{ij}^m = 0$ at $X_{ij}^m = 0$) subject to equations (6.36) and (6.37).

Whatever the chosen approach to the numerical solution of the model, its overall structure should be clear from the above equations. Equation (6.36) will pick up successive multiplier rounds following some initial exogenous change in G_j^m, thus giving the model true input–output multiplier characteristics. Lowry model characteristics may be retained as before by appropriately specifying G_j^m and z_j^{mn}.

An analytical solution to this model may be derived from classical Lagrangian methods. This gives

$$X_{ij}^{m} = B_{j}^{m} \exp[-\mu^{m} c_{ij}^{m}] \tag{6.39}$$

where μ^m is the cost parameter (dual variable) associated with constraint (6.37) and B_j^m is a similar sort of parameter (this time a logarithmic transformation of the dual variable) associated with constraint (6.36). B_j^m can in fact be determined exactly by substituting (6.39) into (6.36), that is, from

$$\sum_{i} B_{j}^{m} \exp[-\mu^{m} c_{ij}^{m}] = \sum_{i'n} z_{j}^{mn} B_{i'}^{n} \exp[-\mu^{n} c_{ij}^{n}] + Y_{j}^{m} \tag{6.40}$$

The question of immediate interest at this point is to ask how strong the similarities are between the two formulations given so far; one a matrix multiplier model with mathematical programming (entropy-maximizing) submodels, and the other a mathematical programming (entropy-maximizing) model with the corresponding multiplier embedded within the constraints.

A comparison of the two models given above (matrix and mathematical programming versions of an input–output based Lowry model) suggests the existence of strong similarities but not of equivalences. This has been confirmed in numerical computations.

Rather than pursuing here a detailed analysis of equivalences and similarities between the models of this and the previous subsection, however, it is intended to postpone such analysis until Chapter 10. We will then not only analyse similarities and equivalences between various representations of input–output based Lowry models, but also pick up a similar equivalence discussion started in Chapter 4, and address the much wider situation that is revealed in the more general modelling literature, namely the formal identification of similarities and equivalences between apparently different methodological approaches for a given modelling purpose.

6.3.3 Further developments

The mathematical programming formulation as an alternative to the matrix representation may be compared with the maximum-entropy representation of the rectangular input–output model multiplier given in the previous chapter as an alternative to existing rectangular input–output multiplier methods. This comparison turns on the fact that in each case a matrix representation has been replaced by a mathematical programming representation. Advantages of the matrix representations are their relative ease in numerical calculations over the mathematical programming representations (linear algebra as opposed to a non-linear programming approach) and their readier ability to monitor individual multiplier rounds giving them a pseudo-dynamic interpretation. (Given the nature of the present non-linear programming model, however, the computational advantage is now far less significant than is usually the case when comparing matrix and programming models, since the latter model above has been shown to have an analytical solution.) Advantages of the mathematical programming representations are their immediate ability to introduce further constraining equations and other optimizing criteria into the objective function, either in addition to or as a replacement for the entropy function (6.38).

The introduction of further constraints will be discussed immediately and is concerned in particular with the specification of land-use density constraints; this remedies an oversimplification of the Lowry model analysis given so far in this chapter.

The inclusion of land-use density constraints in the mathematical programming formulation is a relatively straightforward matter, merely involving the explicit specification of appropriate constraints in addition to (6.36) and (6.37). Such constraints are familiar elsewhere from the use of mathematical programming formulations in land-use models and for the Lowry-based model above could typically take the form:

$$\sum_j d_i^m X_{ij}^m \leqslant L_i^m \tag{6.41}$$

where L_i^m is the total land area available in zone i for activity m, and d_i^m coefficients defining land area requirements of activity m (with an optional zonal subscript). A subset of such constraints could control population densities, and immediate more general refinement to such constraints could be made, though these will not be considered explicitly here.

Thus in order to take account of population and land-use density constraints, the mathematical programming version of the Lowry model is easily amended, requiring only the addition of a constraint such as (6.41) to the formulation already given. Unfortunately, a simple Lagrangian solution is now no longer possible for numerical work.

For the matrix representation, the incorporation of land-use density considerations would appear to be rather less straightforward, since its accomplishment would tend to involve considerable loss of simplicity (and clarity) of the current model equations. At present these equations are built on the (somewhat unsatisfactory) assumption that there is no effective constraint on the ability of sector outputs at particular locations to increase or decrease in immediate response to the demands placed on them. Density considerations, on the other hand, would clearly directly seek to undermine this assumption, by constraining the output of particular sectors at particular locations not to violate given specifications. The general approach needed would be to abandon the form given by equation (6.32), summarizing the whole chain of reactions, but rather to monitor individual rounds of change in sector outputs (pulled through by the linear coefficients in Figure 6.5), and subject these to land-use constraints using the type of method explained in Batty (1971, 1977); Wilson (1974, Chapter 11), and, more recently, Banister (1977) and Williams (1979).

A further oversimplification of the Lowry model analysis of this chapter has been the neglect of work-based retail utilization. This in turn will now be remedied.

The incorporation of work-based retail utilization will be illustrated firstly with respect to the unextended Lowry model (represented in input–output coefficient form in Figure 6.1), from which its development to cover the extended Lowry model given above can follow directly the same pattern. It is a relatively straightforward matter as long as it can be assumed that total sector outputs are directly related to employment in these sectors. In this case the assumptions underlying coefficients d^k in equation (6.7) can again be brought into use, namely that existing employment will induce further retail employment through its demand for retail-sector output. Thus it is possible to define coefficients $g^{R_kR_n}$ denoting retail goods R_k demanded per employee in sector R_n, and hence coefficients $\bar{g}^{R_kR_n}$ defining sector R_n's work-based utilization of retail sector R_k, per unit output of sector R_n. These coefficients would appear in the currently empty retail sector × retail sector partition in Figure 6.1. For an extended Lowry model, such as given in Figure 6.2, coefficients $g^{R_kR_n}$ would appear in addition to the technical coefficients $a^{R_kR_n}$ and $a^{R_kB_n}$ giving cells of the form $(a^{R_kR_n}+\bar{g}^{R_kR_n})$ and $(a^{R_kB_n}+\bar{g}^{R_kB_n})$ in the retail goods rows in Figure 6.2. As well as appearing in the two cited figures, the additional coefficients $\bar{g}^{R_kR_n}$ and $\bar{g}^{R_kB_n}$ can occupy corresponding positions in the full model equations (6.28)–(6.31), thus methodically accounting for work-based retail utilization throughout the input–output representations.

It is finally a further extension to incorporate prior values $\{\bar{X}_{ij}^m\}$ for the $\{X_{ij}^m\}$ terms to be estimated in the mathematical programming models here. This will not be demonstrated explicitly as it would follow identically the corresponding incorporation of prior values $\{\bar{X}^{mnk}\}$ in the rectangular input–output models of the previous chapter. It would provide a base-year weighting for the $\{X_{ij}^m\}$ terms to be projected, and as before would contribute non-redundant information to the model estimation. It will be left as a matter for later investigation to compare rigorously this use of prior values within a (mathematical programming) Lowry model framework to their alternative use within a (matrix) Lowry-based model in a recent paper by Batty and March (1976).

6.4 CLOSING REMARKS

In the introduction to this chapter we cited a recent paper by Williams (1979) which also recognized explicitly the full input–output associations of the Lowry model. It is therefore appropriate to close the present chapter by remarking on the specific developments made in that paper. Its aim was in fact to structure the numerical solution approach of a multi-regional input–output model (of which the Lowry model could readily be shown to be a special case) giving first priority to minimizing computational requirements. The structure of the model turns out to be less elegant and flexible than those presented in earlier sections here, but these possible disadvantages must be weighed against the computational considerations. Finally, it is worth noting that we have above already presented two methodologically different forms of a Lowry or input–output model (a matrix representation and an algebraic representation) and we can now add the approach of Williams (1979) as a further form, which again could give slightly different numerical results. As hinted earlier, a fuller discussion of these considerations is given in Chapter 10.

7. *Embedding theorems and applications*

7.1 INTRODUCTION

The integration of spatial-interaction submodels into a mathematical programming planning framework for the location of facilities is achieved in this chapter by showing how to generate those submodels within an overall optimization setting. First, an embedding theorem is proved which shows that through the use of dispersion terms in the objective function, the spatial-interaction variables in the problem satisfy the usual submodels. Non-linear mathematical programs whose primal and dual versions may be solved by standard methods are obtained. This approach is contrasted with the TOPAZ (the optimal placement of activities in zones) suite due to Brotchie and his collaborators (see Brotchie, 1969), and the method proposed by Mazor and Pines (1973), which seek the maximization of the net benefits of establishing and operating activities in zones minus the interaction costs where the latter are determined from gravity interaction submodels introduced into the optimization framework through iterative linear programming. The results presented in this chapter will be illustrated with two applications to residential and employment activities allocation, and the design of shopping centres, respectively.

7.2 INTEGRATION OF SPATIAL-INTERACTION SUBMODELS WITHIN OVERALL MATHEMATICAL PROGRAMMING FRAMEWORKS

7.2.1 Introduction

Several developments in recent years have established links between programming and spatial-interaction approaches. In a study for the city of Rotterdam (Mazor and Pines, 1973), the following iterative procedure has been used: the mathematical programming problem is first solved with an estimated interaction pattern, then the results are input into the interaction submodel to generate a new interaction pattern, which is fed back again into the mathematical programming problem. This procedure stops when the difference between successive solutions is small, but no convergence proof for this procedure has been provided. Brotchie (1969) has proposed a quadratic programming approach in which the second-order terms in the objective function represent interaction costs. Also, models developed using the program TOPAZ (see, for example, Dickey and Hopkins, 1972; Dickey and Najafi, 1973, based on the work of Brotchie, 1969) use either unconstrained or singly constrained spatial-interaction models and are solved iteratively as a sequence of linear programming problems, but this procedure, although very simple, again does not necessarily converge. Recently, it has been argued (Evans, 1973b; Wilson and Senior, 1974; Senior and Wilson, 1974b) that the mathematical programming and the spatial-interaction approaches are branches of the same family of models. The presentation here is based on papers by Coelho and Wilson (1976) and Coelho, Williams, and Wilson (1978).

The integration of these two approaches is pursued by showing that spatial-interaction patterns may be generated in a mathematical programming framework through the maximization of 'dispersion terms' leading to non-linear problems that may be solved by standard methods. It will be noted here that these 'dispersion terms' will represent under certain assumptions and for specified values of parameters, measures of total consumers' surplus; a comprehensive discussion of this topic was given in Section 3.3. This both contributes further to the study of programming–spatial-interaction relationships, and provides the basis for embedding the spatial-interaction submodel in an optimizing framework. The general results of this section are illustrated in Section 7.3 with an elementary model of residential and employment activities allocation, and in Section 7.4 with a shopping model, but they may also be applied to other fields, particularly in transportation planning, national and regional development, energy modelling, and industrial and building layouts.

7.2.2 Equivalence lemma: adding constraints into the objective function

Let us assume that the functions $f(\mathbf{x})$ and $g_i(\mathbf{x})$, $i = 1, \ldots, m$, where $\mathbf{x} = (x_1, \ldots, x_n)$, have continuous first partial derivatives and that $f(\mathbf{x})$ is a concave function and $g_i(\mathbf{x})$ are either convex or concave functions as required by condition (7.9) below. Then we have:

Lemma: The following problems are equivalent:
Problem 1: Choose $\mathbf{x}$ to maximize

$$f(\mathbf{x}) \tag{7.1}$$

subject to constraints

$$g_i(\mathbf{x}) = b_i \qquad i = 1, \ldots, m \tag{7.2}$$

and

$$\mathbf{x} > 0 \tag{7.3}$$

Problem 2: Choose $\mathbf{x}$ and λ_i, $i = 1, \ldots, p$, to maximize

$$f(\mathbf{x}) + \sum_{i=1}^{p} \lambda_i (b_i - g_i(\mathbf{x})) \tag{7.4}$$

subject to constraints

$$g_i(\mathbf{x}) = b_i, \qquad i = p+1, \ldots, m \tag{7.5}$$

$$\mathbf{x} > 0 \tag{7.6}$$

Proof: The Lagrangian functions of problems 1 and 2 are, respectively:

$$L_1(\mathbf{x}, \boldsymbol{\lambda}) = f(\mathbf{x}) + \sum_{i=1}^{m} \lambda_i (b_i - g_i(\mathbf{x})) \tag{7.7}$$

and

$$L_2(\mathbf{x}, \boldsymbol{\lambda}) = f(\mathbf{x}) + \sum_{i=1}^{p} \lambda_i^{(1)} (b_i - g_i(\mathbf{x})) + \sum_{i=p+1}^{m} \lambda_i^{(2)} (b_i - g_i(\mathbf{x})) \tag{7.8}$$

where $\boldsymbol{\lambda} = (\lambda_1, \lambda_2, \ldots, \lambda_m)$, $\lambda^{(1)} = (\lambda_1, \ldots, \lambda_p)$, $\boldsymbol{\lambda}^{(2)} = (\lambda_{p+1}, \ldots, \lambda_m)$.

It is clear from (7.7) and (7.8) that $L_1 = L_2$, that is, that the Lagrangian functions of both problems are the same. Let us now denote by $(\mathbf{x}^*, \boldsymbol{\lambda}^*)$ the solution of the Kuhn–Tucker conditions for problem 1, and require $L_1(\mathbf{x}, \boldsymbol{\lambda}^*)$ to be a concave function of $\mathbf{x}$. This condition is fulfilled if the functions $g_i(\mathbf{x})$, $i = 1, \ldots, m$, satisfy any of the following conditions (Hadley, 1964):

(a) $g_i(x)$ is concave if $\lambda_i^* < 0$
(b) $g_i(x)$ is convex if $\lambda_i^* > 0$ (7.9)
(c) $g_i(x)$ is linear (so, concave and convex simultaneously).

In this case, it is easily shown that the Kuhn–Tucker conditions for both problems are identical, that is, both problems have the same necessary and sufficient conditions of optimality. Thus, the optimum of problem 1 is also that for problem 2, and vice versa.

This result can easily be extended to relax the non-negativity conditions by using the Lagrange conditions instead of the Kuhn–Tucker conditions. In effect, we have shown how to 'absorb' constraints into the objective function or vice versa.

7.2.3 A theorem for embedding spatial-interaction submodels in an optimisation program

Let us consider a family of spatial-interaction models $\{T_{ij}^k\}$ $(k = 1, \ldots, p)$ such as

$$T_{ij}^k = A_i^k B_j^k x_i^{0k} x_j^{1k} f_{ij}^k(c_{ij}^k) \tag{7.10}$$

where

$$A_i^k = \left[\sum_j B_j^k x_j^{1k} f_{ij}^k(c_{ij}^k) \right]^{-1} \tag{7.11}$$

$$B_j^k = \left[\sum_i A_i^k x_i^{0k} f_{ij}^k(c_{ij}^k) \right]^{-1} \tag{7.12}$$

It is well known that such variables $\{T_{ij}^k\}$ may be generated by a mathematical program, and that one form of this is:

$$\max_{\{T_{ij}^k\}} Z(T_{ij}^k) = -\sum_{ijk} T_{ij}^k \left(\ln \frac{T_{ij}^k}{f_{ij}^k(c_{ij}^k)} - 1 \right) \tag{7.13}$$

subject to

$$\sum_j T_{ij}^k = x_i^{0k} \tag{7.14}$$

$$\sum_i T_{ij}^k = x_j^{1k} \tag{7.15}$$

We now turn to a particular class of mathematical programming problems which arise in allocation problems when the interactions between activities are represented by gravity patterns. In effect, we are going to define an objective function which includes locational costs and benefits, and optimize over variables like x_i^{0k} and x_i^{1k}. Let $\mathbf{x}^{0k} = (x_1^{0k}, \ldots, \mathbf{x}_m^{0k})$ and $\mathbf{x}^{1k} = (x_1^{1k}, \ldots, x_n^{1k})$ and $\mathbf{x}^2 = (x_1^2, \ldots, x_r^2)$ be subvectors of variables forming vector $\mathbf{x}$, ie., $\mathbf{x} = (x_1^{01}, \ldots, x_m^{0p}; x_1^{11}, \ldots, x_n^{1p}; x_1^2, \ldots, x_r^2) = (\mathbf{x}^{01}, \ldots, x^{0p}; \mathbf{x}^{11}, \ldots, \mathbf{x}^{1p}, \mathbf{x}^2)$. Usually, $\mathbf{x}^{0k}$ and $\mathbf{x}^{1k}$ will be interpreted as locational variables, and the vector $\mathbf{x}^2$ is a set of variables which incorporate any additional features of the planning problem. Define the vector $\mathbf{T} = (T_{ij}^k)$ for $i = 1, \ldots, m$, $j = 1, \ldots, n$, and $k = 1, \ldots, p$. We shall then investigate the conditions in which the solution of the mathematical program below will satisfy the gravity pattern defined by the set of spatial-interaction models (7.10)–(7.12). Consider:

Problem 3: Choose $\mathbf{x}$ and $\mathbf{T}$ to maximize

$$Z = f(\mathbf{x}) + h(\mathbf{T}) \tag{7.16}$$

subject to conditions

$$g_h(\mathbf{x}) < b_h \qquad \text{for } h = 1, \ldots, l \tag{7.17}$$

$$\sum_j T_{ij}^k - x_i^{0k} = 0 \qquad \text{for } i = 1, \ldots, m; \quad k = 1, \ldots, p \tag{7.18}$$

$$\sum_i T_{ij}^k - x_j^{1k} = 0 \qquad \text{for } j = 1, \ldots, n; \quad k = 1, \ldots, p \tag{7.19}$$

The following result may be proved.

Theorem: If $f(\mathbf{x})$ is a concave function, $g_h(\mathbf{x})$ (h $= 1, \ldots, p$) are convex functions which, with the linear constraints (7.18) and (7.19) define a non-empty feasible set, and $h(\mathbf{T})$ is defined as follows:

$$h(\mathbf{T}) = -\sum_{ijk} \alpha_k T_{ij}^k \left(\ln \frac{T_{ij}^k}{f_{ij}^k(c_{ij}^k)} - 1 \right) \tag{7.20}$$

where α_k are arbitrary positive constants, then problem 3 has a unique solution which satisfies the spatial-interaction pattern defined by (7.10)–(7.12) for $\{T_{ij}^k\}$.†

Proof: The uniqueness of the solution of problem 3 is due to $h(\mathbf{T})$ being a strictly concave function and therefore (7.16) is also strictly concave. It is well known that in a problem which consists of maximizing a strictly concave function over a convex domain, the optimal solution if it exists is unique (Hadley, 1964). The set of feasible solutions of problem 3 is convex and it has been assumed that it is non-empty, and so problem 3 has a unique solution.

The second statement of the theorem is readily proved using the Lagrangian conditions for problem 3. In fact, if we form its Lagrangian

$$L(\mathbf{x}, \mathbf{T}, \lambda) = f(\mathbf{x}) + \sum_{ijk} \alpha_k T_{ij}^k \left(\ln \frac{T_{ij}^k}{f_{ij}^k(c_{ij}^k)} - 1 \right)$$

$$+ \sum_h \lambda_h^{(1)} (b_h - g_h(\mathbf{x})) + \sum_i \lambda_i^{(2)k} (x_i^{0k} - \sum_j T_{ij}^k)$$

$$+ \sum_j \lambda_j^{(3)k} (x_j^{1k} - \sum_i T_{ij}^k) \tag{7.21}$$

where $\lambda = (\lambda_h^{(1)}, \lambda_i^{(2)k}, \lambda_j^{(3)k})$ is the vector of Lagrange multipliers associated with constraints (7.17)–(7.19) respectively, then differentiating L with respect to T_{ij}^k and equating to zero will give a condition which $(\mathbf{x}, \mathbf{T}, \lambda)$ must satisfy to be a solution of problem 3. Therefore, the optimal solution $(\hat{\mathbf{x}}, \hat{\mathbf{T}})$ of this problem will satisfy the set of equations:

$$\frac{\partial L}{\partial T_{ij}^k} = \alpha_k \ln \frac{T_{ij}^k}{f_{ij}^k(c_{ij}^k)} - \lambda_i^{(2)k} - \lambda_j^{(3)k} = 0 \tag{7.22}$$

which is equivalent to

$$\ln \frac{T_{ij}^k}{f_{ij}^k(c_{ij}^k)} = \frac{\lambda_i^{(2)k}}{\alpha_k} + \frac{\lambda_j^{(3)k}}{\alpha_k} \tag{7.23}$$

that is,

$$T_{ij}^k = \exp\left[\frac{1}{\alpha_k} (\lambda_i^{(2)k} + \lambda_j^{(3)k}) \right] f_{ij}^k(c_{ij}^k) \tag{7.24}$$

Then defining

$$A_i^k = \exp[\lambda_i^{(2)k}/\alpha_k]/x_i^{0k} \tag{7.25}$$

$$B_j^k = \exp[\lambda_j^{(3)k}/\alpha_k]/x_j^{1k} \tag{7.26}$$

† Note that constraints (7.18) and (7.19) are not written in the form

$$\sum_j T_{ij}^k = x_i^{0k} \quad \text{and} \quad \sum_i T_{ij}^k = x_j^{1k}$$

since x_i^{0k} and x_j^{1k} are variables within the problem and not constraints as in the usual spatial-interaction model.

equation (7.24) may be rewritten as

$$T_{ij}^k = A_i^k B_j^k x_i^{0k} x_j^{1k} f_{ij} (c_{ij}^k) \tag{7.27}$$

This means that $(\mathbf{x}, \mathbf{T})$ will satisfy an equation equivaent to (7.10). On the other hand, constraints (7.18), (7.19), and equation (7.27) will imply

$$\sum_j A_i^k B_j^k x_i^{0k} x_j^{1k} f_{ij}^k (c_{ij}^k) = x_i^{0k} \tag{7.28}$$

which gives

$$A_i^k = \left[\sum_j B_j^k x_j^{1k} f_{ij}^k (c_{ij}^k) \right]^{-1} \tag{7.29}$$

and

$$\sum_i A_i^k B_j^k x_i^{0k} x_j^{1k} f_{ij}^k (c_{ij}^k) = x_j^{1k} \tag{7.30}$$

which gives

$$B_j^k = \left[\sum_i A_i^k x_i^{0k} f_{ij}^k (c_{ij}^k) \right]^{-1} \tag{7.31}$$

Thus the proof is complete.

This result also holds when constraints (7.17) are 'greater than' inequalities, or equalities, and when the function $g_h(\mathbf{x})$ is concave or linear, respectively. Hence, constraints (7.17) can take the form

$$\begin{aligned} g_h(\mathbf{x}) &< b_h \qquad h = 1, \ldots, u \\ g_h(\mathbf{x}) &= b_h \qquad h = u+1, \ldots, v \\ g_h(\mathbf{x}) &> b_h \qquad h = v+1, \ldots, l \end{aligned} \tag{7.32}$$

where $g_h(\mathbf{x})$ is a convex, linear or concave function as $1 < h < u$, $u < h < v$, and $v < h < l$, respectively.

The doubly constrained gravity models (7.10)–(7.12) may also be replaced by a singly constrained or any other constrained gravity model without affecting the result asserted by this theorem, provided that constraints (7.18)–(7.19) are changed in problem 3 accordingly. We note that the proof given above is still valid if the α_k are substituted by functions $\alpha_k(\mathbf{x})$ such that $h(\alpha_k(\mathbf{x}), \mathbf{T})$ defined by (7.20) is a strictly concave function of $\mathbf{x}$ and $\mathbf{T}$. Of course, the optimal solution $(\hat{\mathbf{x}}, \hat{\mathbf{T}})$ of problem 3 will itself be a function of the values or functional forms ascribed to α_k, although the spatial-interaction pattern defined by (7.10)–(7.12) is always generated.

7.2.4 Discussion

The first point to note is that the solution does depend on the choice of constants, α_k. If $\alpha_k = 1$ for each k, the results is what we called in Chapter 3 a group entropy model. An interesting case is $\alpha_k = 1/\beta_k$ for each k: in this case, it can be shown that the objective function represents consumers' surplus for this disaggregated case (again, as shown in Chapter 3).

The theorem shows that, for a wide class of optimization problems, the spatial-interaction variables *do* satisfy an entropy-maximizing type of spatial-interaction model. In this sense, therefore, this theorem does show that such spatial-interaction behaviour can be combined with a planning optimization framework determined by the function $f(\mathbf{x})$ and the constraints (7.17). This result is produced by the set of entropy terms which appear in $h(\mathbf{T})$, in equation (7.20). In section 7.2.1, we called these 'dispersion terms', because they produce solutions which are realistic, but suboptimal in the usual linear programming interaction model sense (cf. Wilson and Senior, 1974). We should also note that even if the functions $g_k(\mathbf{x})$ are non-linear, suitable algorithms can be found (Rosen, 1960).

It is also interesting to recall that the dual of problem 3 usually has considerably less variables than the primal, since the variables T_{ij}^k can be eliminated and it determines the primal solution uniquely. This often forms the basis of useful algorithms. In certain cases, and depending on the nature of the planning constraints, the dual program becomes simply an unconstrained minimization problem. There is an example in Champernowne, Williams, and Coelho (1976); and further details are given in Chapter 8 below. Such problems can then be easily solved using standard computational methods for unconstrained optimization.

7.2.5 An example: activity location

To illustrate problem 3 in a land-use context we shall now assign the following definitions to our hitherto abstract variables:

$\mathbf{x}$—activity variables vector
within which
x_i^{0k}—amount of activity X^k in zone i
x_j^{1k}—amount of activity Y^k in zone j
and T_{ij}^k—interaction level between activity X^k in zone i and activity Y^k in zone j.

We do not consider, explicitly, any further planning variables, $\mathbf{x}^2$, in this illustration. We also have:

$f(\mathbf{x})$—allocation net benefit function, i.e. the net benefit of allocating activities spatially according to the pattern $\mathbf{x}$
c_{ij}^k—interaction cost of one unit of interaction between activity X^k in zone i and activity Y^k in zone j
$g_h(\mathbf{x}) < b_h$—set of physical and planning constraints.

Next, we shall assume in order to adopt the notation introduced in Chapter 3 that the deterrence functions $f_{ij}^k(c_{ij}^k) = R_{ij}^k \exp(-\beta_k c_{ij}^k)$ where R_{ij}^k are attractiveness weights (which, when normalized, can also be interpreted as *a priori* probabilities) attached to interaction between activity X^k in zone i and activity Y^k in j, and $\beta_k > 0$ are cost deterrence parameters. If constants α_k are equated to one, then the function $h(\mathbf{T})$ will become

$$h(\mathbf{T}) = -\sum_{ijk} T_{ij}^k \left(\ln \frac{T_{ij}^k}{R_{ij}^k} - 1 \right) - \sum_{ijk} \beta_k c_{ij}^k T_{ij}^k \tag{7.33}$$

which is a linear combination of an entropy function and the interaction costs weighted by the parameters β_k. As noted above, a very important particular case is when the arbitrary constants α_k are set equal to $1/\beta_k$. Then we have

$$h(\mathbf{T}) = -\sum_k \frac{1}{\beta_k} \sum_{ij} T_{ij}^k \left(\ln \frac{T_{ij}^k}{R_{ij}^k} - 1 \right) - \sum_{ijk} c_{ij}^k T_{ij}^k \tag{7.34}$$

It was shown in Chapter 3 that (7.34) is the measure of total consumers' surplus derived from the spatial-interaction demand models (7.10)–(7.12). This measure is particularly meaningful when the variables T_{ij}^k represent demand for travel. Then, problem 3 may be presented as follows: choose the pattern $\mathbf{x}$ which maximizes the allocation net benefits $f(\mathbf{x})$ and the total consumers' surplus $h(\mathbf{T})$ and satisfies the physical and planning constraints (7.17)–(7.19). The values of α_k, of course, would typically be obtained from a calibration of the spatial-interaction models for a time before the planning change; these are then assumed for the present example to retain this constant value for the future situation which is being planned using the optimization method. (It would be possible to relax this assumption, however.) We emphasize that although problem 3 is non-linear it is readily solved through existing computational techniques when $f(\mathbf{x})$ and $g_h(\mathbf{x})$ are 'regular' functions. In particular, if those functions are linear then problem 3 has a convex and separable objective function and wholly linear constraints. A number of methods for solving numerically this class of problems were discussed in Chapter 3. As before, the dual program may be formed, and it will be shown that it generates simpler computational versions than the primal program. This will be discussed in the next section.

Finally, it is interesting to explore briefly the $\beta_k \to \infty$ limit. The results of Evans (1973a) suggest that the minimization of interaction costs within our overall objective function should then coincide with that given by the transportation problem of linear programming associated with constraints (7.18) and (7.19). It is easy to use our formulation of the problem to show that this is indeed the case. In fact, when $\beta^k \to \infty$ the term

$$\sum_k \frac{1}{\beta^k} \sum_{ij} T_{ij}^k \left(\ln \frac{T_{ij}^k}{R_{ij}^k} \right)$$

disappears from the objective function.

7.2.6 The dual program for the general problem

It is easily shown using Wolfe's dual formulation described in Chapter 2 that the dual of the program (7.16)–(7.20) is

$$\min_{\{\mathbf{x}, \mathbf{T}, \boldsymbol{\lambda}\}} U(\mathbf{x}, \mathbf{T}, \boldsymbol{\lambda}) = \mathrm{f}(\mathbf{x}) - \sum_h \lambda_h^{(1)} (g_h(\mathbf{x}) - b_h) + \sum_{ijk} \alpha_k T_{ij}^k \tag{7.35}$$

subject to

$$\frac{\partial f(\mathbf{x})}{\partial x_i^{0k}} < \sum_{h=1}^{p} \lambda_h^{(1)} \frac{\partial g_h(\mathbf{x})}{\partial x_i^{0k}} + \lambda_i^{(1)k} \tag{7.36}$$

$$\frac{\partial f(\mathbf{x})}{\partial x_j^{1k}} < \sum_{h=1}^{p} \lambda_h^{(1)} \frac{\partial g_h(\mathbf{x})}{\partial x_j^{1k}} + \lambda_j^{(3)k} \tag{7.37}$$

$$\frac{\partial f(\mathbf{x})}{\partial x_l^{2}} < \sum_{h=1}^{p} \lambda_h^{(1)} \frac{\partial g_h(\mathbf{x})}{\partial x_l^{2}} \tag{7.38}$$

$$\alpha_k \ln \frac{T_{ij}^k}{f_{ij}^k(c_{ij}^k)} < \lambda_i^{(2)k} + \lambda_j^{(3)k} \tag{7.39}$$

$$\lambda_h^{(1)} > 0 \tag{7.40}$$

where $\boldsymbol{\lambda} = (\lambda_h^{(1)}, \lambda_i^{(2)k}, \lambda_j^{(3)k})$ is the vector of dual variables (or Lagrange multipliers) associated with constraints (7.17)–(7.19) respectively. Since the variables T_{ij}^k are automatically non-negative, the equality will prevail in (7.39). Thus:

$$T_{ij}^k = \exp\left[\frac{1}{\alpha_k}(\lambda_i^{(2)k} + \lambda_j^{(3)k})\right] f_{ij}^k(c_{ij}^k) \tag{7.41}$$

Substituting this last expression in (7.35), the dual program becomes

$$\min_{\{\mathbf{x},\, \boldsymbol{\lambda}\}} U^1(\mathbf{x}, \boldsymbol{\lambda}) = f(\mathbf{x}) - \sum_h \lambda_h^{(1)} (g_h(\mathbf{x}) - b_h)$$

$$+ \sum_{ijk} \alpha_k \exp\left[\frac{1}{\alpha_k}(\lambda_i^{(2)k} + \lambda_j^{(3)k})\right] f_{ij}^k(c_{ij}^k) \tag{7.42}$$

subject to (7.36)–(7.38) and (7.40). Usually, this program has considerably fewer variables than the primal program since the variables $\{T_{ij}^k\}$ have been eliminated. If the objective $f(\mathbf{x})$ and the planning and physical constraints $g_h(\mathbf{x}) < \mathrm{b}_h$ are linear, then the dual program will simply be as follows:

$$\min_{\{\boldsymbol{\lambda}\}} U(\boldsymbol{\lambda}) = \sum_{ijk} \alpha_k \exp\left[\frac{1}{\alpha_k}(\lambda_i^{(2)k} + \lambda_j^{(3)k}\right] f_{ij}^k$$

$$+ \sum_h \lambda_h^{(1)} b_h \tag{7.43}$$

subject to

$$c_i^{(0)k} < \sum_h \lambda_h^{(1)} a_{hi}^{(0)k} + \lambda_i^{(2)k} \tag{7.44}$$

$$c_j^{(1)k} < \sum_h \lambda_h^{(1)} a_{hj}^{(1)k} + \lambda_j^{(3)k} \tag{7.45}$$

$$c_l^{(2)} < \sum_h \lambda_h^{(1)} a_{hl}^{(2)} \tag{7.46}$$

$$\lambda_h^{(1)} > 0 \tag{7.47}$$

where $(c_i^{(0)k}, c_j^{(1)k}, c_l^{(2)})$ and $(a_{hi}^{(0)k}, a_{hj}^{(1)k}, a_{hl}^{(2)})$ are the coefficients of the variables $(x_i^{(0)k}, x_j^{(1)k}, x_l^{(2)})$ in the linear functions $f(\mathbf{x})$ and $g_h(\mathbf{x})$, respectively. The dual program is therefore a convex minimization problem subject to wholly linear constraints, with as many variables as the number of constraints included in the primal program.

7.3 A FIRST APPLICATION: AN ELEMENTARY MODEL OF RESIDENTIAL AND EMPLOYMENT ACTIVITIES ALLOCATION

Let us consider now the planning problem which consists of allocating H households and E jobs to zones $i = 1, \ldots, n$, with land area L_i for zone i. Define variables H_i denoting the number of house holds allocated to zone i, and E_j the number of jobs allocated to j. Then

$$\sum_i H_i = H \tag{7.48}$$

$$\sum_j E_j = E \tag{7.49}$$

and there is a land constraint

$$a^H H_i + a^E E_i < L_i \tag{7.50}$$

where a^H and a^E are the areas assumed here to be required by each household and workplace respectively. We neglect any possible spatial variation in these coefficients for the time being.

Let T_{ij} be the flow of workers from zone i to zone j, and assume that ε is the average number of workers by household. Then the following equations hold:

$$\sum_j T_{ij} - \varepsilon H_i = 0 \tag{7.51}$$

and

$$\sum_i T_{ij} - E_j = 0 \tag{7.52}$$

Assume that the household–workplace transportation interaction may be properly modelled by the spatial-interaction trip-distribution model

$$T_{ij} = A_i B_j (\varepsilon H_i) E_j \exp[-\beta c_{ij}] \tag{7.53}$$

where c_{ij} is the transportation cost from i to j, β is a parameter and, the balancing factors A_i and B_j are given by

$$A_i = \left\{ \sum_j B_j E_j \exp[-\beta c_{ij}] \right\}^{-1} \tag{7.54}$$

$$B_j = \left\{ \sum_i A_i \varepsilon H_i \exp[-\beta c_{ij}] \right\}^{-1} \tag{7.55}$$

to ensure that (7.51) and (7.52) are satisfied.

If p_i^H and p_j^E represent the net benefit of allocating a household to zone i and a workplace to zone j, respectively, and it is assumed that the cost c_{ij} and the net benefits p_i^H and p_j^E, are comparable through suitable definitions (involving perhaps annuities or present values), then a reasonable planning goal may be to select H_i and E_j to maximize

$$Z = \sum_i p_i^H H_i + \sum_j p_j^E E_j - \frac{1}{\beta}\sum_{ij} T_{ij}(\ln T_{ij} - 1) - \sum_{ij} c_{ij} T_{ij} \tag{7.56}$$

subject to constraints (7.48)–(7.52), i.e. to determine the allocation of households and workplaces which maximizes the locational net benefits and the consumers' surplus arising from travel demand. This mathematical programming problem has a very simple structure because all constraints are linear and, as noted earlier, the objective function is a separable, strictly concave function. Thus, it provides (of course, only given these extremely simplified assumptions!) a computationally efficient framework for the allocation of households and workplaces while at the same time simulating their spatial interaction. We emphasize that in the objective function Z the maximization of allocation net benefits and the minimization of travel costs appears alongside the non-linear term

$$\frac{1}{\beta}\sum_{ij} T_{ij}(\ln T_{ij} - 1)$$

which produces the spatial-interaction non-zero variables dispersion, in contrast to the linear programming approach. It also gives a measure of total utility arising from travel demand as discussed previously.

The dual of the concave optimization problem (7.56) subject to the linear constraints (7.48)–(7.52) is to minimize

$$U(\lambda^{(1)}, \lambda^{(2)}, \lambda_i^{(3)}, \lambda_i^{(4)}, \lambda_j^{(5)}) = \sum_{ij} \exp[-\lambda_i^{(4)} - \lambda_j^{(5)} - \beta c_{ij}]$$

$$+ \lambda^{(1)} H + \lambda^{(2)} E + \sum_i \lambda_i^{(3)} L_i \tag{7.57}$$

subject to

$$\lambda^{(1)} + a^H \lambda_i^{(3)} - \varepsilon \lambda_i^{(4)} = P_i^H \tag{7.58}$$

$$\lambda^{(2)} + a^E \lambda_j^{(3)} - \lambda_j^{(5)} = P_j^E \tag{7.59}$$

$$\lambda_i^{(3)} > 0 \tag{7.60}$$

for all i and j, where the dual variables $\lambda^{(1)}$, $\lambda^{(2)}$, $\lambda_i^{(3)}$, $\lambda_i^{(4)}$, and $\lambda_j^{(5)}$ are associated with constraints (7.48)–(7.52), respectively. From the optimality conditions we have

$$T_{ij} = \exp[-\lambda_i^{(4)} - \lambda_j^{(5)} - \beta c_{ij}] \tag{7.61}$$

The dual program and the optimality conditions provide an alternative way of solving the problem above. The number of variables and constraints is considerably smaller in the dual than in the primal program, particularly for n

large. Thus, the dual program will be more efficient than the primal for the purpose of computation, mainly for large-scale models. The approach illustrated here will be further discussed in a broader context in Chapter 8.

7.4 A SECOND APPLICATION: THE OPTIMUM LOCATION AND SIZE OF SHOPPING CENTRES

7.4.1 Introduction

Since the work of Huff (1964) and Lakshmanan and Hansen (1965), spatial-interaction models have been commonly used to describe flows of shopping expenditure. However, there have been only a few attempts to embed this descriptive approach in a framework to seek the optimum location and size of retail facilities. Brotchie (1969) and Dickey and Najafi (1973) proposed a program which minimizes a cost function including interaction variables satisfying a spatial-interaction model. The program is solved iteratively with the spatial-interaction flows re-adjusted in each new situation. Wilson (1974) considered a consumers' welfare maximization criterion subject to the condition that the flows must satisfy a spatial-interaction model. But this generated a complicated non-linear programming problem which could not be solved in practical terms. The presentation here is based on the paper by Coelho and Wilson (1976).

In this section the results obtained above are illustrated by showing that Wilson's approach can be transformed into an optimization problem which can be solved using known non-linear programming methods. We will also show that the program obtained in this way corresponds to the maximization of consumers' surplus.

7.4.2 A consumers' welfare maximization model

The Huff (1964), or Lakshmanan and Hansen (1965), model may be expressed mathematically as:

$$S_{ij} = A_i(e_i P_i)\, W_j^{\alpha} \exp[-\beta c_{ij}] \tag{7.62}$$

$$A_i = 1/\sum_j W_j^{\alpha} \exp[-\beta c_{ij}] \tag{7.63}$$

where the variables are:

S_{ij}— number of trips (†) from residents in zone i to shopping centres j
e_i— the shopping trip rate for residents of zone i
P_i— the population of zone i
W_j— the attractiveness of shopping centre j, which we shall assume to be measured by the size, in appropriate units

† We use trips rather than cash flows as the interaction variables in order to get the measure of consumers' surplus right.

c_{ij}— the cost of travel from zone i to shopping centre j
α, β— parameters
A_i— a balancing factor defined by equation (7.63) to ensure that:

$$\sum_j S_{ij} = e_i P_i \tag{7.64}$$

Now, if we note that

$$W_j^\alpha = \exp[\alpha \ln W_j] = \exp[\beta(\alpha/\beta)\ln W_j] \tag{7.65}$$

then (7.62) can be rewritten as

$$S_{ij} = A_i(e_i P_i)\exp\left[-\beta\left(c_{ij} - \frac{\alpha}{\beta}\ln W_j\right)\right] \tag{7.66}$$

This equation suggests that $\frac{\alpha}{\beta}$ ln W_j may be interpreted as the 'size benefits' of shopping at j which are set against the disutility of travel c_{ij}. Thus, each consumer will try to maximize

$$\frac{\alpha}{\beta}\ln W_j - c_{ij}$$

and a criterion for the optimization of the shopping centres' sizes W_j might be to choose W_j to maximize total consumers' welfare given by (Wilson, 1976a):

$$Z = \sum_{ij} S_{ij}\left(\frac{\alpha}{\beta}\ln W_j - c_{ij}\right) \tag{7.67}$$

subject to equations (7.62) and (7.63) and any other planning constraints. Due to the nature of equations (7.62) and (7.63), this is a rather complicated non-linear mathematical program. In the next section we shall see how this problem may be simplified by using the theorem proved in section 7.2.3 and show that the above criterion is transformed into the maximization of consumers' surplus.

7.4.3 Embedding of the Lakshmanan and Hansen shopping model in an optimization framework

It is easily seen through the theorem proved in section 7.2.3 that the Lakshmanan and Hansen shopping pattern (7.62)–(7.63) may be embedded in Wilson's criterion through the mathematical programming problem which consists of selecting W_j and S_{ij} to

$$\max_{\{W, S\}} = \sum_{ij} S_{ij}\left(\frac{\alpha}{\beta}\ln W_j - c_{ij}\right) - \frac{1}{\beta}\sum_{ij} S_{ij}(\ln S_{ij} - 1) \tag{7.68}$$

subject to the set of linear constraints

$$\sum_j S_{ij} = e_i P_i \tag{7.69}$$

The objective function (7.68) may also be rewritten as follows:

$$\max_{\{\mathbf{W},\,\mathbf{S}\}} = -\frac{1}{\beta}\sum_{ij} S_{ij}\ln\left(\frac{S_{ij}}{W_j^{\alpha}}-1\right)-\sum_{ij} c_{ij}S_{ij} \tag{7.70}$$

We note now that this is the measure of consumers' surplus (group surplus) arising for shoppers when the Lakshmanan and Hansen demand model is satisfied and the attractiveness of the shopping centre j is assumed to be a power function W_j of its size. Thus, Wilson's criterion expressed in (7.67) corresponds to the maximization of consumers' surplus. Alternatively, it has been argued by Pilgrim and Carter (1970) that

$$-\frac{1}{\beta}\sum_{ij} S_{ij}\ln\frac{S_{ij}}{W_j}$$

is a 'measure of locational benefits'. Then (7.70) will represent the locational benefits derived from the shopping centres distribution $\{W_j\}$ minus the total travel costs. Recalling that $CS = U–C$ (where U represents the utility and C the total cost), we will conclude that the 'locational benefits' referred to by Pilgrim and Carter are in fact the utility functions associated with the demand model (7.62) and (7.63).

Any additional planning constraints, say,

$$g_k(\mathbf{x}) = b_k \qquad k = 1, \ldots, p \tag{7.71}$$

where $\mathbf{x} = (\mathbf{W}, \ldots)$ may also be added to the optimization program above. When these constraints are linear, the model reduces to the optimization of a non-linear function subject to linear constraints. So, if the consumers' surplus maximization criterion is accepted as a planning goal, then the model (7.69)–(7.71) is a sound workable method for computing the W_j, the location and size of retail facilities.

It is also interesting to observe that when β tends to infinity, the objective function (7.70) reduces to the minimization of travel costs and therefore the model (7.69)–(7.71) becomes a simple transportation problem (cf. Wilson and Senior, 1974).

The dual program associated with this model depends on the nature of the planning constraints. For simplicity, let us assume that the planning constraints are linear, as follows:

$$\sum_j a_{kj}W_j = b_k \qquad k = 1, \ldots, p \tag{7.72}$$

Then, the dual program is

$$\min_{\{v,\lambda,\mathbf{S},\mathbf{W}\}} U = \left[\frac{1}{\beta}\sum_{ij} S_{ij}+\sum_i v_i\,(e_iP_i))\right]+\left[-\frac{\alpha}{\beta}\sum_{ij} S_{ij}+\sum_k \lambda_k b_k\right] \tag{7.73}$$

subject to the optimality conditions

$$S_{ij} = W_j^{\alpha}\exp[-\beta(v_i+c_{ij})] \tag{7.74}$$

$$W_j = \frac{\alpha}{\beta}\sum_i S_{ij}/\sum_k a_{kj}\lambda_k \tag{7.75}$$

where v_i and λ_k are the Lagrange multipliers associated with constraints (7.69) and (7.72), respectively. Thus, using (7.74) and (7.75) to replace S_{ij} in (7.73), the dual program becomes the unconstrained minimization problem:

$$\min_{\{v,\lambda,W\}} U = \frac{1}{\beta}\sum_{ij} W_j \exp[-\beta(v_i+c_{ij})] + \sum_i v_i(e_i P_i) - \sum_{kj} a_{kj}\lambda_k W_j + \sum_k \lambda_k b_k \tag{7.76}$$

The solution of (7.76) will determine the spatial-interaction variable S_{ij} uniquely through (7.74), but the size of the dual mathematical program (7.76) is smaller than that of the primal model (7.69), (7.70), and (7.71).

In this section, we will mainly concentrate on the theoretical developments of the model above following its primal version. We remark, however, that the number of variables in the primal will increase very quickly with disaggregation. This is a barrier for any practical work based on it which is not very easy to overcome with the present stage of development of computer programs for non-linear mathematical optimization problems. The size difficulties, though, are much less acute if the dual formulation is used. In the next chapter, land-use and transportation models involving the dual program formulation will, then, be explored.

7.4.4 Disaggregation and dynamics

The shopping model introduced above is too aggregated to be of help in the complex problem of planning retail facilities. We will now present a number of developments of the model and the associated optimization program which take us nearer to reality.

First, we note that the cost of developing a shopping centre depends on its site, because of varying land costs and many other factors such as topography or municipal regulations concerning the size and shape of the buildings in the zone. Thus, it is quite normal to add to the objective function (7.68) the development costs of the retail facilities. Let p_j be the development cost per unit area of facilities in zone j. Then (7.68) may be replaced by

$$\max_{\{S_{ij}, W_i\}} Z = -\frac{1}{\beta}\sum_{ij} S_{ij}\ln\frac{S_{ij}}{W_j} - \sum_{ij} c_{ij}S_{ij} - \sum_j p_j W_j \tag{7.77}$$

In this case, we are subtracting the development costs which are a component part of the definition of the overall producers' surplus. Another direction of development is through disaggregation of the interaction model variables. Let us first disaggregate by type of goods. Define S^g_{ij} as the number of service trips by residents in zone i to shopping centre j for the purchase of goods of type g, W^g_j the size of the shopping centre in j for goods of type g, and e^g_i the service trip rate for residents in zone i for the purchase of goods of type g; then the retail model (7.62)–(7.63) is replaced by

$$S^g_{ij} = A^g_i(e^g_i p_i)(W^g_j)^{\alpha^g}\exp(-\beta^g c_{ij}) \tag{7.78}$$

$$A^g_i = 1/\sum_j (W^g_m)^{\alpha^g}\exp(-\beta^g c_{ij}) \tag{7.79}$$

where the parameters α^g and β^g are also assumed to vary with the type of good g. In this case, the consumer surplus function is

$$CS = -\sum_{ijg} \frac{1}{\beta^g} S_{ij}^g \ln \frac{S_{ij}^g}{(W_j^g)^{\alpha^g}} - \sum_{ijg} c_{ij} S_{ij}^g \tag{7.80}$$

and therefore the objective function (7.77) may be replaced by

$$\max_{\{S_{ij}^g, W_j^g\}} Z = -\sum_{ijg} \frac{1}{\beta^g} S_{ij}^g \ln \frac{S_{ij}^g}{(W_j^g)^{\alpha g}} - \sum_{ijg} c_{ij} S_{ij}^g - \sum_{jg} p_j^g W_j^g \tag{7.81}$$

where p_j^g is the development cost per unit area for goods g in site j.

It is well known that the type of goods that are available in a shopping centre are intimately related to the level of hierarchy of the centre, as defined by Christaller (1933) in his central-place theory. Let us now assume that g, as the order of a type of goods, represents the level of hierarchy of the retail centre. The above model may be retained in all aspects, but now interpreted as disaggregated by hierarchy. In order to adapt the model to the principles of the central-place theory it is necessary, however, to add further details. It is commonly accepted that each level of hierarchy of retail centres has lower and upper size limits related to the population they are supposed to serve. Let W^g and $\bar{W}^g$ be the lower and upper size bounds for centres of level of hierarchy g, and define a zero-one variable δ_j^g which takes the value one if a centre of level of hierarchy g exists in zone j. Then, the following sets of constraints may be added to the model above:

$$W_j^g < \bar{W}^g \delta_j^g \tag{7.82}$$

$$W_j^g > W^g \delta_j^g \tag{7.83}$$

$$\sum_j \delta_j^g = \mu^g \tag{7.84}$$

$$\delta_j^g = \{0, 1\} \tag{7.85}$$

where μ^g is the total number of centres of level of hierarchy g. Constraints (7.82) and (7.83) ensure that the shopping centre size W_j^g is between the fixed limits and (7.84) is the total number of centres required in hierarchy g. μ^g may be defined by planning considerations, in which case it is assumed fixed, or it may be varied so that the sensitivity of the solution to the total number of retail centres in that level of hierarchy can be investigated. Such sensitivity analysis may then help to find the optimal value of μ^g.

If fixed set-up costs q_j^g, for the building of a retail centre of level of hierarchy g in site j, in addition to those varying with size, p_j^g, are defined, then (7.81) will be replaced by

$$\max_{\{S_{ij}^g, W_j^g, \delta_j^g\}} Z = -\sum_{ijg} S_{ij}^g \ln \frac{S_{ij}^g}{(W_j^g)\alpha^g} - \sum_{ijg} c_{ij}^g S_{ij}^g - \sum_{jg} p_j^g W_j^g - \sum_{jg} q_j^g \tag{7.86}$$

The optimization problem which consists of maximizing (7.86) subject to constraints (7.71), (7.82)–(7.85), and

$$\sum_j S_{ij}^g = e_i^g P_i \tag{7.87}$$

may not be solved directly by the computer programs referred to above since it includes zero–one variables δ_j^g. However, the solution may be obtained with the help of those programs using a branch-and-bound technique by fixing the values of δ_j^g in each iteration appropriately. We note also that this model may be included in the family of location-allocation models (Scott, 1969 and 1971; Sheppard, 1974) with the demand-side benefits measured by consumers' surplus.

Disaggregation by type of person and transportation mode may similarly be added to the model. Let S_{ij}^{kng} be the number of service trips by person type n, travelling by mode k from the residence zone i to the shopping centre j of hierarchic level g; P_i^n the number of type n persons in zone i; e_i^{ng} the service trip rate for centres of level of hierarchy g for type n residents in zone i; and c_{ij}^k the cost of travel from the zone i to j by mode k. The retail model then becomes:

$$S_{ij}^{kng} = A_i^{ng} (e_i^{ng} p_n) (W_j^g)^{\alpha^g} \exp(-\beta^{ng} c_{ij}^k) \tag{7.88}$$

$$A_i^{ng} = 1/\sum_{jk} (W_j^g)^{\alpha^g} \exp(-\beta^{ng} c_{ij}^k) \tag{7.89}$$

where α^g and β^{ng} are parameters. The consumer's surplus is now given by

$$CS = -\sum_{ijkng} \frac{1}{\beta^{ng}} S_{ij}^{kng} \ln \frac{S_{ij}^{kng}}{(W_j^g)^{\alpha^g}} - \sum_{ijkng} c_{ij}^k S_{ij}^{kng} \tag{7.90}$$

Thus, the complete model in this case could be:

$$\max_{\{S_{ij}^{kng}, W_j^g, \delta_j^g\}} \quad Z = CS - \sum_{jg} p_j^g w_j^g - \sum_{jg} q_j^g \delta_j^g \tag{7.91}$$

subject to constraints (7.82)–(7.85), the linear constraints

$$\sum_{jk} S_{ij}^{kng} = e_i^{ng} P_i^n \tag{7.92}$$

and the planning constraints (7.71).

A dynamic version of the model can also be obtained easily. To explain the principles of this development, consider the model without any disaggregation; this can always be added in a straightforward way. Let $t = 1, 2, \ldots, T$ denotes times. Then, if $S_{ij}(t)$ is the number of service trips by residents in zone i to shopping centre j at time t, $e_i(t)$ is the service trip rate for residents in zone i at time t, $P_i(t)$ is the population of zone i at time t, and $\Delta W_j(t)$ so that $W_j(t) = W_j(t-1) + \Delta W_j(t)$ is the increment of retail facilities area in zone j at the same period, a suitable dynamic model is:

$$\max_{\{S_{ij}(t), \Delta W_j(t)\}} \quad Z = \sum_t - \left(\frac{1}{\beta(t)} \sum_{ij} S_{ij}(t) \ln \frac{S_{ij}(t)}{W_j(t)} + \sum_{ij} c_{ij}(t) S_{ij}(t) + \sum_j p_j(t) \Delta W_j(t) \right) \tag{7.93}$$

where $c_{ij}(t)$ and $p_j(t)$ are the costs at time t, suitably discounted, subjected to the linear constraints

$$\sum_j S_{ij}(t) = e_i(t)P_i(t) \tag{7.94}$$

and the usual planning constraints (7.71). Further dynamical questions are pursued in Chapter 9.

It has been assumed implicitly until now throughout the chapter that a global optimization of the retail facilities was being sought. However, it is a simple task to adapt the previous models to seek a marginal modification in the stock of retail facilities. Suppose that W_j^0 is the present 'size' of the retail facilities in shopping centre j and denote by ΔW_j its increment due to a marginal alteration in retail facilities stock. Then the total attractiveness of shopping centre j is

$$W_j = W_j^0 + \Delta W_j \tag{7.95}$$

The consumers' surplus will now be evaluated replacing W_j by $W_j^0 + \Delta W_j$ in (7.70) and then the remaining argument is as before.

Finally, we note that the possibility of renewal may also be incorporated in this model by further small adjustments.

In the literature, as we noted earlier, we usually find the variables S_{ij} defined as flows of expenditure because planners are usually interested in the estimation of turnovers in shopping centres. The S_{ij} variables have been defined in this chapter as trips in order to construct the consumers' surplus generated in the transportation system. However, turnover can always be estimated by multiplying the number of service trip destinations by the average expenditure per trip.

We recognize that, in both the disaggregated and the dynamic versions of the procedure, appropriate data are more difficult to obtain. It is useful, however, to indicate the principles on which such developments are based so that the planner can design an optimization problem which meets his own requirements *and* data limitations as effectively as possible.

7.5 CONCLUDING CONSIDERATIONS

In this chapter it has been shown how to embed spatial-interaction submodels in a mathematical programming framework, through the maximization of a non-linear function that under certain assumptions represents consumers' surplus.

Some authors (Mazor and Pines, 1973) have suggested that spatial interaction could be embedded in linear programming models by an iterative procedure in which the output of the optimization problem is used as input to the interaction submodel, then a 'new' interaction pattern is generated and this is fed back into the linear programming problem, and so on. But no convergence proof for this procedure has been presented.

It has been shown in Section 7.2 that the integration of these submodels is accomplished through the maximization of a 'dispersion' term that leads to a strictly concave and separable objective function, which can be solved by standard computer methods.

The aim of the examples given in Sections 7.3 and 7.4 has been to illustrate some concepts developed previously and also to demonstrate that it is possible conceptually and in practical terms to construct mathematical programming models to optimize the location of residential, service, and employment activities, with flows that are consistent with the spatial-interaction submodels which have been in common use since the mid-1960s. The criterion adopted here has been the maximization of consumers' surplus, though it would also be easily extended, as emphasized earlier, to involve other welfare components such as the maximization of producers' surplus. This will be pursued further in the next chapter.

Clearly, these results can also be useful in many other fields where interaction models can be applied, such as national and regional development, transportation planning, energy modelling, and industrial and building layouts. Some such further examples will be presented in another context in Chapter 10.

8. *Optimal design and activity location*

8.1 INTRODUCTION

In Chapter 4 a family of activity allocation models incorporating an economic base mechanism and founded upon a behavioural theory of spatial interaction was presented. The extension of this approach to the design of land-use and transport plans follows naturally from its mathematical programming basis and evaluation properties as noted previously (Sections 4.4 and 4.5), and this topic will be pursued here.

The importance of the design phase in the land-use planning process is self-evident. Many authors have emphasized this point in the past. For example, Schlager (1965) remarked that 'plan design lies at the heart of planning process'. Undoubtedly, one of the most relevant tasks of the urban and regional planners is the invention or creation of alternative land-use plans, the evaluation of the social costs and benefits of each plan, and the identification of the 'non-inferior' ones out of all feasible courses of action.

Until recently this operation has been mainly accomplished using intuitive methods. In essence it was considered more like an architectural problem. But the rapid urbanization characteristic of recent times has increased the size and complexity of the problems that nowadays confront planners. This point was explicitly made by Britton Harris (1961), who wrote: 'As a result of technological advance and social change, the size and complexity of our urban concentrations has grown enormously. Their function and growth patterns now surpass the intuitive understanding and powers of normative reduction of any single individual.'

The intuitive approach to the design problem has a number of inherent limitations. A first one is associated with the inability of the human mind to deal with a large number of conflicting goals and interacting relationships. The possibility that intuition, by missing many intermediate steps and secondary effects, will lead to erroneous decisions, cannot be disregarded. Another limitation results from the fact that intuition has no feedback mechanisms to guarantee 'the self-correction of internal inconsistencies', as noted by Ben-Shahar *et al.* (1970). These shortcomings of the traditional intuitive approach have led to the exploration of recent advances in operational research and computer science to replace some of the tedious aspects of non-mechanized search in the design of land-use and transport plans, and for the generation of more effective solutions.

In recent years several linear programming approaches have been developed within overall planning frameworks. Examples may be found in the work of Schlager (1965), Ben-Shahar *et al.* (1969), Farnsworth *et al.* (1969), Young (1972), Jenkins and Robson (1974), and Ripper and Varaya (1974), among others. In terms of computational convenience the advantages of linear programming are considerable and do not need elaboration. The assumptions of linearity in the formulation of the objective function and the constraints do, however, impose important restrictions in the representation of design problems as emphasized by Parry-Lewis (1969) and others. Perhaps the most important limitation relates to the failure to incorporate into the design framework realistic spatial-interaction models such as the class of gravity models, discussed by Wilson (1970), which are characterized by a finite elasticity parameter. Some form of non-linearity thus seems an inevitable feature of an acceptable planning representation.

Relatively few non-linear programming approaches to the design problem have reached an operational stage. Notable examples of those which have, are the planning frameworks developed by the TRANSLOC group in Stockholm described by Lundqvist (1975), and the TOPAZ (the optimal placement of activities in zones) suite widely applied in Australia and elsewhere by Brotchie and his collaborators; see, for example, Brotchie—(1969) and Sharpe, Brotchie, and Ahern (1975). In the Swedish model the objective function is written as the sum of an interaction cost indicator, expressed as a regional average distance travelled, and a spatial-congestion index defined in terms of a capital-to-land ratio. The building stock allocated over a planning period is determined by quadratic programming (although the combination of activity location with a network design process, which is a broader feature of the TRANSLOC model, leads to a highly non-linear mixed integer program).

The allocation process in TOPAZ is essentially a modification and extension of the quadratic assignment program developed by Koopmanns and Beckmann (1957). The objective function consists of the benefits and costs of establishing and operating activities in the zones, and the interaction costs are determined from a doubly constrained gravity model. Solution of the constrained optimization problem is by iterative linear programming.

In his comments on the Swedish model and TOPAZ, Broadbent (1975) has referred to the need to formulate the program and benefit indicators fully consistent with the spatial-interaction model assumed to underpin the model framework in such a way that: 'travel production and consumption across space through spatial interaction are documented consistently between the entities in the model.' This is an aspect that has often been neglected in the process of designing and evaluating land-use plans which partially explains the failure of some modelling attempts to date in preparing alternative courses of action using mathematical models. In the absence of a clear understanding of the way in which people will react to land-use design alternatives, their consistent evaluation is not feasible and therefore the process of selecting the 'best' or 'most preferred' plan becomes meaningless. An attempt is made in this chapter to address this question by developing a family of mathematical programming models for the design of land-use plans founded upon a representation of spatial-interaction behaviour resulting from, and indeed supporting, the activity system. This is achieved by assuming rational choice behaviour for the 'spatial actors' in a random-utility frame of reference which is also assumed to underpin the surplus components that are integrated in the welfare maximization criterion. The formalism of random utility and the theory of choice behaviour that it uses it described in Chapter 4. The focus here is on the formulation of a range of design models using the language of mathematical programming which are consistent with the individual choice behaviour theory implicit in the representation of spatial interaction adopted throughout this work. The optimization approach is also characterized by the integration of both phases of generation and evaluation of the alternative plans through a unique operation that simultaneously ensures the internal consistency of the plans with respect to the physical, social, and planning constraints and the selection of the most preferable one by reference to the performance criteria. This can be contrasted with the simulation approach in which the outcomes of predetermined courses of action are predicted by means of a model representation of the land-use system. The generation of alternative courses of action, the evaluation of their consequences, and the selection of one or more preferred options are operations that are external to the simulation method. Therefore, the pure simulation models perform only a partial role in the planning process, whereas the optimization approach attempts to combine the four major steps concerning the creation of land-use and transport plans, as follows:

(i) the generation of alternative courses of action consistent with the physical, social, and planning constraints which have previously been identified explicitly;
(ii) the prediction of outcomes conforming to the behaviour of the spatial locators;
(iii) the computation of the costs and benefits associated with each alternative;
(iv) the selection of the 'best' alternative by reference to the welfare criterion.

However, the existence of plan components involving very difficult or even impossible quantification tasks prevents the 'ideal' generation of the 'most preferred' land-use and transport plan in 'one-shot operation' by means of an optimization framework. In certain cases, it will be appropriate in order to measure the impact of non-quantifiable factors to generate alternative solutions by varying specific parameters of the model. In that sense, an optimization model is also a simulation tool. The set of non-inferior alternatives generated through this procedure can then be subject to a multiple-objective evaluation analysis integrating further elements that might eventually not have been included in the model structure.

As a final comment on the family of optimization models to be discussed below, we would like to emphasize the irreplaceable role of the 'human-planner' in the overall process of planning. As Chadwick (1971, p. 273) notes, 'we must not despise the commonplace—and yet beautifully, wonderfully complex—attributes of planners as human beings, for planning is essentially a man–machine system, and man as operator as well as decision maker is an essential component'.

8.2 SPATIAL-INTERACTION MODELS, LOCATIONAL SURPLUS, AND PLAN DESIGN

8.2.1. The benefit functions

In general terms the benefit functions associated with a given distribution of activities A is usually written in the form

$$B = B_0 + B_1 + B_2 \tag{8.1}$$

in which B_0 contains terms independent of the spatial configuration of A, B_1 depends on zonal properties, while B_2 embraces all interaction terms expressing the mutual dependence of activities over space through the flows of people or goods between them. For the class of location problems that is discussed here, the term B_0, common to all configurations, is irrelevant and will be omitted. On the other hand, the term B_1 typically contains establishment costs or benefits associated with the configuration of activities in each zone, while B_2 is frequently dominated by a transport-cost component. A somewhat more careful discussion of the latter term is needed, however. It has been pointed out by several authors (Quarmby and Neuberger, 1969; Neuberger, 1971; and Williams, 1976 and 1977) that transportation cost alone is not an appropriate constituent of the interaction terms in the (dis-) benefit functions, as it ignores the preferences which give rise to trip dispersion. There is the well-known paradox of a city crossed by a river for which the 'most preferred' alternative, if the cost-minimization criterion is adopted, consists of the splitting of the city along the river with the activities clustered around two centres, one in each margin. This is due to the very simple fact that the cost-minimization criterion does not take into account the increase of accessibility arising from the very existence of a bridge across the river. The cost-minimization criterion is for this reason inconsistent with the basis upon which travel decisions are made. The appropriate criterion is, rather, the maximization of group surplus which was explicitly defined in Chapters 3 and 4.

8.2.2 The locational surplus maximization model: an overview

We will define a set of planning models in this chapter which are based on the *maximization* of a welfare function LS—the locational surplus function—of the following general form:

$$LS = \{\text{interaction benefits} - \text{establishment costs}\}$$

subject to a series of constraints relating to: the gravity-type representation of the spatial-interaction pattern between land-use activities which link land supply to travel demand; the consistency and economic base conditions (of section 4.2.1); and any other planning or 'problem oriented' constraints. For the reasons expounded in Chapters 3 and 4, the group surplus function GS is the appropriate measure of the interaction benefits. This representation of the interaction benefits will, by virtue of the embedding theorem proven previously (of section 7.2.3), imply that a gravity pattern for the spatial interaction between land-use activities is automatically generated.

The modelling approach for the design of land-use plans described in this chapter will be characterized by the following five main features which have already been encountered in the formulation of the group surplus model:

(i) the foundation of the spatial-interaction submodels and associated benefit measures within the framework provided by the random utility theory which offers a behavioural background based on rational choice for the 'most probable' states of the land-use system;
(ii) the formulation of the model in terms of a mathematical optimization program that allows great flexibility for handling constraints;
(iii) the derivation of an overall benefit measure—the locational surplus—by means of the objective function of the mathematical program;
(iv) the generation of the most probable spatial-interaction pattern consistent with the information available and the dispersion of preferences underpinning the population through the process of surplus maximization;
(v) the exploitation of duality to reduce the dimensionality of the programs for numerical purposes.

8.3 A FAMILY OF DESIGN PROBLEMS AND MODELS

8.3.1 The general design model framework

A range of design problems and models will be formally considered in this section which are derived within the general framework

$$\text{maximize } \{\text{total locational surplus}\}$$

subject to:

(i) consistency conditions (or accounting constraints);
(ii) economic base relations;
(iii) 'market clearing conditions';
(iv) planning constraints;
(v) non-negativity conditions on planning and trip variables.

A variety of situations encountered in the formulation of land-use and transport plans can be represented in this way. Five specific examples will be outlined, including: a long-term equilibrium design problem; a housing reallocation model; a service activity reallocation model; a housing and service activity redistribution model; and an employment redistribution model.

8.3.2 The long-run equilibrium design problem

Consider the long-run equilibrium design problem which involves selection of the housing stock and configurations of basic and service employment to maximize overall locational surplus LS for a population P or a corresponding total E^B for basic employment, determined through the multipliers in the economic base system. This design process is expressed through the following mathematical optimization model:

$$\max_{\{\mathbf{T},\mathbf{S},\mathbf{H},\mathbf{E}^B,\mathbf{E}^S\}} LS = -\frac{1}{\beta^w}\sum_{ij} T_{ij}\left(\ln\frac{T_{ij}}{\bar{W}_i^R}-1\right)-\sum_{ij} T_{ij}c_{ij}^w$$
$$-\sum_k \frac{1}{\beta^{Sk}}\sum_{ij} S_{ij}^k\left(\ln\frac{S_{ij}^k}{\bar{W}_j^{SK}}-1\right)-\sum_{ijk} S_{ij}^k c_{ij}^{S^k}$$
$$-\sum_i m_i^H H_i - \sum_{jk} m_j^{S^k} E_j^{S^k} - \sum_j m_j^B E_j^B \tag{8.2}$$

subject to

(i) the consistency condition

$$\sum_j T_{ij} - a^{(1)k}\sum_j S_{ij}^k = 0 \tag{8.3}$$

(ii) the economic base relation

$$\sum_i T_{ij} - \sum_k a^{(2)k}\sum_i S_{ij}^k - b^{(2)}E_j^B = 0 \tag{8.4}$$

(iii) the market clearing conditions

$$\sum_j T_{ij} - a^{(3)}H_i = 0 \tag{8.5}$$

$$\sum_i S_{ij}^k - a^{(4)k}E_j^{S^k} = 0 \tag{8.6}$$

(iv) the supply constraint on total basic employment

$$\sum_{ij} T_{ij} - \sum_k a^{(2)k}\sum_{ij} S_{ij}^k = b^{(2)}E^B \tag{8.7}$$

(v) a land-use capacity constraint

$$\delta^H H_i + \sum_k \delta^{S^k}E_i^{S^k} + \delta^B E_i^B \leqslant L_i \qquad i = 1, \ldots, N \tag{8.8}$$

(vi) and the non-negativity conditions on the planning and interaction variables

$$\mathbf{H} \geqslant 0,\ \mathbf{E}^{S^k} \geqslant 0,\ \mathbf{E}^B \geqslant 0,\ \mathbf{T} \geqslant 0,\ \mathbf{S} \geqslant 0 \tag{8.9}$$

It is assumed for the sake of simplicity that the final costs of establishing the urban 'commodities' $\{H_i\}$, $\{E_j^{S^k}\}$, and $\{E_j^B\}$ are constant and equal to $\{m_i^H\}$, $\{m_j^{S^k}\}$, and $\{m_j^B\}$ respectively. The conditions for relaxing this 'constant returns to scale' assumption and the further significance of the m-parameters in a location-choice setting will be discussed later.

The land-use capacity constraint (8.8) is expressed in terms of the density coefficients δ^H, δ^{S^k}, δ^B, and may be rewritten as

$$\sum_j T_{ij} + \sum_k c^{(5)k}\sum_j S_{ij}^k + b^{(5)}E_i^B \leqslant c^{(5)}L_i \tag{8.10}$$

where $a^{(5)k}$, $b^{(5)}$, and $c^{(5)}$ are suitably defined transformations of the δ's, and L_i is the area of zone i.

The dual program corresponding to (8.2)–(8.10) becomes

$$\min_{\{v,\gamma,\sigma,\rho,\xi,\eta\}} U = \frac{1}{\beta^w} \sum_{ij} \bar{W}_i^R \exp\left[-\beta^w \left(\sum_k v_i^k + \gamma_j + \xi_i + \rho + \sigma_i + c_{ij}^w \right) \right]$$

$$+ \sum_k \frac{1}{\beta^{S^k}} \sum_{ij} \bar{W}_j^{S^k} \exp[\beta^{S^k}(a^{(1)k} v_i^k + a^{(2)k}\gamma_j + a^{(2)k}\rho - \eta_j^k - a^{(5)k}\sigma_i - c_{ij}^{S^k})]$$

$$+ \rho b^{(2)} E^B + \sum_i \sigma_i c^{(5)} L_i \tag{8.11}$$

subject to the following constraints:

$$b^{(2)}\gamma_j - b^{(5)}\sigma_j \leqslant m_j^B \qquad j = 1, \ldots, N \tag{8.12}$$

$$a^{(3)}\xi_i \leqslant m_i^H \qquad i = 1, \ldots, N \tag{8.13}$$

$$a^{(4)k}\eta_j^k \leqslant m_j^{S^k} \qquad j = 1, \ldots, N, \; k = 1 \ldots, K \tag{8.14}$$

The dual variables $\mathbf{v}$, $\boldsymbol{\Upsilon}$, $\boldsymbol{\sigma}$, $\boldsymbol{\rho}$, $\boldsymbol{\xi}$, and $\boldsymbol{\eta}$ are associated with the constraints (8.3), (8.4), (8.10), (8.7), (8.5), and (8.6), respectively.

The dual inequality (8.12), for example, expresses that, at optimality, in any zone the surplus arising from basic employment allocation cannot exceed the sum of the land-use 'rent' and the cost of supplying the basic employment.

The primal and dual programs defined above express the residential-services-transportation users' benefit maximization subject to the equilibrium and planning constraints. A special case arises when the marginal cost coefficients $\{m_i^H\}$ and $\{m_j^{S^k}\}$, are constant over space, and corresponds to the supply of housing and services which 'follow' demand, as in the original Lowry long-run equilibrium configuration. Here the basic employment distribution is the major input component for the generation of the residential and service activities allocation. The land-use design program consists of seeking the distribution of basic employment which optimizes the total social benefit measure subject to the series of constraints defined above (section 4.10.2).

The application of the design framework to several 'short-run' models is now outlined.

8.3.3 A housing allocation model

The optimal plan for housing allocation is one of determining the spatial distribution of housing stock **increments** ΔH_i (positive or negative), defined relative to a base distribution $\{H_i^{(0)}\}$, which maximizes locational surplus subject to an adapted set of constraints. We shall assume that the distribution of basic employment $\{E_j^B\}$ is known, and that the service employment $\{E_j^{S^k}\}$ will 'follow demand' through the allocation process described above. The appropriate planning program may now be written

$$\max_{\Delta\mathbf{H},\mathbf{T},\mathbf{S}} LS = -\frac{1}{\beta^w}\sum_{ij} T_{ij}\left(\ln\frac{T_{ij}}{\bar{W}_i^R}-1\right)-\sum_{ij} T_{ij}c_{ij}^w$$
$$-\sum_k \frac{1}{\beta^{S^k}}\sum_{ij} S_{ij}^k\left(\ln\frac{S_{ij}^k}{\bar{W}_j^{S^k}}-1\right)-\sum_{ijk} S_{ij}^k c_{ij}^{S^k}$$
$$-\sum_i m_i^H \Delta H_i \tag{8.15}$$

subject to the consistency and economic base equations; the market clearing conditions

$$\sum_j T_{ij} - a^{(3)}\Delta H_i = a^{(3)}H_i^{(0)} \tag{8.16}$$

the stock capacity constraint

$$\sum_i \Delta H_i = \Delta H \tag{8.17}$$

which may be written

$$\sum_{ij} T_{ij} = a^{(6)}(H^{(0)}+\Delta H) \tag{8.18}$$

where $H^{(0)}$ and ΔH are the initial housing stock and the total for reallocation, respectively; and any other planning constraints. While it is possible to distinguish between demolition, construction, and renewal by defining appropriate variables, these refinements will not be considered explicitly here.

The dual model can be written

$$\min_{\{\mathbf{v},\boldsymbol{\gamma},\boldsymbol{\xi},\boldsymbol{\tau}\}} U = \frac{1}{\beta^w}\sum_{ij}\bar{W}_i^R \exp\left[-\beta^w\left(\sum_k v_i^k+\gamma_j+\xi_i+\tau+c_{ij}^w\right)\right]$$
$$+\sum_k \frac{1}{\beta^{S^k}}\sum_{ij}\bar{W}_j^{S^k}\exp[\beta^{S^k}(a^{(1)k}v_i^k+a^{(2)k}\gamma_j-c_{ij}^{S^k})]$$
$$+\sum_j \gamma_j b^{(2)}E_j^{B(0)}$$
$$+\sum_i a^{(3)}\xi_i H_i^{(0)}+\tau a^{(6)}(H^{(0)}+\Delta H) \tag{8.19}$$

subject to

$$\xi_i \leqslant \frac{m_i^H}{a^{(3)}} \tag{8.20}$$

where $\mathbf{v}$, $\boldsymbol{\gamma}$, $\boldsymbol{\xi}$, and τ are the dual variables associated with the constraints (8.3), (8.4), (8.16), and (8.18), respectively. In the case where no sign restriction is placed on ΔH_i, equation (8.20) becomes an equality, the variable $\boldsymbol{\xi}$ can be eliminated from the program, and the dual becomes an unconstrained minimization problem (unless other planning restrictions involve inequalities).

This type of model can be of use in the design of housing renewal programs, slum clearance projects, residential developments, density regulations, etc.

8.3.4 A service activity reallocation model

A similar type of model operating on the service sector may be of use in the planning of shopping centres, service facility location, and 'accessibility-to-services' related contexts. Assume that the housing and basic employment distributions are known and that an adjustment in the location of service facilities is sought subject to the criterion of overall locational surplus maximization. The design program becomes

$$\begin{aligned}\max_{\{\Delta \mathbf{E}^S,\, \mathbf{T},\, \mathbf{S}\}} \quad LS = & -\frac{1}{\beta^w}\sum_{ij} T_{ij}\left(\ln\frac{T_{ij}}{\bar{W}_i^R}-1\right)-\sum_{ij} T_{ij}c_{ij}^w \\ & -\sum_k \frac{1}{\beta^{S_k}}\sum_{ij} S_{ij}^k\left(\ln\frac{S_{ij}^k}{\bar{W}_j^{S_k}}-1\right)-\sum_{ijk} S_{ij}^k c_{ij}^{S^k} \\ & -\sum_{jk} m_j^{S^k}\Delta E_j^{S^k}\end{aligned} \tag{8.21}$$

subject to equations (8.3), (8.4). and

$$\sum_j T_{ij} = a^{(3)} H_i^{(0)} \tag{8.22}$$

$$\sum_i S_{ij}^k - a^{(4)k}\Delta E_j^{S^k} = a^{(4)k} E_j^{S^k(0)} \tag{8.23}$$

$$\sum_{ij} S_{ij}^k = a^{(4)k}(E^{S^k(0)}+\Delta E^{S^k}) \tag{8.24}$$

$$\mathbf{T} \geqslant 0,\ \mathbf{S} \geqslant 0$$

Here $\{H_i^{(0)}\}$ is the known housing stock, $\{E_j^{S^k(0)}\}$ is the initial distribution of service employment, and $\{\Delta E_j^{S^k}\}$ is the incremental distribution which is the controllable variable. The dual program may be found, as before, and is of unconstrained form if no sign is placed on the variation of $\Delta E_j^{S^k}$.

8.3.5 A housing and service activity redistribution model

Consider now the situation in which the configuration of basic employment is known and an increment or change in the distribution of housing and service activities is sought to improve overall accessibility at a minimial cost. This condition may arise, for example, if more industries are to be erected or demolished in known locations, generating therefore a new configuration of basic employment for which housing and service activity reallocation adjustments are required. This problem may be formulated using the criterion of locational surplus maximization, as follows:

$$\max_{\{\Delta \mathbf{H}, \Delta \mathbf{E}^S, \mathbf{T}, \mathbf{S}\}} \quad LS = -\frac{1}{\beta^W}\sum_{ij} T_{ij}\left(\ln\frac{T_{ij}}{W_i^R}-1\right)-\sum_{ij} c_{ij}^W T_{ij}$$
$$-\sum_k \frac{1}{\beta^{S^k}}\sum_{ij} S_{ij}^k\left(\ln\frac{S_{ij}^k}{W_j^{S^k}}-1\right)-\sum_{ijk} S_{ij}^k c_{ij}^{S^k}$$
$$-\sum_{jk} m_j^{S^k}\Delta E_j^{S^k} \tag{8.25}$$

subject to the consistency equation (8.3), the economic base conditions (8.4), and market clearing relations

$$\sum_j T_{ij} - a^{(3)}\Delta H_i = a^{(3)} H_i^{(0)} \tag{8.26}$$

$$\sum_i S_{ij}^k - a^{(4)k}\Delta E_j^{S^k} = a^{(4)k} E_j^{S^k(0)} \tag{8.27}$$

$$\mathbf{T} \geqslant 0, \qquad \mathbf{S} \geqslant 0 \tag{8.28}$$

Restrictions on the sign of ΔH_i and $\Delta E_j^{S^k}$ can be considered to distinguish between new developments, renewal programs, and clearing projects. The dual model is found as before.

8.3.6 An employment redistribution model

A similar model is generated when the stock of housing is assumed known and a change in the configuration of basic and service employment is sought. This corresponds to the situation in which new residential developments, housing renewal and clearing projects have altered the pattern of distribution of the population and the construction of industrial estates and shopping centres is considered to accommodate the shift in the access to labour. Again, primal and dual models are easily formulated, as before, employing the locational surplus maximization framework.

8.4 SOME FURTHER EXTENSIONS AND COMMENTS

8.4.1 Introduction

In the previous section a number of models for the design of land use and transportation plans were considered under the assumption of 'constant returns to scale' in the supply of housing, services, and employment commodities. A general supply–demand equilibrium framework using the concept of locational surplus maximization will be outlined in section 8.4.2. We shall also add several comments on a number of additional issues concerning the locational surplus approach. These are:

(i) The validity of the locational surplus criterion in the 'real-world' context (cf. section 8.4.3)
(ii) The interpretation of the parameter β (cf. section 8.4.4)
(iii) The dual variables and the generalization of the Hansen accessibility index (cf. section 8.4.5)
(iv) The access to markets and suppliers for industrial firms (cf. section 8.4.6)
(v) The linear programming limiting case (cf. section 8.4.7)
(vi) The size of the primal and dual design programs and the relaxation of the 'constant returns to scale' assumption (cf. section 8.4.8)

8.4.2 A general supply/demand spatial equilibrium framework

A further step towards a more general framework may be taken by introducing explicitly the supply side of the market through the addition of specific producers' surplus terms accompanying the provision of the stock $\mathbf{H}$, $\mathbf{E}^{S^k}$, and $\mathbf{E}^B$. An overall locational surplus function LS may now be defined as:

$$\begin{aligned} LS = & -\frac{1}{\beta^W}\sum_{ij} T_{ij}\left(\ln\frac{T_{ij}}{W_i^R}-1\right)-\sum_{ij} c_{ij}^W T_{ij} \\ & -\sum_k \frac{1}{\beta^{S^k}}\sum_{ij} S_{ij}^k\left(\ln\frac{S_{ij}^k}{W_j^{S^k}}-1\right)-\sum_{ijk} c_{ij}^{S^k} S_{ij}^k \\ & -\sum_i \int_0^{H_i} \psi_i(t)\mathrm{d}t-\sum_{jk}\int_0^{E_j^{S^k}} \phi_j^k(t)\mathrm{d}t \\ & -\sum_j \int_0^{E_j^B} \Omega_j(t)\mathrm{d}t \end{aligned} \tag{8.29}$$

in which ψ_i, ϕ_j^k, and Ω_j are zone-dependent supply functions in the basic, housing, and service sectors. The symmetric role played by demand and supply variables in (8.29) is emphasized by rewriting LS equivalently as follows:

$$\begin{aligned} LS = & \sum_{ij}\int_0^{T_{ij}}\left(-\frac{1}{\beta^W}\ln\frac{t}{W_i^R}-c_{ij}^W\right)\mathrm{d}t \\ & +\sum_{ijk}\int_0^{S_{ij}^k}\left(-\frac{1}{\beta^{Sk}}\ln\frac{t}{W_j^{Sk}}-c_{ij}^{S^k}\right)\mathrm{d}t \\ & -\sum_i\int_0^{H_i}\psi_i(t)\mathrm{d}t-\sum_{jk}\int_0^{E_j^{S^k}}\phi_j^k(t)\mathrm{d}t \\ & -\sum_j\int_0^{E_j^B}\Omega_j(t)\mathrm{d}t \end{aligned} \tag{8.30}$$

The commonly accepted criterion of social welfare maximization corresponds to the optimization of the locational surplus function LS that represents the 'area' between the demand and supply curves associated with the housing, services, and basic employment activities net of the transportation costs. A general land-use supply/demand spatial equilibrium model is now provided by the maximization of *LS* subject to the consistency and economic base relations (8.3) and (8.4) respectively, and to the market clearing conditions (8.5) and (8.6); other planning constraints such as density regulations, capacity restrictions, etc., may of course be readily added.

8.4.3 The locational surplus criterion in the real-world context

The maximization of the locational surplus measure *LS* expresses the optimization of overall accessibility/attractiveness for the consumers and the minimization of the supply costs to accommodate the demand for housing, services, and employment. If perfect competition, which implies a perfect knowledge of the market opportunities by the spatial actors, could be assumed, then the aim of social welfare maximization underpinning the *LS* modelling framework would be automatically fulfilled (see, for example, Samuelson, 1952; and Beckmann, 1968, p. 92). Yet, in the failure of these conditions the maximization of locational surplus will remain a valid social criterion for guiding public-sector intervention either in a market or state-controlled economy. Thus, the *LS* maximization approach is, we believe, suitable for a large class of 'real-world' land-use planning problems.

8.4.4 A note on the β parameters

It is well known in transportation planning theory that the mean trip cost is inversely related to the value of the model parameter β. It might therefore be implied that any change in the mean trip costs of the spatial-interaction submodel in the *LS* approach would inevitably be accompanied by alterations in the value of the parameters β. Of course, this is not necessarily true. In effect, it has been emphasized previously in the context of the random utility derivation of the spatial-interaction model that the parameter β is inversely related to the standard deviation of the probabilistic distribution representing the dispersion of preferences over the population P (cf. section 3.2.3). The parameters β of the spatial-interaction submodels are therefore measures of the trade-off over the population P between transport costs and the attributes associated with the opportunities at given locations. The parameters are determined by the intrinsic preferences of locators which do not depend directly on the specific location of activities. In the absence of any further evidence on the variation of the trade-off with spatial configuration, the forecasting assumption of constant β's may be adopted. Thus, in summary, by assuming the spatial-interaction submodels founded upon the random utility framework for spatial location choice, the mean trip costs may be envisaged as primarily dependent on the spatial configuration of activities, that is, any change in the mean trip costs does not necessarily imply a change in the dispersion of preferences and will, as in conventional model contexts, be explained by alterations in the transport system and spatial configuration of activities.

8.4.5 The dual variables and the generalization of the Hansen accessibility index

The relation between the dual variables and the Hansen-type accessibility indices is readily established. Let us first consider the production-constrained gravity model

$$S_{ij} = A_i P_i W_j \exp[-\beta c_{ij}] \tag{8.31}$$

which is generated by the program

$$\max_{\{S\}} -\frac{1}{\beta} \sum_{ij} S_{ij}\left(\ln \frac{S_{ij}}{W_j} - 1\right) - \sum_{ij} c_{ij} S_{ij} \tag{8.32}$$

subject to

$$\sum_{ij} S_{ij} = P_i \tag{8.33}$$

The optimality conditions of this latter program are

$$S_{ij} = W_j \exp[-\beta(v_i + c_{ij})] \tag{8.34}$$

which can be identified with the model, as discussed before, by writing

$$A_i = \exp[-\beta v_i]/P_i \tag{8.35}$$

On the other hand, it is well known that the Hansen accessibility index

$$X_i = \sum_j W_j \exp[-\beta c_{ij}] \tag{8.36}$$

is inversely related to the balancing factor A_i. Thus we have

$$X_i = A_i^{-1} = P_i \exp[\beta v_i] \tag{8.37}$$

which expresses the relationship between the dual variables v_i and the Hansen's indices. Next, consider the doubly constrained gravity model written as follows:

$$T_{ij} = \exp[-\beta(v_i + \gamma_j + c_{ij})] \tag{8.38}$$

subject to the trip end constraints

$$\sum_j T_{ij} = P_i \tag{8.39}$$

$$\sum_i T_{ij} = E_j \tag{8.40}$$

where P_i and E_j represent zone population and employment variables, respectively. Hansen-type indices measuring the accessibility of zone i to employment and of zone j to labour are readily provided by

$$X_i^E = \sum_j E_j \exp[-\beta(\gamma_j + c_{ij})] = P_i \exp[\beta v_i] \tag{8.41}$$

$$X_j^L = \sum_i P_i \exp[-\beta(v_i + c_{ij})] = E_j \exp[\beta \gamma_j] \tag{8.42}$$

as transformations of the dual variables v_i and γ_j, respectively. This procedure may clearly be extended to define generalized Hansen-type accessibility/attractiveness indices, directly from the dual program solution of more general locational surplus models. Thus, for example, the following indices may be considered for the long-run equilibrium model (8.2)–(8.9):

(i) accessibility to service activity sector k

$$X_i^{Sk} = P_i \exp[\beta^{S^k}(a^{(1)k} v_i^k)] \tag{8.43}$$

(ii) accessibility to service employment

$$X_i^S = P_i \exp[\beta^W(\sum_i v_i^k)] \tag{8.44}$$

(iii) access to labour in service activity sector k

$$X_j^{Lk} = E_j^{Sk} \exp[-\beta^{S^k} a^{(2)} \gamma_j] \tag{8.45}$$

(iv) overall access to labour

$$X_j^L = E_j \exp[\beta^W \gamma_j] \tag{8.46}$$

where P_i, $E_j^{S^k}$, and E_j are the population, service employment, and total employment internally defined by the model.

8.4.6 Some comments on industrial location, accessibility, and plan design

In the general design model considered above, households seek to maximize access to home-based activities and retailers (we have assumed) maximize access to purchasing power, an objective achieved by following 'aggregate demand'. The factors influencing industrial location which must be incorporated into the criterion function require, however, a more explicit discussion. Many factors influence the location of particular firms, but it is found that in many contexts, for example, (see the Tyne and Wear Plan, 1973) three factors predominate: accessibility to labour; access to markets and supplies; and land availability (physical constraints).

In urban development studies, access to labour is usually simulated through some form of labour potential operationalized through traditional Hansen-type indices. The access to labour is introduced directly into the locational surplus criterion through the dual variable Υ, and maximization of LS with respect to the control $\mathbf{E}^B$ will thus simultaneously perform the required function of maximizing accessibility of employment to workers.

With regard to access to markets and supplies, two problems emerge: to gauge the relative importance of each to firms and to determine the importance of external and internal access with reference to a given study area. For the exporting sector we shall take, in the absence of specific inter-industry relations, the access to external markets as the dominant factor. (In the Type and Wear Plan, 1973, labour availability, access to markets, and land availability, the three principal factors were weighted in the proportions 0.45: 0.30: 0.25.) In the urban simulation model, discussed by Turner (1974), and the Tyne and Wear model (1973), the process of basic employment location involved an index defined in terms of zonal access to external transport routes and facilities as a surrogate measure for external market accessibility.

To incorporate this notion into the programming model we shall define a similar index as follows:

$$\mu_j = M_j \sum f_{jl} c_{jl} \tag{8.47}$$

in which f_{jl} is the proportion of commercial goods leaving the study area by exit l from zone j in the base year, and M_j is the volume of goods produced for export per unit employment at j. A suitable aggregate sub-objective for industrial location now becomes

$$\underset{\mathbf{E}^B}{\text{minimize}} \sum_j E_j^B \mu_j \tag{8.48}$$

The term m_j^B in equation (8.2) may now conveniently be viewed as comprising of a set-up cost and the accessibility component defined above (reduced to a unit time measure). That is,

$$m_j^B = \bar{m}_j^B + \varepsilon M_j \sum_l f_{jl} c_{jl} \tag{8.49}$$

where ε is a criterion weighting factor.

8.4.7 The linear programming limiting case

It has been shown by Evans (1973b) that the doubly constrained gravity model converges to the linear programming transportation problem when the β-parameter tends to infinity (cf. section 2.4.2). A similar result for the locational surplus model is straightforwardly obtained. Note that when the parameters β tend to infinity, the non-linear terms of the form $\frac{1}{\beta}\xi \ln \xi$ in the locational surplus function will tend to zero. Thus, it is clear that the solution of the model (8.2)–(8.9) will converge to that of the linear program

$$\underset{\{\mathbf{T},\,\mathbf{S},\,\mathbf{H},\,\mathbf{E}^S,\,\mathbf{E}^B\}}{\min} \quad LS = \sum_{ij} c_{ij}^W T_{ij} + \sum_{ijk} c_{ij}^{S^k} S_{ij}^k$$
$$+ \sum_i m_i^H H_i + \sum_{jk} m^{S^k}{}_j E^{S^k}{}_j + \sum_j m_j^B E_j^B \tag{8.50}$$

subject to the same set of constraints. Computational evidence for this behaviour has been obtained and will be presented elsewhere (see also Coelho and Williams, 1978). This result shows that the linear programming land-use design models of Schlager (1965), Ben–Shahar *et al.* (1969), and others are limiting cases of the more general locational surplus approach discussed here.

8.4.8 Additional remarks on the programming representation

We would like here to re-emphasize the dramatic differences between the size of primal and dual design program. For example, if the planning area is divided into 50 zones and **three** service employment sectors are considered, then the 'long-run' primal design model formulated above has 10 250 variables and 451 constraints, while the dual program has only 451 variables, 50 constraints and upper bounds in 200 variables. Of course, the solution of the dual determines uniquely the primal outcome through the optimality conditions, and the interaction flows can be extracted directly from the dual variables as, for example, in equations (4.95) and (4.96).

It is also worth while to stress that the primal model is a concave and separable maximization problem subject to linear constraints, whereas the dual is a convex minimization problem with linear constraints (in some situations the dual may be simply an unconstrained convex minimization problem). These particular classes of non-linear problems can be solved by a number of existing methods such as the projected-gradient method, the convex simplex method, the variable reduction method, and the reduced gradient method proposed by Rosen (1960), Zangwill (1967), McCormick (1970a and 1970b), Wolfe (1965), and Faure and Huard (1965).

Finally, it is appropriate to add a comment on the relaxation of the assumption of linear cost forms in the objective function, involving the constants $\mathbf{m}^H$, $\mathbf{m}^S$, and $\mathbf{m}^B$—that is, the assumption of no economies or diseconomies of scale. In fact, it is important to emphasize that as long as the locational surplus function *LS* remains concave no difficulties with local optima will be encountered. Suitable adjustment to the dual must of course be made within the Rockafellar or Wolfe frameworks. However, if cost functions are chosen which cause this concavity condition to be violated, local optima will in general exist, and we cannot speak of a dual program in the conventional sense.

8.5 A BRIEF NOTE ON MULTIPLE GOALS

The method for the design of land use and transport plans expounded in this work is essentially based on the maximization of a single objective—the locational surplus function—which measures economic efficiency taking into account the benefit arising from accessibility and attractiveness. The evaluation procedure underpinning the approach is characterized by the cost–benefit analysis framework in which the benefits accruing to consumers from accessibility and attractiveness are evaluated in a consistent manner with the basis in which location decisions are taken.

It can also be argued that this approach may be used to study the impact of a number of other goal strategies. In particular, the goal of transport and supply cost minimization has been shown to correspond to a limiting case of the locational surplus maximization approach. Other goal strategies may be inserted into the modelling framework by varying the nature of the cost components and including alternative evaluation indicators. The above-mentioned generalized Hansen accessibility and attractiveness indicators provide a specific example of how additional information for a more general and comprehensive evaluation framework can be generated.

It is well established that many practical planning problems require the consideration of mutually conflicting and non-commensurable objectives. Pollution externalities, social effects, and environmental preservation are examples of important values that need to be considered in land-use design which cannot be converted into money units without very substantial controversy. The occurrence of these plan evaluation conflicts in the urban and regional field among others, has given rise to an intensive research for new methods of 'vector optimization' or 'multiobjective programming' in the last decade. The locational surplus method, by virtue of its mathematical programming representation, is naturally fitted for integration in such multiple-goal evaluation procedures. The discussion of those techniques falls, however, outside the scope of this chapter. Comprehensive surveys of multiobjective programming methods are provided by Cohen and Marks (1975) and Nijkamp (1977).

8.6 CONCLUDING COMMENTS

In this chapter a framework for the design of land use and transport plans involving a mathematical programming representation has been proposed. The latter consists of the maximization of a locational surplus function subject to equilibrium and planning constraints. A relevant feature of this approach, as we have argued earlier, is that the process of maximizing the benefit measure associated with travel and location choices will generate trip distribution submodels consistent with the gravity-type patterns commonly observed in urban areas. This may be regarded as an improvement on the mathematical optimization models based upon the linear programming frameworks proposed by Schlager (1965) and Ben-Shahar *et al.* (1969).

Another characteristic of this approach is its foundation upon a behavioural model of location choice which is derived within the framework of random utility theory. This enables the definition of evaluation measures which are fully consistent in behavioural terms with the representation of spatial interaction underpinning the approach. A further characteristic of the *LS* maximization method is the exploitation of duality for reducing the dimensionality of the mathematical programs when a numerical solution is required. This is undoubtedly of primary importance for model implementation purposes.

The criterion embedded in the *LS* framework involves exclusively benefits and costs expressed in monetary units and therefore it must be viewed as representing just a component in a more comprehensive multidimensional evaluation setting which will clearly integrate other non-pecuniary elements that also contribute to the overall welfare of 'spatial actors'. Bearing that in mind, the model-building technique outlined in this chapter based on locational surplus maximization is, we believe, a useful tool for land use and transportation planning.

9. *Optimization and dynamical systems theory*

9.1 INTRODUCTION

Most of the models presented so far in this book are static: any forecasting uses are on a comparative static basis, and nowhere is time treated explicitly. In this chapter, we relate optimization theory to the modern theory of dynamical systems and show how many of the model structures presented hitherto can be embedded (in a new way—cf. Chapter 7) in dynamical frameworks, but also that the dynamical theory gives new insights into comparative static approaches.

Most familiar models are based on strong equilibrium assumptions, and the equations of the model usually enable an equilibrium state of the system represented by the model to be calculated. Dynamical analysis is important in a number of respects. Firstly, it provides tools for the analysis of the *stability* of such equilibrium points. Secondly, by focusing on equilibrium points as a function of values of model parameters, it enables multiple equilibria to be identified (which corresponds to the surface of possible equilibrium points in phase space—including both state variables and parameters in this space—being folded). Thirdly, it provides the framework for building more general dynamical models which do not rely on equilibrium assumptions; so transient effects can be modelled, or the behaviour of systems which are far from equilibrium.

An optimizing model, in the broadest terms, takes the form:

$$\min_{\{\mathbf{x}\}} f(\mathbf{x}, \boldsymbol{\alpha}) \tag{9.1}$$

subject to

$$\mathbf{g}(\mathbf{x}, \boldsymbol{\alpha}) > 0 \tag{9.2}$$

(for state variables $\mathbf{x}$ and parameters $\boldsymbol{\alpha}$).

When the relations (9.2) are present, the problem is constrained; otherwise, it is unconstrained and totally specified by (9.1). As we saw in Chapter 2, the constrained problem can be written as an unconstrained problem in a larger number of variables if a vector of Lagrangian multipliers is introduced and the problem written as

$$L = f(\mathbf{x}, \alpha) + \lambda . \mathbf{g}(\mathbf{x}, \alpha) \tag{9.3}$$

in terms of a Lagrangian L.

The comparative statics of such models usually involves situations in which small changes in the parameters lead to small changes in the state variables **x**—the whole process of change is smooth. However, these ideas can be extended by those of catastrophe theory: the equilibrium surface can be folded and we will see that this introduces new kinds of system behaviour even into a comparative static analysis: discrete change (jumps), hysteresis effects, and divergence. These ideas will be introduced in Section 9.2 below.

The next step in the argument is to consider embedding the static model in a more dynamic framework. If states of the system away from equilibrium are investigated, the simplest differential equations which govern the return to equilibrium for a system such as (9.1)–(9.2) or equivalently (9.3), are

$$\dot{\mathbf{x}} = -\frac{\partial L}{\partial \mathbf{x}} \tag{9.4}$$

together with

$$\dot{\boldsymbol{\lambda}} = -\frac{\partial L}{\partial \boldsymbol{\lambda}} \tag{9.5}$$

These equations can be modified in various ways, but we leave the detail of that until later. The signs of the derivatives can be used to investigate the stability of the equilibrium points (the first-order conditions) and we will see later that the second-order conditions—second derivatives or Hessian matrices as appropriate—can be used to identify certain critical points. A non-equilibrium trajectory of the system can be constructed simply by solving the equations, possibly by numerical methods as analytical solutions are often difficult to obtain. The simplest system is when the original optimization problem is unconstrained, as in (9.1) when the differential equations would be

$$\dot{\mathbf{x}} = -\frac{\partial f}{\partial \mathbf{x}} \tag{9.6}$$

and the stable equilibrium points are then (some of) the solutions of

$$\frac{\partial f}{\partial \mathbf{x}} = 0 \tag{9.7}$$

Such systems are known as gradient systems (because the return to equilibrium after a disturbance is along a gradient, defined by (9.7), towards an equilibrium point). These can be connected directly to the topic of catastrophe theory mentioned earlier. The more general constrained systems can also be treated as gradient systems (though in a larger number of variables, or, if the constraints are treated separately, as involving new kinds of catastrophes relative to the basic elementary ones).

The most general dynamical system of all can be written canonically in the form

$$\dot{\mathbf{x}} = \mathbf{f}(\mathbf{x}, \boldsymbol{\alpha}) \tag{9.8}$$

where in this case, in general, it is not a gradient system. However, modern dynamical systems theory has much to say about the bifurcation properties of equilibrium points of such systems, and we will pursue this topic in Section 9.3 below.

Finally, we should mention a more obvious connection of optimization theory to dynamic models—control theory. This will be pursued briefly in relation to an example in a later section.

9.2 CATASTROPHE THEORY

Consider a gradient system with equilibrium states defined by (9.1) (which are the solutions of (9.7)) and whose dynamics are governed by differential equations of the form (9.6). We will not consider constraints at this stage, but will return to them later. Suppose that there are n state variables in the vector $\mathbf{x}$ and k parameters $\boldsymbol{\alpha}$. The possible equilibrium states then form a surface, or manifold, in $(n+k)$-dimensional space. The behaviour of the system over time is described by trajectories over this manifold. Thus, comparative statics is concerned with the consequences of parameter change, specified exogenously; the model of a dynamical system is completed if differential equations are added for the $\mathbf{x}$ variables, as previously indicated, and also, at an even deeper level of explanation, for the $\boldsymbol{\alpha}$ parameters.

Catastrophe theory is about the possible shapes of the equilibrium manifold for particular functions f. The theory was developed by the French mathematician René Thom and his main result is that, for a wide class of functions, the number of types of shapes of equilibrium surfaces is relatively small. However, although we will try to give the flavour of these results, the main interest at the present time in urban modelling lies in the introduction and awareness of new modes of system behaviour. Before pursuing this, however, we will try to give the flavour of some of the results developed by Thom (1975), Zeeman (1977), and others—see for example, Amson (1975), Chillingworth (1975, 1976), and Poston and Stewart (1978) for elementary accounts.

Thom's first concern in the initial development of catastrophe theory was *structural stability*, of functions, and of related aspects of the real world modelled by functions. An idea of the basic concept can be given by the following example. Consider the function x^3 as plotted in Figure 9.1(b). This is structurally *unstable*

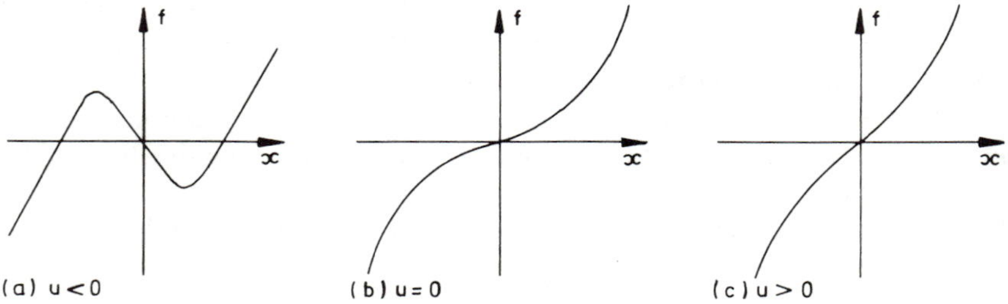

Figure 9.1 Plots of $y = x^3 + ux$ for different u

in the following sense: if it is perturbed slightly, for example by the addition or subtraction of a small term ux, then its behaviour in the neighbourhood of $x = 0$ undergoes a qualitative change. This can be seen on the other parts of Figure 9.1, which distinguishes the cases $u < 0$ and $u > 0$. The question can then be asked: what is the simplest way to make such a function stable? The answer is to add the term ux, and then to note that the function $x^3 + ux$ is, typically, structurally stable: for a typical non-zero value of u, the shape of the curve does not change under small perturbations.

x^3 is said to have a degenerate, or structural singularity at $x = 0$. $x^3 + ux$ is known as an *unfolding* of this singularity and it can be shown that, in a topological sense, this is an universal unfolding: all degenerate singularities of the same order have this form.

This can be presented more formally as follows. The stationary points of

$$f(x) = x^3 \tag{9.9}$$

are given by the solutions of

$$\frac{\partial f}{\partial x} = 3x^2 = 0 \tag{9.10}$$

and there is obviously only one:

$$x = 0 \tag{9.11}$$

At this point, we also note that

$$\frac{\partial^2 f}{\partial x^2} = 6x = 0 \tag{9.12}$$

which shows that the singularity is degenerate: it is neither a maximum nor a minimum. It can easily be checked that a progression through the same steps shows that $x^3 + ux$ does not have any degenerate singularities except when $u = 0$. Such a point is a critical value of the parameter u. For other values of u, the function either has one minimum (for $u < 0$ as depicted on Figure 9.1(a)) or no minimum at all (for $u > 0$ as shown on Figure 9.1(c)). This information can be plotted graphically as in Figure 9.2. In this case, the catastrophe occurs at the

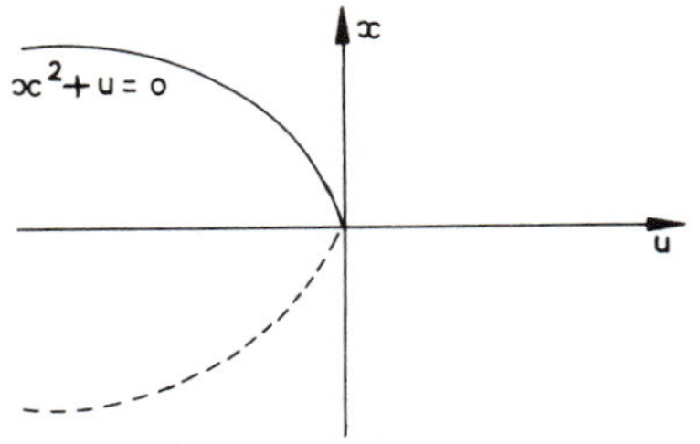

Figure 9.2 Plot of $x^2 + u = 0$: the fold catastrophe

critical point in parameter space, $u = 0$, when the minimim disappears, and this corresponds to the changing values of u shown in Figure 9.1.

What do higher-order degenerate structural singularities look like? The simplest form is represented by higher-order power functions like x^n for $n = 4, 5, 6, \ldots$ .The universal unfolding of these singularities involves the addition of terms analogous to ux, but more of them. In general, we look at the singularities of functions like

$$f(x, u_1, u_2, \ldots) = x^m + u_1 x^{m-2} + u_2 x^{m-3} + \ldots + u_{m-2} x \tag{9.13}$$

of which $x^3 + ux$ can now be seen to be a special case. The highest-order degenerate singularity obviously occurs at the point

$$(u_1, u_2, \ldots, u_{m-2}) = (0, 0, 0, \ldots, 0) \tag{9.14}$$

in parameter space, but there turn out to be interesting lower-order structural singularities in *regions* of parameter space. The x^3 example, as we shall see more formally shortly, illustrates the *fold* catastrophe. The next most complicated unfolds x^4 and involves two parameters u_1 and u_2 (and hence a three-dimensional—(x, u_1, u_2)—space), and this generates the *cusp* catastrophe. The $(u_1, u_2) = (0, 0)$ singularity is the cusp point, and on cusp-shaped curves in (u_1, u_2) space, there occurs a family of folds.

The mathematical importance of Thom's theorems stems from the following observation: as the parameters vary in (9.13), the right-hand side can be viewed as a truncated Taylor expansion of a family of functions—one function for each 'point' in parameter space. What he shows is that for up to four control parameters, the number of forms of structural singularity are, at worst, of very limited types. For example, functions spanned by $x^3 + ux$ can only have folds; functions spanned by $x^4 + u_1 x^2 + u_2 x$ can only have cusp points and folds. Because of the wide class of functions spanned by truncated Taylor expansions (9.13), Thom's main theorem says that the forms of singularity of a wide class of functions of one, two, three, or four parameters can only take this form.

Two important qualifications must be added at this stage. The theorem applies *locally*, in the neighbourhood of the degenerate singularity; and secondly, the illustrations using power functions employ a canonical form—other functions of the same number of parameters have the same form, at worst,† of structural singularities in the topological sense that they can be transformed into the

† More generally, it can be stated that most functions are one of a relatively limited number of types:

(1) they have no structural singularities, and these are the most common;

(2) they have Morse, or regular, singularities: they can be transformed into $x_1^2 + x_2^2 + \ldots - x_r^2 - x_{r+1}^2 - \ldots$ which means that their stationary values are maxima, minima, or saddle points;

(3) they have degenerate singularities, and in many cases these are of a limited number of types, by Thom's theorem, depending on the co-rank and co-dimension of the singularity. Essentially, co-rank measures the order of degeneracy and co-dimension the number of parameters needed to provide a universal unfolding. Thom's theorem is concerned with singularities up to co-rank 2 and co-dimension 4.

canonical form. However, the transformations may be very difficult in practice. We will comment further on the practical uses of Thom's theorem shortly.

Probably the best framework for the introduction of further concepts is an example. Rather than take the simplest, which is the fold, we take the cusp catastrophe as this provides a richer illustration of the associated concepts. As we have seen, this involves the universal unfolding of the singularities of x^4, of co-rank 1 and co-dimension 2. Suppose the possible equilibrium states of a system are the minima of a function $f(x, u_1, u_2)$ given by

$$f(x, u_1, u_2) = x^4 + u_1 x^2 + u_2 x \tag{9.15}$$

The stationary values are given by

$$\frac{\partial f}{\partial x} = 4x^3 + 2u_1 x + u_2 = 0 \tag{9.16}$$

The cubic equation (9.16) can have one or three real roots and the condition for the existence of three real roots is

$$(-\tfrac{1}{3}u_1)^3 > (\tfrac{1}{2}u_2)^2 \tag{9.17}$$

which also implies

$$u_1 < 0 \tag{9.18}$$

By squaring each side of equation (9.17), we can see that the boundary of the region (of one real root or three) is defined by

$$4u_1^3 + 27u_2^2 = 0 \tag{9.19}$$

Some further manipulation shows that $\dfrac{\partial^2 f}{\partial x^2}$ is zero on this curve (which includes the point (0,0,0) in (x, u_1, u_2)-space) and hence that the function has degenerate singularities along the curve.

The second derivative can be used to identify the minima: in the case of three real roots, two are minima; and in the case of the single real root, that turns out to be a minimum. The surface represented by equation (9.16) can be plotted as shown on Figure 9.3. This is a very important figure: it is sufficiently rich to illustrate the set of concepts which are the basis of applied catastrophe theory.

The curves which join in the cusp point are plotted in the (u_1, u_2) plane. It can now be seen that for values of u_1 and u_2 inside the region, the surface of possible equilibrium states is multivalued—in fact with two possible stable values with the middle sheet added for geometrical completeness but representing unstable states. It is also immediately clear that the curves which form the boundaries of this region are critical values of (u_1, u_2): if the system is moving in a trajectory whose projection in the (u_1, u_2) plane crosses one of these curves, then in some cases there can be a jump from the upper sheet to the lower, or vice versa. However, this is anticipating some results which are best discussed more generally and will be picked up again shortly.

The region inside the curves is known as the *conflict set* (because the system can

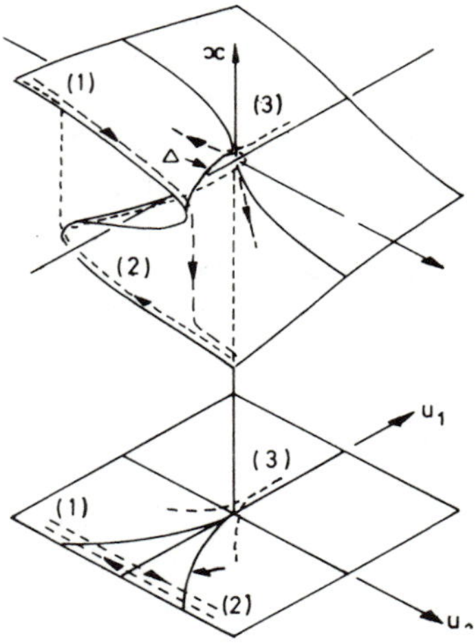

Figure 9.3 The cusp catastrophe equilibrium surface

be in more than one state) or sometimes the *catastrophe set*. The set of points on the curves themselves form the bifurcation set, since it is at these values of u_1 and u_2 that jumps can take place. The u_1 and u_2 variables are sometimes called control variables (although they are not always used in this way in the usual sense of 'control'), and so the (u_1, u_2) plane is known as the *control plane*, or more generally as the *control manifold* (since if more than two control variables are involved, it has a correspondingly higher dimension).

A good way of showing in more detail the way in which the cusp catastrophe surface has been assembled is to use a diagrammatic device due to Zeeman, which is shown in Figure 9.4. This shows the control manifold, a typical trajectory on the manifold, and a set of plots of f against x for typical points on that trajectory. This shows clearly how there is one possible minimum outside the critical region and two inside it, and that one of the minima crashes into the maximum to form a point of inflexion (and hence the degenerate singularity which produces the folds) as a trajectory crosses the bifurcation curve from the outside.

This is a useful point to tie up some of the mathematics used previously in relation to these diagrams, especially Figure 9.3. We noted earlier that at critical points (which we now recognize as the bifurcation set) degenerate singularities occur, and at such points, the second derivative of the objective function vanishes: $\frac{\partial^2 \mathbf{f}}{\partial x^2} = 0$. We are considering f and x also to be functions of the control variables, in this case u_1 and u_2. The implicit function theorem shows that

$$\frac{\partial u_1}{\partial x} = -\frac{\partial f'}{\partial x} \bigg/ \frac{\partial f'}{\partial u_1} \tag{9.20}$$

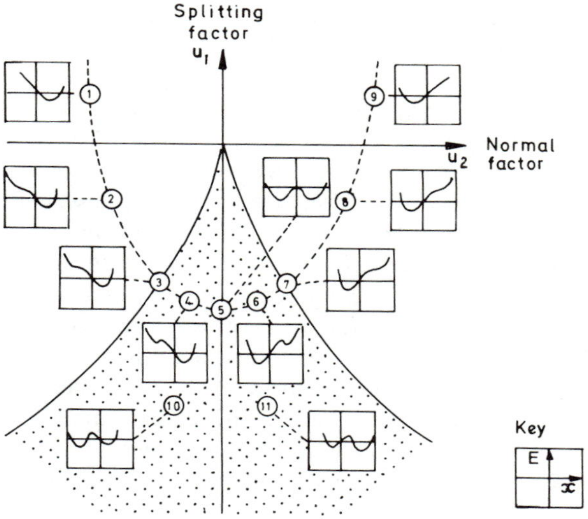

Figure 9.4 The (u_1, u_2) control plane for the cusp catastrophe with (E, x) plots for different (u_1, u_2) values

and

$$\frac{\partial u_2}{\partial x} = -\frac{\partial f'}{\partial x} \bigg/ \frac{\partial f'}{\partial u_2} \tag{9.21}$$

where

$$\frac{\partial f'}{\partial x} = \frac{\partial^2 f}{\partial x^2} \tag{9.22}$$

and we have already seen that this vanishes. This shows, therefore, that

$$\frac{\partial u_1}{\partial x} = \frac{\partial u_2}{\partial x} = 0 \tag{9.23}$$

and hence that at critical points (structural singularities or bifurcation points), the tangent to the equilibrium surface in Figure 9.3 is perpendicular to the control manifold. This result applies more generally whatever numbers of dimensions are involved. It also shows that a jump near a bifurcation point can be caused by a change in any of the control variables.

We now have set up the equipment needed to see the possible modes of 'interesting' behaviour for a system governed by the cusp catastrophe equilibrium surface. We can now return to Figure 9.3 and examine the trajectories marked (1), (2), and (3) in turn. The projections of these trajectories are shown as dashed curves in the control manifold. In case (1), the trajectory crosses the bifurcation curve for the first time without any jump taking place, but when it crosses the second curve, the system must make a discrete jump to a new state. Within the conflict set, where there are two possible states, an additional rule must be supplied to say which state the system will be in at any given time. In this case we have assumed what is called the 'perfect delay convention': the system stays on a particular sheet as long as possible. (The alternative is known as the Maxwell convention, which states that the system always accepts the lowest minimum, and in this case any jumps take place along the axis which bisects the cusp curves.) Note that for jumps to take place, we must have $u_1 < 0$ and for this reason u_1 is known as the 'splitting factor' which determines whether jump behaviour is included or not; the u_2 variable is known as the 'normal factor' and generally shows the effect of the state variable of monotonic change in u_2.

The second kind of 'new' behaviour is illustrated by trajectory (2): this shows that, with the perfect delay convention again assumed to operate, since any jump takes place at the second crossing of the bifurcation curve, when a process is reversed, the jump, of course, takes place at a different point. If we look at a section of this perpendicular to the (x, u_2) plane as in Figure 9.5, we get what can be recognized as a hysteresis curve.

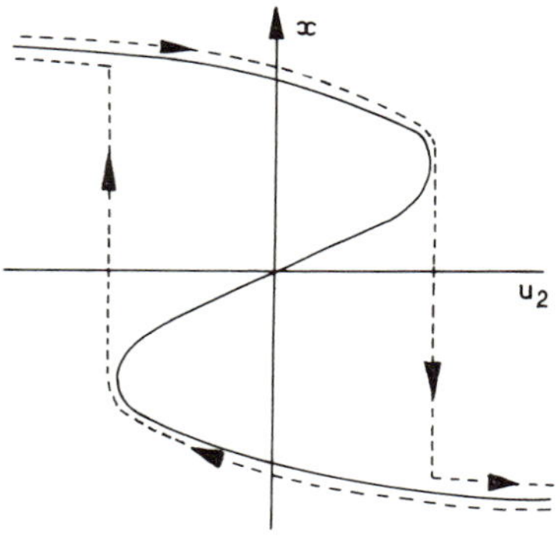

Figure 9.5 Hysteresis effects

Finally, consider trajectory (3): in this case it approaches the cusp point from above, and it is clear that a small variation in its position will take it smoothly either to the upper sheet or the lower, and this behaviour is known as divergence.

The fold catastrophe can be given the same formal treatment as the cusp, though in this case the results are simpler and are already clear from Figure 9.1. The equivalent of the critical cusp curves is the single point at which the minimum disappears. There is no alternative state for the system to jump to—it simply 'disappears', though we shall see in later applications that alternative states can be added to models based on the fold.

Thom's main theorem can now be put another way (bearing in mind that the typical function is not in canonical form but is transformable into such a form): if a family of functions is parameterized by a single control variable, then the only possible structural singularities are folds. A two-parameter function can have structural singularities which are at worst folds or cusps. As we then progress to degenerate singularities of higher co-rank and co-dimension, new and more complicated structures emerge. However, it is a feature of catastrophe theory that each higher level retains the possibility of the forms of structural singularity which have been identified at lower levels.

This is not the place to present in detail the form of the remaining catastrophes. There are now many available sources for this information, including the references cited earlier in this chapter. However, Table 9.1 lists Thom's seven elementary catastrophes, of which we have discussed two in some detail—the fold and the cusp. These also suffice for the application to be presented in this chapter.

Table 9.1 The seven elementary catastrophes

Name	End variables	Control variables	Potential function
Fold	1	1	$\frac{1}{3}x_1^3 + u_1x_1$
Cusp	1	2	$\frac{1}{4}x_1^4 + \frac{1}{2}u_1x_1^2 + u_2x_1$
Swallowtail	1	3	$\frac{1}{5}x_1^5 + \frac{1}{3}u_1x_1^3 + \frac{1}{2}u_2x_1^2 + u_3x_1$
Hyperbolic umbilic	2	3	$\frac{1}{3}x_1^3 + \frac{1}{3}x_2^3 + u_1x_1x_2 - u_2x_1 - u_3x_2$
Elliptic umbilic	2	3	$\frac{1}{3}x_1^3 - \frac{1}{2}x_1x_2^2 + \frac{1}{2}u_1(x_1^2 + x_2^2) - u_2x_1 - u_3x_2$
Butterfly	1	4	$\frac{1}{6}x_1^6 + \frac{1}{4}u_1x_1^4 + \frac{1}{3}u_2x_1^3 + \frac{1}{2}u_3x_1^2 + u_4x_1$
Parabolic umbilic	2	4	$\frac{1}{2}x_1^2x_2 + \frac{1}{4}x_2^4 + \frac{1}{2}u_1x_1^2 + \frac{1}{2}u_2x_2^2 - u_3x_1 - u_4x_2$

When more than four parameters, or control variables, are involved, the forms of structural singularity cannot be classified in the same straightforward way as for the elementary catastrophes. Indeed, so far, relatively little progress has been made in classifying them at all. Since the number of independent variables is often very large for locational or transport problems, this does mean that catastrophe theory as such is not always very useful. We will return to this point shortly. However, one further form of catastrophe should be mentioned which is important in application and for which work is in progress on classification (Poston and Stewart, 1978): these are the so-called constraint catastrophes. It has long been known that in mathematical programming, especially with linear constraints, the optimum solution occurs at a corner of the feasible region. This means that if one of the parameters of the problem changes, there can be a 'jump' from one corner of the feasible region to another. We have seen that in catastrophe theory, jumps occur on the boundary of a region in parameter space (i.e. on the control manifold) which divides a multivalued part of the equilibrium surface from a single-valued part. This kind of multivaluedness is created by constraints as illustrated in Figure 9.6. The intersection of the curve and the axis

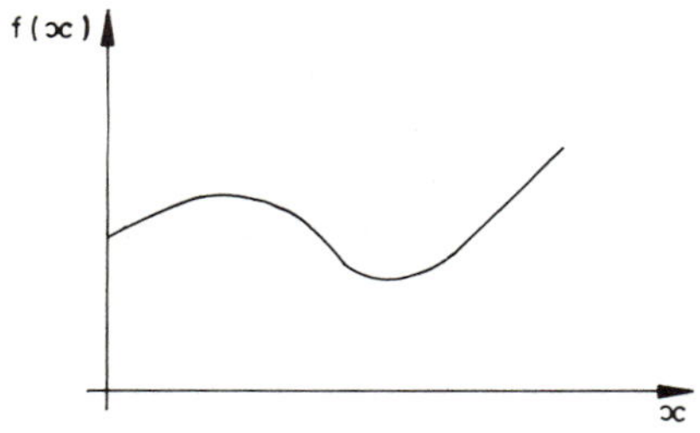

Figure 9.6 Local minimum created by the constraint $x > 0$

shows that a local minimum has been created on the axis by the constraint $x > 0$.

Finally, it is useful to comment in broad terms on the uses of catastrophe theory in applications. We will see in the next section how it can be linked closely to the development of models involving differential equations. Meanwhile, we note that it can be used at two levels: first, it makes us aware of the possibility of new types of system behaviour. We can then seek to recognize these types in empirical work and we can try to build such behaviour explicitly into our models where appropriate. At a higher level, we can apply the theorems of catastrophe theory: we can say that if a function contains two parameters, its structural singularities are at worst folds and cusps and we can use this result. There are two drawbacks to this in practice, both of which have been mentioned earlier: firstly, the theorem applies locally, in the neighbourhood of the highest-order structural singularity present; and secondly, the typical function for applied work is not of course in canonical form and would need to be transformed for the cusp surface to be identified in its correct orientation for the particular problem. In the examples presented in this chapter, therefore, the applications take the first style: we seek to build models which contain this new kind of behaviour.

9.3 BIFURCATION AND DIFFERENTIAL EQUATIONS

We saw in Section 9.1 how the behaviour in time of a system is usually represented by a set of differential equations, and in part of Section 9.2, we looked briefly at the kinds of equations which arise from gradient systems in the context of catastrophe theory. In this section, we look more broadly at differential equations but with the new emphasis on bifurcation of solutions. We then see that, although many of these results have been known in a specific way for a long time, we can now use the modern theory (of which the previous section forms a small part) to set them in a broader context. We then see that jump behaviour in non-gradient systems (and indeed more complicated kinds of bifurcation behaviour) can be identified.

A set of simultaneous differential equations can always be written in first-order form (by introducing a new variable defined by $y = \dot{x}$ if $\ddot{x}$ appears, and so on) and so typically can be taken as

$$\begin{aligned}\dot{x}_1 &= f_1(x_1, x_2, \ldots, x_n)\\ \dot{x}_2 &= f_2(x_1, x_2, \ldots, x_n)\\ &\vdots\\ \dot{x}_n &= f_n(x_1, x_2, \ldots, x_n)\end{aligned} \tag{9.24}$$

or in vector form as

$$\dot{\mathbf{x}} = \mathbf{f}(\mathbf{x}) \tag{9.25}$$

The analysis of such equations usually turns first on the existence (or otherwise) and stability of equilibrium points which are the solutions of

$$\dot{\mathbf{x}} = \mathbf{f}(\mathbf{x}) = 0 \tag{9.26}$$

(and hence the obvious connection with the discussion of the previous section and catastrophe theory): the equilibrium surfaces are the same when the derivatives arise from a potential function—that is, when $f_j(\mathbf{x}) = \partial L/\partial x_j$ for some L). And then, secondly, for non-equilibrium positions, the system behaviour, the solutions of the differential equations, are usually described by trajectory sketching in phase space (though this has obvious limitations, at least in a geometrical sense when the number of dimensions is high). The stability of equilibrium points is revealed by the sign of $\dot{\mathbf{x}}$ in the neighbourhood of such a point and this can be exhibited, as we shall see shortly, by the form of trajectories near the point. In addition to equilibrium behaviour (and perhaps return to stable equilibrium points following a disturbance), there are a number of other 'regular' types of solution which we now summarize briefly.

The main types of solution to differential equations can be summarized as follows:

(i) Stable equilibrium points, which means that trajectories in the neighbourhood lead into them and so they are called attractors. They are illustrated in a two-dimensional phase space in Figure 9.7.

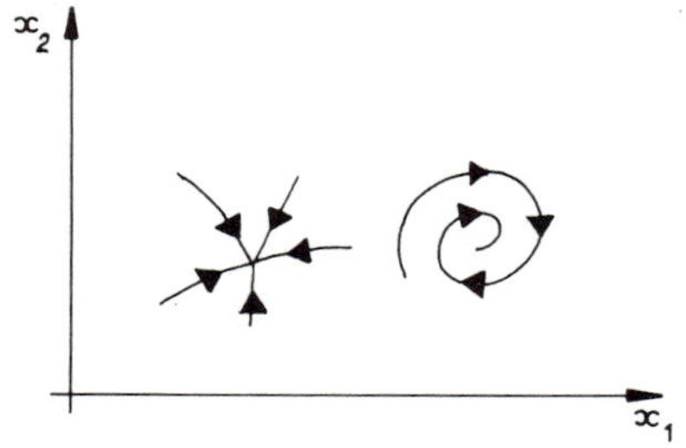

Figure 9.7 Examples of stable equilibrium points in phase space

(ii) Unstable equilibrium points. Trajectories avoid these and so they are known as repellers (as in Figure 9.8). They are not usually considered to be observable states for real systems.

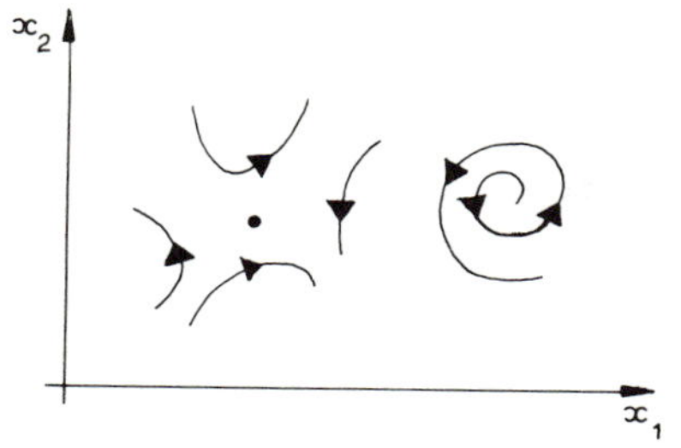

Figure 9.8 Examples of unstable equilibrium points in phase space

(iii) Some equilibrium points are saddle points. Formally, they are unstable but the (two) trajectories (curves in two dimensions) which pass through them play a special role. They are known as separatrices and they separate phase space into distinct regions: trajectories on either side of a separatrix typically go to different stable equilibrium points (see Figure 9.9).

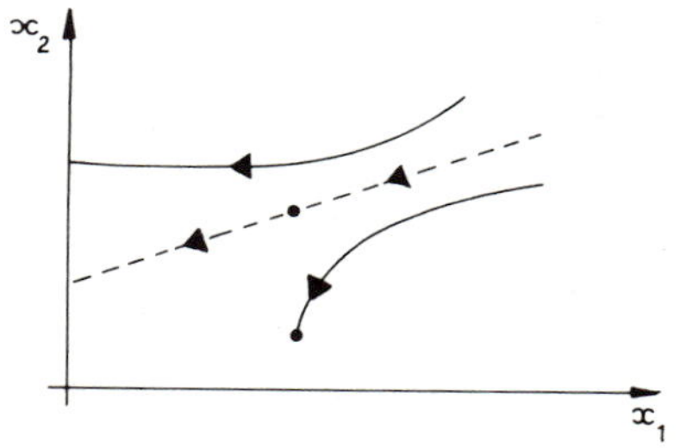

Figure 9.9 A separatrix

(iv) Solutions may be periodic and two kinds can be distinguished at this stage: those with closed orbits in phase space (as in Figure 9.10) and those which are asymptotically close to an orbit, which are known as limit cycles (see Figure 9.11). The latter type are typically stable, while closed-orbit solutions are structurally unstable as we will see with an example shortly.

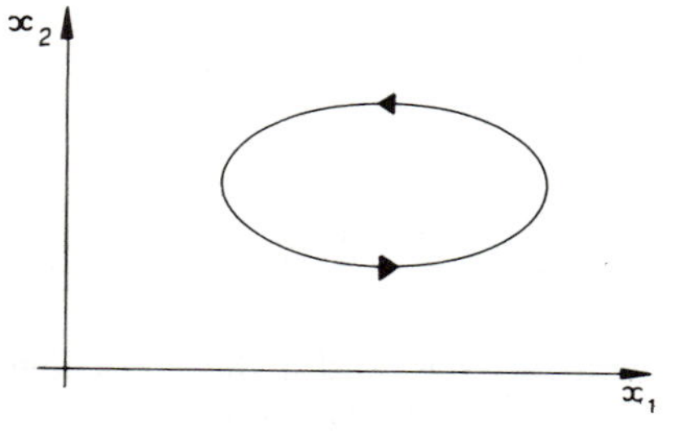

Figure 9.10 A closed orbit

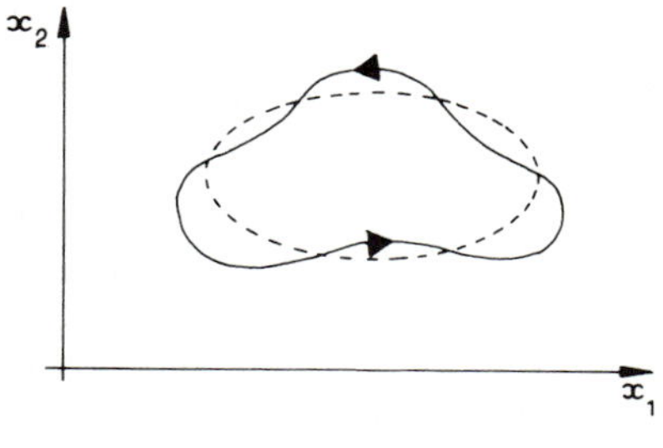

Figure 9.11 A limit cycle

(v) There are two further kinds of oscillation, which are periodic but with either decreasing or increasing amplitude. These are known as convergent or divergent oscillations respectively and are illustrated in phase-space terms as spirals in Figure 9.12 and in more familiar time plots in Figure 9.13.

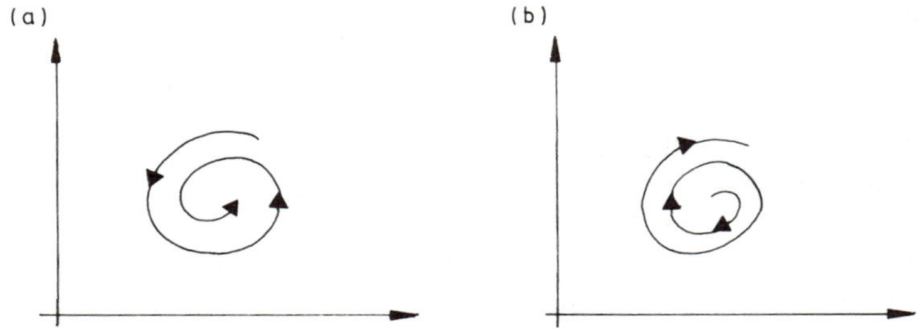

Figure 9.12 (a) convergent oscillation; (b) divergent oscillation

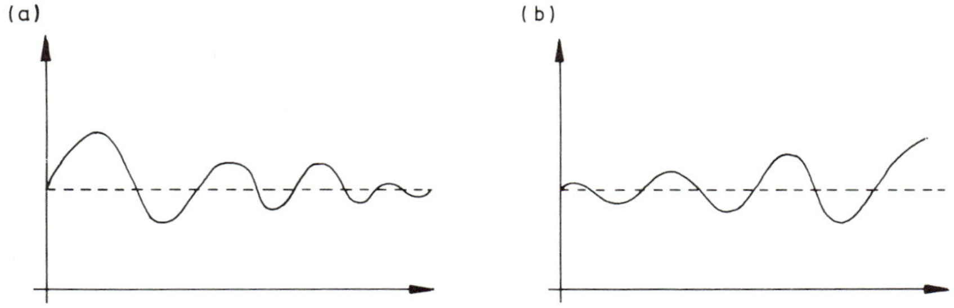

Figure 9.13 The two types of oscillation on time plots

(vi) The behaviour of solutions may be fundamentally irregular, and this is known as chaotic behaviour.

This range of behaviour of solution can be illustrated very easily in any standard textbook of differential equations. Here we are more interested in the transition from one type of solution to another as some system parameters change. Bifurcation in the theory of differential equations can now be seen as the 'jump' (in a more general sense than in the catastrophe theory context) from one kind of solution to another when some parameter of the system (and hence of the equations) changes. We can show the presence of parameters in the equations systems explicitly as a vector $\boldsymbol{\alpha}$:

$$\dot{\mathbf{x}} = \mathbf{f}(\mathbf{x}, \boldsymbol{\alpha}) \tag{9.27}$$

We are therefore interested in critical values of $\boldsymbol{\alpha}$ at which bifurcation of solutions takes place. A critical value at which a stable equilibrium solution disappears may then correspond to a fold in catastrophe theory, for example. But there may also be more complicated transitions—as from a stable solution to a periodic one (the

Hopf bifurcation). We illustrate the basis of these ideas with a number of simple (and mostly well-known) examples which then form the basis of some applications in later sections.

We can begin by connecting to the idea of structural stability introduced in the previous section. A well-known equation which produces closed orbit periodic behaviour is the equation

$$\ddot{x} = -ux \tag{9.28}$$

This is structurally unstable in the sense that if a term $\varepsilon\phi(x)$ is added, for arbitrarily small values of ε, then the solution becomes one of the convergent or divergent oscillation—something which is qualitatively different.

An example of more direct application later is provided by the well-known Lotka–Volterra equations, presented here in a manner which follows Maynard-Smith (1974). A pair of such equations are

$$\begin{aligned} \dot{x}_1 &= (a - bx_1 - cx_2)x_1 \\ \dot{x}_2 &= (e - fx_1 - gx_2)x_2 \end{aligned} \tag{9.29}$$

and the equilibrium point is the intersection of the two straight lines (ignoring for the time being points at which $x_1 = 0$ or $x_2 = 0$)

$$\begin{aligned} \dot{x}_1 &= 0 = a - bx_1 - cx_2 \\ \dot{x}_2 &= 0 = e - fx_1 - gx_2 \end{aligned} \tag{9.30}$$

These are plotted for two cases in Figure 9.14. The difference between the two cases is that the relative slopes of the lines are reversed. An examination of the trajectories near the equilibrium point in the two cases shows that one is stable and one is unstable, thus illustrating a critical value of the parameters (in this case the coefficients which determine the slopes of the lines) at which there will be a flip from stable to unstable or vice versa.

With the signs of the coefficients used in that particular illustration, the equations represent two species competing for resources. If e is considered to be negative and f negative, with g zero, then the equations represent a prey–predator system, with the x_1 population as the prey and the x_2 population as the predator. This then illustrates another point. Such a system is often presented as having a periodic solution, and indeed it has in the case where we have the further condition $b = 0$ (which means there is no limit on the growth of the prey population from crowding and such phenomena). But for any value of b which is non-zero (though perhaps arbitrarily close to zero), the solutions are either stable or unstable equilibrium *points* with corresponding non-equilibrium trajectories, and so this provides another example of bifurcation and also of structural instability of the periodic solution to the equations when $b = 0$.

There are more general non-linear versions of these equations (as discussed for

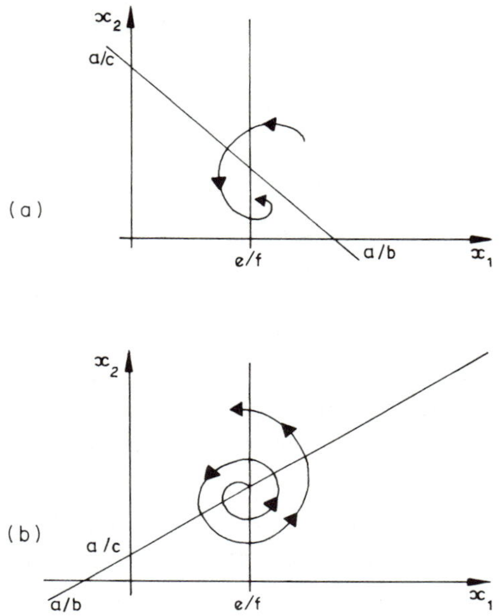

Figure 9.14 Solutions to Lotka–Volterra equations: (a) stable, (b) unstable

example by Hirsch and Smale, 1974) which are usually written in the form

$$\dot{x}_1 = u_1(x_1, x_2)x_1$$
$$\dot{x}_2 = u_2(x_1, x_2)x_2 \tag{9.31}$$

The equilibrium points are then the intersections of the curves

$$\dot{x}_1 = u_1(x_1, x_2) = 0$$
$$\dot{x}_2 = u_2(x_1, x_2) = 0 \tag{9.32}$$

together with the axes

$$x_1 = 0$$
$$x_2 = 0 \tag{9.33}$$

This leads to more complicated bifurcation possibilities and will be used to illustrate a problem in locational analysis in a later section.

We have seen that bifurcation can arise with rather simple looking equations (such as (9.29)) though even these were non-linear, and it turns out that non-linearity is the basic feature in differential equations which generates bifurcation. We give another illustration of this in a context which occurs for many systems, including both locational and transport analysis: the basic growth equation. (And this is a special case of (9.31) above). The logistic growth equation is taken as

$$\dot{x} = u(D - x)x \tag{9.34}$$

where D is a measure of capacity and u is another parameter. However, D can itself be a function of x and the equation becomes:

$$\dot{x} = u(D(x) - x)x \tag{9.35}$$

and the non-linearities thus introduced into the solution of

$$u(D(x) - x)x = 0 \tag{9.36}$$

then generate interesting bifurcation behaviour. However, it is more interesting to proceed directly to the simultaneous equation form of these equations, where we have capacities D_i for each variable x_i, and with each of these, in principle, a function of all the state variables. So the full system is

$$\begin{aligned} \dot{x}_1 &= u_1(D_1(x_1, x_2, \ldots, \alpha_1, \alpha_2, \ldots) - x_1)x_1 \\ \dot{x}_2 &= u_2(D_2(x_1, x_2, \ldots, \alpha_1\ \alpha_2, \ldots) - x_2)x_2 \\ &\vdots \\ \dot{x}_n &= u_n(D_n(x_1, x_2, \ldots, \alpha_1, \alpha_2, \ldots) - x_n)x_n \end{aligned} \tag{9.37}$$

In effect, we have introduced feedbacks between the individual equations and allowed for the possibility of competition (e.g. for resources as we saw in a simpler ecological example earlier). We have also shown the function D_i to be functions of sets of parameters $\boldsymbol{\alpha} = (\alpha_1, \alpha_2, \ldots)$. To analyse the bifurcation properties of such equations, therefore, we start to look for the equilibrium points which are the solutions of the non-linear simultaneous equations

$$\begin{aligned} D_1(x_1, x_2, \ldots, \alpha_1, \alpha_2, \ldots) &= x_1 \\ D_2(x_1, x_2, \ldots, \alpha_1, \alpha_2, \ldots) &= x_2 \\ &\vdots \end{aligned} \tag{9.38}$$

together with the equations representing the axes (which show that there are likely to be many equilibrium points with some combination of the x_i's zero—which connects these problems to the location-allocation problem).

Typically, the solution of these equations (as functions of the parameters $\boldsymbol{\alpha}$) leads to rich bifurcation behaviour, illustrated formally by the branching of equilibrium points in Figure 9.15. We will take this up more specifically in the context of an example in location analysis below.

9.4 SUMMARY: ZEEMAN'S SIX STEPS FOR MODELLING

A convenient way of summarizing the argument so far and taking it forward is to relate it to six possible stages in the modelling of dynamical systems suggested by Zeeman (1977). Consider a dynamical system described by state variables **x** and control variables, or parameters, **u**. Zeeman's steps are directed towards

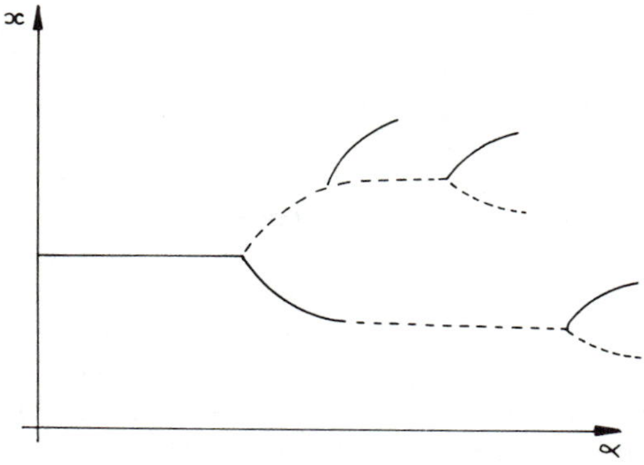

Figure 9.15 Bifurcation of equilibrium x-values against a parameter

catastrophe theory and extensions of it, but are easily seen to be applicable more broadly, as we indicate at times in parenthesis in the summary of the steps below:

(i) Study the structural singularities of the objective function. (In effect this is a way of extending traditional comparative statics and could also be taken as applying to studying the structural singularities of any equilibrium surface—even for a non-gradient system.)
(ii) Specify the 'fast' dynamics: this involves writing down the differential equations for the x-variables assuming the parameters are fixed. These then indicate the trajectories of 'return to the equilibrium surface' after a disturbance.
(iii) Specify the differential equations for the **u**-variables—the 'slow' dynamic. In effect, this then charts the trajectory to be followed on the equilibrium surface in (**x**, **u**)-space. This step is described by Zeeman as 'modelling development'.
(iv) Allow for feedback between the fast and slow dynamics. This is the kind of thing we built into the equations (9.37) by making the functions D_i depend on all the **x** variables and on all the parameters—in that notation, $\boldsymbol{\alpha}$. To complete that particular system, we would also have to add differential equations for the $\boldsymbol{\alpha}$-variables.
(v) Recognize the existence of 'noise' in either the state or the control variables. This is important in two ways: it recognizes the stochastic nature of the world which is being modelled; but by incorporating fluctuations, we will see later how new patterns of order can be generated by the presence both of bifurcation properties and fluctuations. The former can amplify the latter.
(vi) In the final step, explicit note can be taken of diffusion phenomena. This is potentially important in the long run for locational and transport modelling for obvious reasons. In mathematical terms, it takes us into the realm of differentiating with respect to spatial variables as well as time, and hence to simultaneous partial differential equations.

It is clear from the form of presentation of these steps that they provide a sensible practical way of tackling a model-building exercise, not least because they involve increasing mathematical difficulty. It is also implicit in this account that there are some difficult model design questions in relation to definition of variables, but the steps give us some insight into how these can be tackled. For example, there often seems to be a degree of arbitrariness in deciding for a particular system or problem which should be state variables and which parameters (or control variables). The analysis above suggests that this is at least partly a matter of the speed of response of the part of the system represented by the particular variables—fast implies state variables and slow implies control variables. This, however, obviously has to be sorted out in relation to other notions of dependence (state) and independence (control) in the traditional applied mathematical way and with notions of control which are often necessary for planning applications.

9.5 EXAMPLES IN LOCATIONAL ANALYSIS: RETAIL CENTRES AND URBAN STRUCTURE

9.5.1 The model to be used

Many examples of urban and regional models have been presented earlier in this book. One which is a part of the previously much-discussed Lowry model, but which has its own life as the Huff–Lakshmanan–Hansen model, involves modelling the flows of cash from residences to retail centres. This is a well-known model and will be presented briefly in this subsection. The point of using it as an example is to show how our new techniques help us to address new and difficult questions in old contexts. The model is usually stated as

$$S_{ij} = A_i e_i P_i W_j \exp[-\beta c_{ij}] \tag{9.39}$$

where the balancing factor is

$$A_i = 1/\sum_k W_k \exp[-\beta c_{ik}] \tag{9.40}$$

to ensure that

$$\sum_j S_{ij} = e_i P_i \tag{9.41}$$

S_{ij} is the flow of cash from residences in zone i to shops in zone j; A_i a balancing factor; e_i the per capita expenditure by residents of zone i; P_i the population of zone i; W_j a measure of the attractiveness of the shops in zone j, and here this will be measured by size; c_{ij} is the cost of travel from i to j; and α and β are parameters.

The usual use of this model is to assume a given set of W_j's, written as an array $\{W_j\}$, perhaps as a plan, or as a set of existing assignments with one or two modifications. The model will then give a prediction of $\{S_{ij}\}$ for this particular

$\{W_j\}$, and of course, given everything else. In particular, if we have the flows S_{ij}, then we can calculate the revenue attracted to the shops in each zone in turn. If this is denoted by D_j, then

$$D_j = \sum_i S_{ij} \tag{9.42}$$

It is obviously useful to be able to calculate this for each j and for any change in such a way that the effect of (e.g. the competition of) other centres is taken into account, and this has always been one of the attractions of this model. This is a good and proper use of the model, and one which will continue to be appropriate in many circumstances. What we want to do here, however, is to investigate some new types of theoretical question mainly to attempt to gain some insight into why the W_j are as they are and the conditions under which new patterns could evolve. In other words, by adding hypotheses about the determination of the W_j, we can attempt a dynamical analysis of their structure.

One reason why these questions are relatively new is that much of urban modelling has been concerned with the distribution of activities and flows taking the structure either as a given existing structure (and certainly that for model calibration purposes) or more likely as above, with the structure as a given plan involving some modifications of the existing situation. Possibly one of the reasons for this is that it has proved difficult to find the techniques to model structural development and change, and so it may be particularly useful that the relatively new techniques of dynamical systems theory should have become available.

9.5.2 Hypotheses for centre size: embedding a static model into a dynamic framework

In order to be able to investigate and model change in $\{W_j\}$, we need to add some hypotheses—formally another n equations, if there are n zones—which determine $\{W_j\}$. One way to do this is to hypothesize (for a suitable system, of course) that suppliers of shops determine the quantity supplied in such a way as to balance revenue and costs. Thus, if k is the cost per unit size of supplying and running shops (which we will assume to be independent of j for simplicity for the time being), then this hypothesis is equivalent to adding the n equations

$$D_j = kW_j \tag{9.43}$$

In principle, equations (9.39), (9.40), (9.42), and (9.43) can now be solved for $\{W_j\}$. We could substitute for D_j in (9.43) from (9.42); then for S_{ij} from (9.39) using the A_i from (9.40). This, as we will see later, produces some highly non-linear equations for $\{W_j\}$, but ones which can be solved numerically and about which we can obtain considerable theoretical insight.

It turns out in this case, probably because of the fairly highly constrained nature of the problem, that there are equivalent formulations for calculating $\{W_j\}$ even though the hypotheses appear to be different. For example, there are at least two exactly equivalent mathematical programming formulations, one based on the maximization of consumer's surplus (Coelho and Wilson, 1976), the other on maximizing log accessibility (Leonardi, 1978). However, it is more convenient to take these up in the next chapter where we pursue a more general treatment of these kinds of equivalences.

The next step in this preliminary argument on hypotheses of $\{W_j\}$ behaviour is to proceed to Zeeman's second step (which is largely as far as we are able to go) and to hypothesize differential equations in $\{W_j\}$. The simplest way of doing this is to say the centre sizes will grow if it is profitable to expand and will decline otherwise. Profitability is implicitly defined by the terms in equation (9.43): profit is $D_j - kW_j$, so that for some suitable constant ε, we can write

$$\frac{dW_j}{dt} = \dot{W}_j = \varepsilon(D_j - kW_j) \tag{9.44}$$

and, of course, equation (9.43) can now be seen as the equilibrium condition for this set of simultaneous differential equations. In fact, the evolution of these equations illustrates Bellman's notion of the embedding of an essentially static equilibrium model within a dynamic framework so that a new range of questions can be explored. We can see the nature of the equations more explicitly if we follow through the substitution procedure suggested earlier for solving the equilibrium equations. First substitute for D_j, then for S_{ij} and then for A_i within that expression. Equations (9.44) can then be written more explicitly as

$$\dot{W}_j = \varepsilon\left[\sum_i \frac{e_i P_i W_j^\alpha \exp[-\beta c_{ij}]}{\sum_k W_k^\alpha \exp[-\beta c_{ik}]} - k W_j\right] \tag{9.45}$$

and can be seen to be very non-linear equations in $\{W_j\}$. It is the non-linearity, as noted in a general context in Section 9.3, which produces interesting bifurcation properties. Note also that these equations bear a family relationship to the logistic growth equation also presented in that section. The difference is that a factor W_j is missing from the right-hand side of (9.45). However, this has no direct effect on the possible equilibrium states and their stability: it merely affects the rate and form of the system's returns to equilibrium after a disturbance (or from a non-equilibrium position, such as the starting position in a new town). These equations also turn out to be a special case of equations (9.31), which turn up in ecology.

9.5.3 A dynamical analysis of equilibrium points

We can now proceed to get some insight into the nature of the equilibrium points and to use the differential equations to explore their stability. The main points will be presented here, with more detail being available in two papers and another book: Harris and Wilson (1978), Wilson and Clarke (1978), and Wilson (1981).

To find the equilibrium points we have to solve equation (9.43) with the related substitutions. This can be accomplished numerically in a variety of ways. One method, which connects to the example presented in Chapter 7, is to set up an equivalent mathematical program, so this dynamic analysis has strong connections to programming ideas. For the present we are more interested in theoretical insight. This can be achieved by the use of the following simple trick: consider D_j as a function of all the W_k's, and in particular of course, of W_j; it can be written $D_j(W_1, W_2, \ldots, W_j, \ldots)$. If we explore the nature of D_j as a function of W_j we can then look at the intersection of this curve with the straight line (9.43) and this gives the equilibrium points. We can use a geometrical argument about the shape of this curve relative to the line to learn something about the nature of the equilibrium points. In full, D_j as a function of W_j can be written

$$D_j = \sum_i \frac{e_i P_i W_j^\alpha \exp[-\beta c_{ij}]}{\sum_k W_k^\alpha \exp[-\beta c_{ik}]} \tag{9.46}$$

We seek to sketch the shape of this curve by looking at the derivatives of D_j with respect to W_j for varying W_j. Note from differentiating (9.42) that

$$\frac{\partial D_j}{\partial W_j} = \sum_i \frac{\partial S_{ij}}{\partial W_j} \tag{9.47}$$

so that first we must concentrate on the derivative of S_{ij} with respect to W_j. It can be shown (with some considerable manipulation) that

$$\frac{\partial S_{ij}}{\partial W_j} = \frac{\alpha S_{ij}}{W_j}\left[1 - \frac{S_{ij}}{e_i P_i}\right] \tag{9.48}$$

and

$$\frac{\partial^2 S_{ij}}{\partial W_j^2} = \frac{2\alpha^2 S_{ij}}{W_j}\left(\frac{S_{ij}}{e_i P_i} - 1\right)\left(\frac{S_{ij}}{e_i P_i} - \frac{(\alpha - 1)}{2\alpha}\right) \tag{9.49}$$

The derivatives have been presented in this form since $S_{ij}/e_i P_i$ can vary, in principle, between 0 and 1 and so we can see both signs of first and second derivatives and their limiting properties for very small or very large S_{ij}. The main results can be summarised as follows:

(i) As $W_j \to \infty$, $S_{ij} \to e_i P_i$ and so equation (9.48) shows that

$$\frac{\partial S_{ij}}{\partial W_j} \to 0 \tag{9.50}$$

(ii) As $W_j \to 0$, the term $S_{ij}/e_i P_i$ can be neglected and $\partial S_{ij}/\partial W_j$ has a factor $W_j^{\alpha-1}$. Thus (recalling that S_{ij} has a factor W_j^α)

$$\frac{\partial S_{ij}}{\partial W_j} \to \begin{cases} 0 & \text{if } \alpha > 1 \\ \text{finite} & \text{if } \alpha = 1 \\ \infty & \text{if } \alpha < 1 \end{cases} \tag{9.51}$$

(iii) As $W_j \to 0$, $S_{ij} \to 0$ and so we see from (9.49) that

$$\frac{\partial^2 S_{ij}}{\partial W_j^2} \to \begin{cases} < 0 \text{ if } \alpha < 1 \\ \phantom{<}0 \text{ if } \alpha = 1 \\ > 0 \text{ if } \alpha > 1 \end{cases} \tag{9.52}$$

(iv) As $W_j \to \infty$, then

$$\frac{\partial^2 S_{ij}}{\partial W_j^2} \to 0 \tag{9.53}$$

(v) For intermediate values of S_{ij}, we must examine the terms $\frac{\alpha - 1}{2\alpha}$ in the last factor of (9.49). Its behaviour is plotted in Figure 9.16. This shows that for $0 < S_{ij}/e_i P_i < 1$ we can distinguish

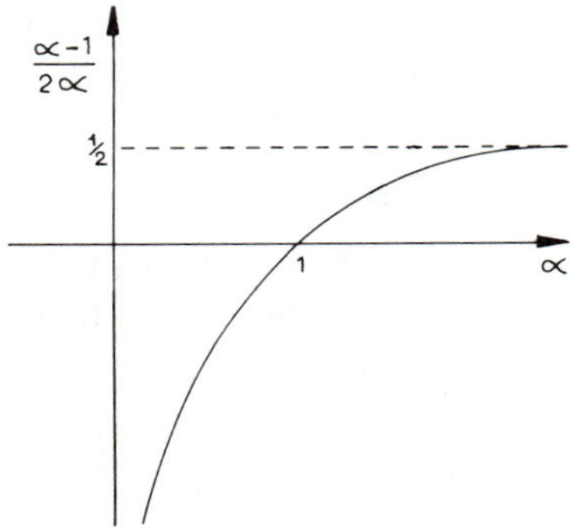

Figure 9.16 Plot of $f(\alpha) = (\alpha - 1)/2\alpha$

(a) $\alpha < 1$:

$$\frac{\partial^2 S_{ij}}{\partial W_j^2} < 0 \text{ for whole range} \tag{9.54}$$

(b) $\alpha = 1$:

$$\frac{\partial^2 S_{ij}}{\partial W_j^2} > 0 \text{ for whole range} \tag{9.55}$$

(c) $\alpha > 1$:

$$\frac{\partial^2 S_{ij}}{\partial W_j^2} \begin{cases} > 0 \text{ for } \dfrac{S_{ij}}{e_i P_i} < x & (9.56) \\ = 0 \text{ for } \dfrac{S_{ij}}{e_i P_i} = x & (9.57) \\ < 0 \text{ for } \dfrac{S_{ij}}{e_i P_i} > x & (9.58) \end{cases}$$

where

$$x = \frac{\alpha - 1}{2\alpha} \tag{9.59}$$

Thus $\frac{\partial S_{ij}}{\partial W_j}$ increases up to x, has a point of inflexion and decreases after that.

The information summarized under the five points above allows us to sketch S_{ij} as a function of W_j, for the different α values, as shown in Figure 9.17. Thus, in spite of the complicated looking behaviour in terms of W_j, the sketches of the S_{ij} curves are fairly simple. The argument for D_j is just slightly more complicated. Because the D_j curve is the sum of S_{ij} curves, it can easily be seen that for the $\alpha < 1$ and $\alpha = 1$ cases, the curves sum 'to the same shape'. In the $\alpha > 1$ case, the curves

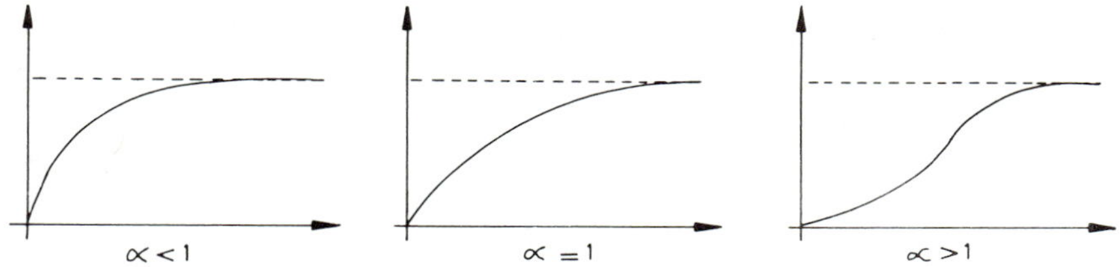

Figure 9.17 Shopping flow revenue–size curves for different α values: (a) $\alpha < 1$; (b) $\alpha = 1$; (c) $\alpha > 1$.

are still most likely to sum to the same shape, but because the points of inflexion of the S_{ij} curves occur at different values of W_j, it is possible that there will be a series of points of inflexion on the D_j curve. The results of these analyses are collected together on Figure 9.18.

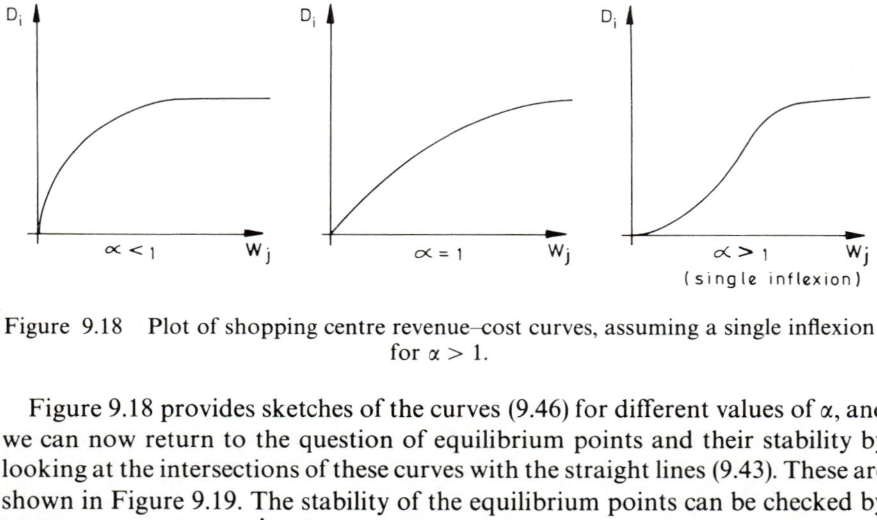

Figure 9.18 Plot of shopping centre revenue–cost curves, assuming a single inflexion for $\alpha > 1$.

Figure 9.18 provides sketches of the curves (9.46) for different values of α, and we can now return to the question of equilibrium points and their stability by looking at the intersections of these curves with the straight lines (9.43). These are shown in Figure 9.19. The stability of the equilibrium points can be checked by looking at the sign of $\dot{W}_j$ (from (9.44)) at either side of a point (and this can be done easily in a geometrical way as indicated in Figure 9.20 for the two main cases). On the basis of this, we see that the origin is a stable equilibrium point for $\alpha < 1$, but not for $\alpha \geqslant 1$ and that the points marked W_j^A are stable while those

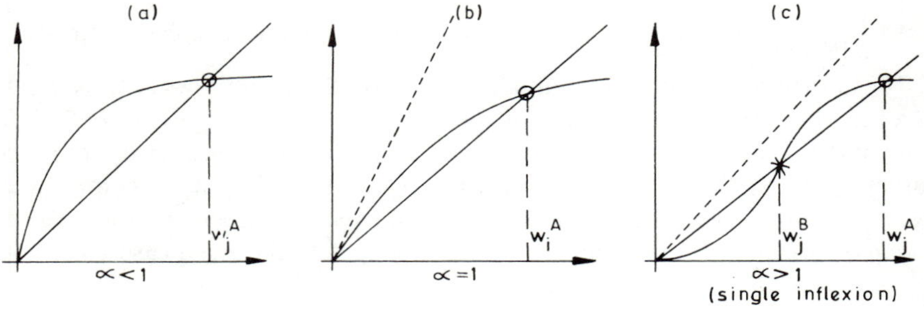

Figure 9.19 Revenue–cost curves and intersections (or non-intersections) with the centre-cost line

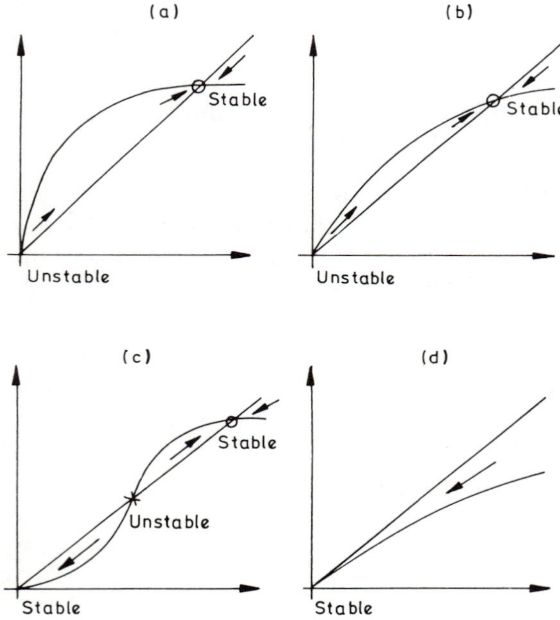

Figure 9.20 Investigations of the stability of equilibrium solutions

marked with a superscript B are unstable. An interesting particular observation is that for $\alpha < 1$ this suggests that there are no stable equilibrium configurations which include zeros. The most interesting case, however is $\alpha > 1$ (and similar considerations apply for $\alpha = 1$) because it is clear from Figure 9.19 that if the slope of the cost line, k, is increased sufficiently, then it need not intersect the revenue curve except at the origin. In fact, for low values of k, it is clear that there will be an intersection and that W_j^A is relatively larger (which is also common sense, since this is the case where the unit cost of supply is cheaper). Consider k increasing smoothly from such a low value: W_j^A decreases smoothly until the critical point is reached at which the cost line merely touches the revenue curve rather than intersecting it. Beyond that slope, the only possible stable equilibrium value is zero. This information can be collected together graphically by plotting W_j^A against k as in Figure 9.21 and we immediately recognize it as an example of the fold catastrophe, but with an additional state added—the zero equilibrium solution which is always possible. We will take this discussion further in a more general account of criticality for this problem in section 9.5.5 below.

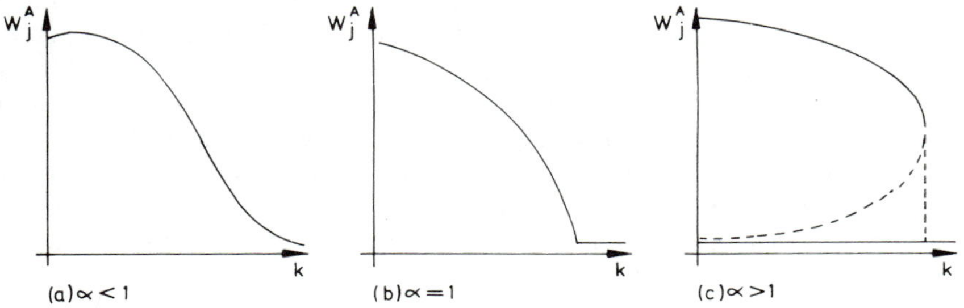

Figure 9.21 Equilibrium size plotted against the k-parameter: (a) $\alpha < 1$; (b) $\alpha = 1$; (c) $\alpha > 1$

We can deduce one more interesting point from this line of argument: if we focus on α rather than k, we see that $\alpha = 1$ is a critical point. For $\alpha < 1$, we have no jump behaviour with respect to k and for $\alpha > 1$, we do. This suggests that $\alpha = 1$ is a cusp point and that α is acting as a splitting parameter. We will see in section 9.5.5 below, however, that there are critical values of α and β at which there are fold-type jumps, exactly as for k. This means that they each also act as normal parameters. This in turn implies that, while we have identified behaviour which is representative of the cusp catastrophe in the parameters α and k, the representation involved is not canonical since α is not a pure splitting factor. Later, we also refer to the fact that since another parameter is involved, β, then we might also expect even higher-order catastrophes of the swallowtail kind. Indeed, the argument generalizes further to include even higher-order catastrophes and this takes us into the realms of generalized catastrophe theory and away from Thom's elementary examples.

9.5.4 Differential equations and bifurcation for this example

We can get some new kinds of insight by considering the differential equations (9.44), written out in full as (9.45), to be special cases of those in (9.31) of Section

9.3 (though for n zones instead of for two). In this case, they could be written as

$$\begin{aligned}\dot{W}_1 &= u_1(W_1, W_2, \ldots, W_n)W_1 \\ \dot{W}_2 &= u_2(W_1, W_2, \ldots, W_n)W_2 \\ &\vdots \\ \dot{W}_n &= u_n(W_1, W_2, \ldots, W_n)W_n\end{aligned} \tag{9.60}$$

Hirsch and Smale (1974) discuss the case in ecology where the W_j's would be species competing for a fixed supply of resources. The retailing equations satisfy the same kinds of conditions—essentially because they represent centres competing for a fixed supply of consumers—and so some of these same general results apply. For example, typically, there is a single stable equilibrium point at which all the variables are non-zero, and there are many others which contain one or more zeros. This leads to another kind of interesting bifurcation property as can be illustrated in the two-zone case in Figure 9.22. Here, the stable equilibrium points are B and P, and Q is a saddle point. The two trajectories through the saddle point divide this phase space into two regions: above them, all trajectories go to B; below, to P. The new kind of bifurcation arises as follows: if the system is in state P but is disturbed to a point near the saddle-point trajectories—the separatrix—then if it stays on the P-side, it returns to P, but if it crosses it, a jump to B will be observed, and this means the jump to zero of one centre (or, in the ecological case, the extinction of a species). This also connects to the other kind of bifurcation. We have just seen that if a parameter such as k changes through a critical point, then the value of W_j in a particular zone j can jump to zero. This, in effect would be the disappearance of the stable equilibrium point P on Figure 9.22.

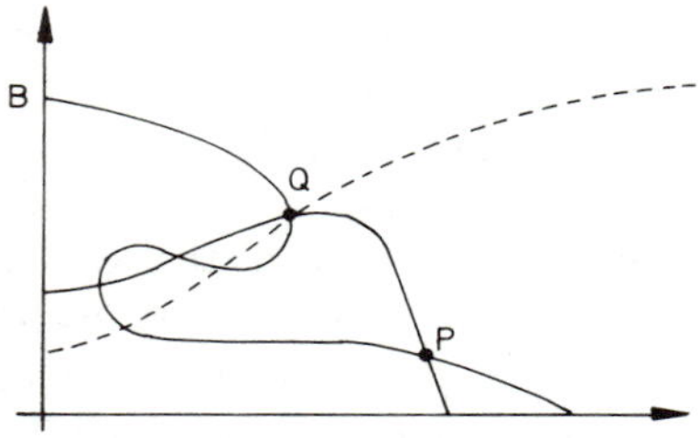

Figure 9.22 Equilibrium points and separatrix for the general model

9.5.5 Zonal criticality and retail structure

We now extend the argument of section 9.5.3 to look at criticality in a zone with respect to all the parameters of the model. We saw there, through Figure 9.21, that there is a critical value of k and that this occurs when the cost line touches the revenue curve as shown in Figure 9.23, line (2), for the $\alpha > 1$ case. The other lines show the intersection (1) and the existence of the stable equilibrium point and the

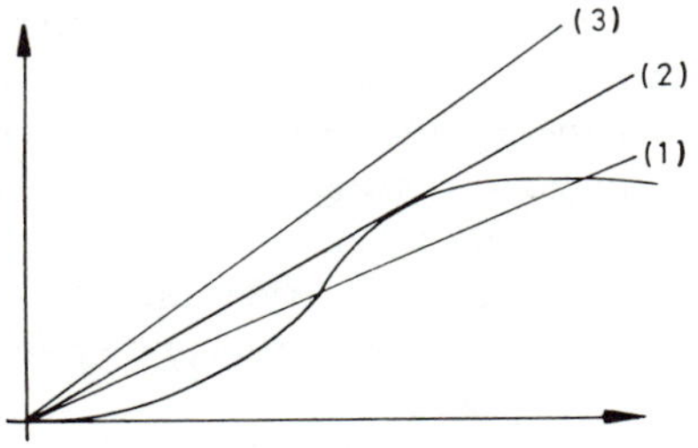

Figure 9.23 Equilibrium points, including the critical case

non-intersection (3) respectively. We can call the slope of the line which touches the curve k_j^{crit} (and note that it is j-dependent because it depends on many features of the particular zone, such as the shape of the revenue curve for that zone, even though the parameter k has been assumed to have a system-wide value). However, it is also clear that, if any other parameter changes when the zone is in a critical state, then there would be a jump to either intersection or non-intersection. For example, if α or β change, then the position of the revenue curve changes as indicated in Figure 9.24. The parameter k only *appears* to play a special role because it is so easily interpretable as the slope of the cost line. In fact, when the zone is in a critical state, it is critical with respect to all parameters (which includes all the independent variables in this case, such as e_i, P_i, and c_{ij}, as well as other terms which are more usually referred to as parameters).

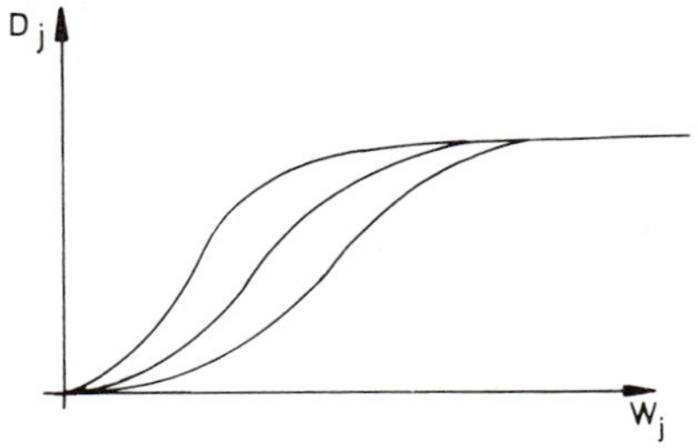

Figure 9.24 Shifting revenue–size curves for different α and β values

To illustrate this notion further, suppose all the independent variables are fixed except for k, α, and β. In the critical state, we can write a critical parameter set as $(k_j^{\text{crit}}, \alpha_j^{\text{crit}}, \beta_j^{\text{crit}})$ and these trace out a three-dimensional surface of zonal criticality. The actual trajectory of a set of zonal parameters will be a curve in this three-dimensional space and we can now see that jumps could take place any time such a curve intersects that surface. It is a substantial research question to determine the shape of that surface. We have seen that the singularities are folds in (W_j, k)-

space, as in Figure 9.21, and an exactly analogous argument shows the existence of folds in (W_j, α)-space and (W_j, β)-space. This is the sort of result which may be expected from catastrophe theory since the equilibrium positions of $\{W_j\}$ can be seen to arise from a maximization problem which can be set up in Lagrangian form with k, α, and β as parameters. Catastrophe theory then tells us that, for this three-parameter case, the worst kind of structural singularities are folds (which we have seen), cusps (which we have also seen), and swallowtails (which we have not seen yet).

This kind of criticality, involving all the parameters simultaneously, arises out of the sort of mathematics already presented in summary form in Section 9.2 and equations (9.20)–(9.23) for the cusp case. It can also be shown that the argument can be extended in the same way to incorporate the other independent variables which we are currently assuming fixed—$\{e_i\}$, $\{P_i\}$, and $\{c_{ij}\}$ (see Wilson, 1981, for a more detailed account).

9.5.6 Simultaneous zonal variation

The argument so far has enabled us to gain some insight into zonal criticality and hence into identifying conditions under which retail centre development is likely or not at a particular location. The weakness in this argument is that we have assumed that, in some sense which itself turns out to be difficult to specify, that centre development in all other zones is 'fixed'. In other words, when we focus on W_j, we assume a fixed backcloth of $\{W_k\}$, $k \neq j$. There are two kinds of problem which we need to take further here. Firstly, how do we specify what 'fixed backcloth' means; and secondly, how do we progress towards handling the much more complicated and realistic problem that the backcloth is changing and evolving simultaneously?

We begin with the problem of fixing the backcloth. If it is literally fixed, then the difficulty which arises is that

$$W = \sum_j W_j \tag{9.61}$$

changes, and hence so does k, since

$$\sum_j D_j = \sum_i e_i P_i = k \sum_j W_j \tag{9.62}$$

and hence

$$k = \frac{\sum_i e_i P_i}{\sum_j W_j} \tag{9.63}$$

This shows that k decreases as W_j (and hence W) increases, and so instead of (9.43) being a straight line, it now becomes a curve of the form shown in Figure

9.25. However, this does not change the argument too much: the basic forms of intersection with the revenue curve and still the same and stability conditions do not essentially change.

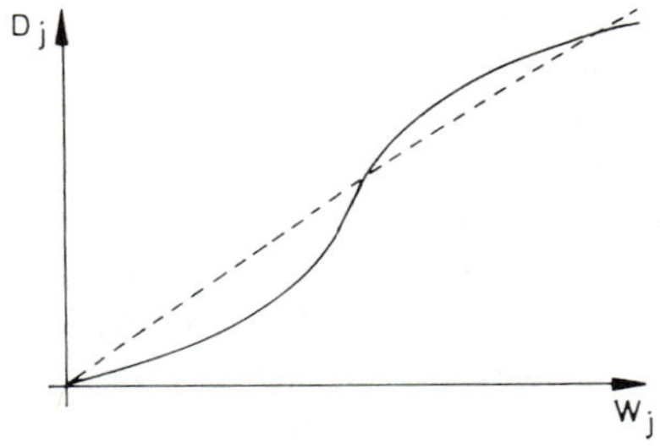

Figure 9.25 The cost-line becomes a cost-curve

A second possibility is to keep W fixed. This involves making some assumptions about how the W_k, $k \neq j$, change when W_j changes, and has the unfortunate consequence of producing silly answers when $W_j \to W$ since then all the W_k tend to zero and the revenue curve has an unrealistic shape. A better possibility is to take sets of W_k which are known equilibrium points, and it is particularly interesting to take these in the neighbourhood of a known jump, and to use these as the fixed W_k while W_j is allowed to vary. In this case, only the portion of the curve near to a jump point may be considered relevant, as sketched in Figure 9.26. Another possibility in the same spirit is to use actual data for the W_k. An important research task is clearly the identification of real jumps and their interpretation in terms of this, or related mechanisms. A preliminary example of this kind of work is presented by Wilson and Clarke (1978).

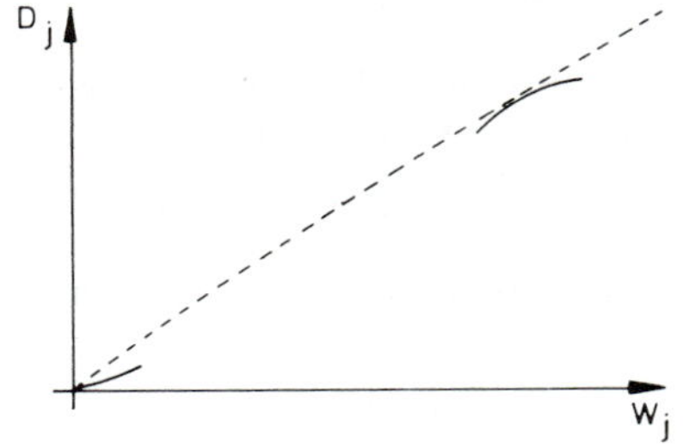

Figure 9.26 Criticality using a small section of the revenue–size curve

The extension of this kind of analysis to cope with simultaneous zonal variation is, in effect, an analysis of what happens in a particular zone with a changing backcloth. A number of comments therefore follow from our earlier argument.

First, recall that if there is an equilibrium point at which all the W_j are non-zero, then it is unique. This implies that each zone must then be in a 'development-possible' state simultaneously with all the others. This suggests, intuitively, that there is a region of space for which this is true and that this is probably a 'whole' and 'closed' region. There will be a larger such region for a particular zone when some of the other zones are allowed to have zero W_k's. In this case, the regions (in the W_k-part of the space) may overlap.

The interpretation of development in particular cases is obviously very different and currently relies on a series of numerical experiments. It has been shown by a number of authors (Harris, 1965; White, 1977; and Wilson and Clarke, 1978, for example) that high α values and low β values tend to favour patterns with a few large centres, and vice versa. In the former case, this means that there are more zones which are on the 'development-impossible' side of criticality and we can interpret the development of and any changes in the pattern in terms of the concept of criticality already presented but with, for each zone, the remaining W_k's now being treated as parameters.

There is a further element in this kind of argument. In reality, the particular pattern which develops may depend on chance events: the particular risk-taking of one entrepreneur in one location, for example. If a big centre is established in this way, then it will have a major impact on all future development. But we may then still find that the parameters of the model determine elements of pattern in a broader sense: perhaps the distribution of centre sizes and their average spacing, if not the detailed spatial pattern.

The next complication involves drawing residential development into the argument (as a function of job location, say) and then seeing how this interacts with the retail field. This involves, in effect, a new approach to central-place theory and will be taken up in the next section.

9.5.7 Interacting fields: the addition of residential location

We illustrate the argument of this section with a very simple residential location model. This shows the principle involved, but clearly to make a realistic model much more disaggregation and detailed specification of mechanisms is required. This is presented elsewhere (Wilson, 1981).

The simplest form of residential location model has people being located around workplaces, say as

$$T_{ij} = B_j W_i^{\text{res}} E_j \exp[-\mu c_{ij}] \tag{9.64}$$

where W_i^{res} is the residential attractiveness factor, T_{ij} is the number of people who work in j and live in i, E_j is the number of jobs in j, c_{ij} is the usual generalized cost of travel from i to j, and μ is a suitable parameter. The balancing factor, B_j, is given in the usual way as

$$B_j = 1/\sum_k W_k^{\text{res}} \exp[-\mu c_{kj}] \tag{9.65}$$

There are a number of different features, clearly, between residential location and retail location, but we can make some interesting points if we pursue an exactly analogous argument in the first instance. First, we need a measure of residential structure and we take number of houses in i, H_i, and we take this as a measure of attractiveness. This topic should obviously be dealt with in a much more complicated way, but we reserve that for later and for elsewhere. Then, an analogous differential equation for H_i would be

$$H_i = \rho(P_i - qH_i) \tag{9.66}$$

where ρ and q are suitable constants. P_i functions here as the demand for housing at i, and q something like the cost per house. P_i is given by

$$P_i = \sum_j T_{ij} \tag{9.67}$$

and so this incorporates the effect of competition of other places in exactly the same way as happens for the shopping centres. The equilibrium conditions are

$$P_i = qH_i \tag{9.68}$$

An exactly analogous analysis to the shopping-centre case would then show some zones where housing development would take place, and some where it would not. The immediate point to be noted is that P_i is a function of $\{E_j\}$, and $\{E_j\}$ in turn is a function of $\{W_j\}$—as the creator of retail employment. $\{W_j\}$, as we already know from the previous analysis, is a function of $\{P_i\}$. So, in this case, we have two strongly coupled sets of differential equations and the corresponding possibilities for bifurcation behaviour will be that much more complicated. At the very least, jumps in any W_j will cause jumps in $\{P_i\}$, and vice versa. The strength of the coupling in the models as presented will depend on how much of total employment is represented by retail employment. But, of course, the links are likely to be much stronger, as will be apparent as soon as the residential location model in particular is refined.

We conclude this section with a review of the kinds of refinements which are needed, though the space is not available to present the detail. First, we note that both the retail model and the residential location model need to be disaggregated. In, the first case, this mainly involves distinguishing different kinds of goods and services. In the residential location case, this has two aspects: distinguishing both person types, especially in relation to purchasing power (or its equivalent for other tenure groups) for housing and house types. The service sector and housing stock variables can then be seen as the basic variables of central-place theory and these methods have a contribution to make towards the development of a dynamic theory in that field. This in effect involves the use of a different representation of the spatial system as a basis. These ideas are presented in more detail elsewhere (Wilson, 1978b).

It is clear that disaggregated models introduce very much more richness into the dynamic behaviour of the system. It will be possible for example to introduce competitive, or cooperative, behaviour between groups and this will introduce new non-linearities in the system of equations and hence new forms of bifurcation.

Bearing in mind the need for disaggregation, we can none the less explore problems with the models, and perhaps especially the residential location model which can be seen even in its aggregate form. Consider again the differential equation (9.66). We can identify a number of complicating factors. The parameter q, for example, represents the cost of housing and will be a function of location, i, and a very complicated function at that. It will itself depend on things like density and hence on H_i, and this will introduce new non-linearities into the equation.

In equation (9.66), we took H_i as a measure of W_i^{res}. This is one field where there is scope for a great deal of research. It is certainly the case that residential attractiveness is made up of a large number of factors, say of the form:

$$W_i^{\text{res}} = X_{1i}^{\gamma_1} X_{2i}^{\gamma_2} X_{3i}^{\gamma_3} \ldots \tag{9.69}$$

where the γ's are suitable parameters. Some of the X_{ki}'s will then be the variables which appear directly in differential equations. For example, one might be H_i, another might be L_i^{res}—the amount of land in i taken up by residential development. Some of the X_{ki}'s may be 'memory' variables which keep a record of earlier developments of a kind which influence future attractiveness. Examples of these might be density and amount of open space. Memory variables of this type are constructed from other variables which do appear in differential equations or from intrinsic characteristics of zones. Then, an outcome of this kind of research would be to replace equation (9.66) by equations in a wider range of variables which took their place in a composite attractiveness factor.

There is one very special problem with residential location: there is an obvious bound on the consumption of land for residential development in a zone. This is not recognized in equation (9.66) at present. There are various ways of coping with this. Firstly, we could do more work on the term q and make it, in effect, a complicated production function of other variables. For example, when H_i rises to a point where densities are so high that further building becomes prohibitively expensive, then this fact would be registered by an increasing value of q, and $\dot{H}_i$ would become negative in equation (9.66). An alternative procedure (which is in effect an approximate form of the previous method) would be to take a constant or simple value of q up to a certain value of H_i which represented a maximum density and then let it be zero after that. This sort of method might work well in a simulation model.

We make one final point in relation to equation (9.66) directly, and then some more general points. This relates to the P_i term. This represents demand for housing in i (at this stage with the usual over-simple assumption of one worker per household for ease of exposition). When the model is disaggregated, this term will have to become much more complicated: it will represent the demand for a type of house at a location by different types of people, and this will have to include terms representing the interactions of these different groups and so again new non-linearities will be introduced.

The more general points are to do with time development. First, we know that there must be a lot of inertia in residential development: once a particular house or group of houses is built, then unless something very unusual happens, it has a life of perhaps sixty years and often very much longer. Further, the development is often lumpy: houses are not typically built singly in such a way that a continuous variable differential equation might be a good representation. This suggests that we have to look much more to simulation models in which this process can be represented directly. There is also an accounting problem which differentiates residential modelling from retailing: the structures, the houses, last much longer than the assignments of people to them. This means that the dynamic assignment of people to the stock should also be represented by difference equations rather than the allocation equations (9.64) and (9.65). This yet again will lead to the introduction of new non-linearities and new bifurcation properties. The kinds of accounting frameworks involved have been presented elsewhere (Wilson, 1974, Chapter 11).

Finally, it may be interesting to try new tacks altogether for residential modelling and not to work directly by analogy with the retailing case. A good example of this is the work of the Brussels school which we present in more detail in the next section. The essence of this is to use the population variables directly in the differential equations rather than to model structure variables. Perhaps the main difficulty with this method is that the structural variables, as we noted in another context, change much more slowly than some of the population variables and this could lead to problems if they are the internal variables of a 'fast' equation and are themselves changing. In effect, some assumption is needed rather like that of the original Lowry model: that residential structural development takes place in response to, and always balances with, population change by location. However, we take this up in more detail in the next section, which also serves to illustrate some new techniques even if the detail of the model mechanisms can be criticized.

9.5.8 Order from fluctuations: the work of the Brussels school

There is a school of research, mainly in physics and chemistry and to a lesser extent biology, which is associated with the name of Ilya Prigogine in Brussels (see, for example, Nicolis and Prigogine, 1977) and whose main ideas are summarized in the phrase 'order from fluctuations'. The essence of the concept is that for a wide variety of material systems, when they are in states which are far from equilibrium, a variety of forms of ordered structures can evolve. Such systems are characterized by some kind of driving force—a source of energy or matter, say—which maintain them away from their equilibrium position; and also by non-linear interaction terms. As will be clear from Section 9.3, such equations have complicated bifurcation properties and these are at the basis of the structures which evolve and the transformations between different kinds of structures. Fluctuations drive such systems from one kind of state to another. These ideas have now been applied by Allen *et al.* (1978) to urban modelling and we can get some interesting new ideas by exploring the basis of their models.

The basic differential equations are similar (and in one application identical) to equations (9.37) presented earlier. They are identical in the interurban application. Here, we concentrate on the intraurban application which is more similar to the other examples presented in this chapter. The city is characterized, as is customary, by a set of variables describing the levels of economic and population activity. Both interactions between components (people or organizations) and interactions across space are considered—first separately and then combined in a simple experimental model. If one of the variables is x_i, then the form of differential equation used is

$$\dot{x}_i = \varepsilon\left[D\frac{F_i}{\sum_k F_k} - x_i\right] \tag{9.70}$$

This is similar in character to the logistic equation (9.34) but because it has the factor x_i omitted, the pattern of growth from some starting point will be different. However, the equilibrium points (defined by $\dot{x}_i = 0$, which is not quite the same sense of equilibrium used by Prigogine when he speaks of 'far from equilibrium') are the same in each case. The functions F_i determine the share of the activity which goes to a particular group or location (or both, as *i* may be a subscript list which covers both a component type and a location—replace *i* by, say, *ij* or *ik*, for example). The equilibrium points are, of course:

$$x_i = D\frac{F_i}{\sum_k F_k} \tag{9.71}$$

and these take the form of the sort of model equations we have used earlier—for example, in the retailing model, provided suitable choices are made for the functions F_i. Allen *et al.* (1978) do, of course, give great attention to the different forms which this function can take for different parts of their models, but essentially they are still within the usual tradition of urban modelling.

The main point we have to learn for the present discussion is that there are new ways of handling these kinds of differential equations which recognize that the system may not be in equilibrium (even in our sense) and that the use of the fluctuations concept is itself a useful computational device. The model equations are 'solved' in simulation fashion: from some initial position (and some given values of 'driving' variables, such as the demand for exports from outside the city), the model is progressed through a series of time increments with the equations in difference equation form. If the driving terms did not change, and these are of course analogous to the control parameters of the previous section, then the system would simply progress to its equilibrium state. If they do change, however, then new kinds of structure can emerge through bifurcation. What remains to be added is the role of fluctuations. In this particular model, they are incorporated as follows: at each step, the calculated value of the variables, say x_i^A, is modified to a new value x_i^N given by

$$x_i^N = x_i^A(1-E) + Ex_i^A G_i \tag{9.72}$$

where E is a parameter and G_i is a random number between 0 and 1. The fluctuations play one major role: they test the stability of the solution which has been achieved so far. If the system is near to a bifurcation point, then a small fluctuation can lead to a jump to a new kind of trajectory in phase space. It is also reasonably argued that this is a representation of real-world processes since systems are subject to random shocks, particularly at the coarse level of resolution at which they are being modelled.

9.6 EXAMPLES IN TRANSPORT ANALYSIS: MODAL CHOICE AND NETWORK STRUCTURES

9.6.1 Modal choice and structural dynamics

There is one sense in which we can follow by analogy with Section 9.5 and this is made clear by the ecological analogy which has been discussed briefly in section 9.5.3 above. In the same way in which the suppliers of shopping centres compete for a fixed supply of customers, the providers of transport modes are also competing for a fixed supply of patrons. But this analogy also shows up one crucial difference: in the shopping case, we were modelling a structural variable, W_j. In the modal choice case, it is more difficult to identify such a variable clearly, though we do grapple with this problem shortly. The state of the system is usually described by the proportion travelling by each mode and this is analogous more to the revenue variable D_j than to the structural variable W_j.

The first point to note is that, since $D_j = kW_j$, the ecological differential equations could have been stated in terms of D_j rather than W_j and similar conclusions could have been drawn. This means that, for two modes, we can apply this argument to equations

$$\begin{aligned} \dot{M}^1 &= G^1(u^1, u^2)u^1 \\ \dot{M}^2 &= G^2(u^1, u^2)u^2 \end{aligned} \tag{9.73}$$

where M^k is the proportion travelling by mode k and the G^k are suitable functions. The Hirsch and Smale argument then tells us that we can expect to find one stable solution where both M^k's are non-zero, and another where one of the variables is zero, and that those will be separated in phase by a pair of trajectories through a saddle point forming a separatrix.

The main practical point to note from this analysis is that the stability results may be of some importance if the situation described by the model is anything like realistic. It would then be important to know where the separatrix is so that the system can be brought back to the 'co-existence' side if it is thought desirable to maintain both modes.

The uniqueness theorem about the all-non-zero point, or the co-existence point, states that such a point is unique if it exists. What we discovered in the retail-centre analysis was that in some circumstances it may not exist. These deeper insights can only be obtained if we can define a suitable structural supply variable to model and look at the interplay of supply and demand. In the modal choice case, this is much more difficult. The supply variables are represented in the models by the measures of generalized cost of travel by different modes—or at least by components of such costs. And the matter is made particularly complicated by the fact that the values of some crucial variables, such as travel time, are affected by demand and through very complex relationships which depend on the way many bundles of traffic are distributed on the underlying network. The analysis of modal evolution is not helped by the fact that when the public transport mode is mainly made up of buses, their travel times and performance are usually determined through an interaction with private cars on the road. What we attempt below, therefore, is an investigation of structural dynamics for modal supply, particularly with respect to public transport, but in a much simplified model.

The first complication arises from the measure to be used of 'supply'. Ideally, we should focus on the levels of service on particular routes and the contribution these make to inter-zonal travel performance through the usual assignment procedure. However, we go instead for a global measure of supply, say as the total annual expenditure on the public transport system, which we call E. We show in broad terms how this affects inter-zonal travel possibilities, but we are making the assumption that the public transport authority optimally allocates its global expenditure to supply services within its district. The revenue derived will be a function of expenditure and will be written $R(E)$. We then assume that the authority has to balance its books (though we consider later and explicitly a possible subsidy) so that the differential equation which governs expenditure is

$$\dot{E} = \rho[R(E) - E] \tag{9.74}$$

where ρ is a constant.

The argument can be taken further with any transport distribution modal choice model which is thought suitable. To illustrate this example, suppose that

$$T_{ij}^{kn} = A_i B_j O_i D_j \exp[-\beta^n c_{ij}^k] \tag{9.75}$$

where T_{ij}^{kn} is the number if trips from i to j by mode k by persons of type n, O_i and D_j are trip productions and attractions respectively, c_{ij}^k is the generalized cost of travel from i to j by mode k, and A_i and B_j are the usual balancing factors. β^n is a set of parameters. Assume there are two modes—car and public—and two person types, those with car available and those not. The latter are obviously captive to the public transport system. Let the generalized cost of travel be given by

$$c_{ij}^k = a_1 t_{ij}^k + m_{ij}^k + x_{ij}^k \tag{9.76}$$

where t_{ij}^k is travel time and m_{ij}^k money cost (which is fares in the case of public transport); x_{ij}^k represents any other relevant contribution to generalized cost; a_1 is a constant representing the value of time. Then the revenue attracted to the public transport system is given by

$$R(E) = \sum_{ijn} T_{ij}^{2n} m_{ij}^2 \tag{9.77}$$

The analytical incompleteness of the presentation so far is evident from the fact that E appears on the left-hand side of this equation but not on the right. The values of m_{ij}^2 will be influenced by it, but they are separate policy variables. The main variables to be determined by it are t_{ij}^2 as these reflect the level of service to be provided: the greater the 'investment' in both running cost and discounted capital cost terms, the smaller will be t_{ij}^2 and the correspondingly greater will be T_{ij}^2 and hence revenue.

The existence of stable solutions is determined by the shape of the $R(E)$ curve as a function of E. Stable solutions occur when this curve crosses the $R = E$ line from above. It is likely that there will be at least one solution at a relatively low level—assuming that car ownership levels are high—(of both E and service provision) which involves catering for only the captive ridership. Whether others exist will depend on the public-mode production function and the nature of the costs of travel by competitive modes. In particular cases, it would be possible to add more detailed assumptions and to construct the curves involved. Our previous analyses of the retailing case then suggest that interesting bifurcation properties may be present. It is difficult to take the analysis much further in a general way because the non-linearities involved are much more complicated than those of the retail case owing to the different nature of the underlying non-linearities which are rooted in the network structure and the way in which bundles of trips share routes which are provided.

9.6.2 Modal choice and hysteresis

In this subsection, we illustrate another way of applying catastrophe theory. If we hypothesize that modal choice is a function of two parameters or control variables, say u_1 representing a 'habit' factor and u_2 representing cost difference between two modes ($u_2 = c_1 - c_2$, say, for modes 1 and 2 between some pair of destinations), then we can say that catastrophe theory tells us that the shape of the equilibrium surface takes cusp catastrophe form. Let us return, therefore to Figure 9.3. Take the state variables x as representing the modal choice decision, the top sheet being taken as mode 1 and the lower mode 2.

Trajectories such as (1) and (2) then represent the effects of changing the relative costs of the two modes. u_1 has an obvious interpretation as a splitting factor in the following sense: if it is positive, this represents no habit, and the traveller chooses the mode which is cheapest. If it is negative, then the traveller only jumps to the new mode after some lag. If the change is reversed, then the jump back in this later situation does not take place at the same point. In other words, there is an hysteresis effect.

This mode of use of catastrophe theory is in many ways a dangerous one. It is not possible to appeal to the theorems to assert that this model is adequate—even if there are only two parameters involved. It is for this reason that in all the other applications in this chapter we have been concerned to represent the mechanisms of change as part of the model and then to interpret jumps after this analysis in terms of the shapes of the catastrophe theory equilibrium surfaces. Indeed, we have argued that bifurcation in general is usually more important. However, in this case, causal theoretical investigation using the theory is suggestive of a new effect—hysteresis—and it turns out that there is some empirical evidence for this as presented by Blase (1979). It is also true, of course, that this effect can be predicted from other theoretical approaches—compare, for example, Wilson (1976) and Goodwin (1977). So the recommendation which follows from this analysis is that the theory may at least be useful for suggesting new ideas (and there are examples in other disciplines where its use is crucial—there is no other way of doing it), but that for the types of examples here, the effects should then be built more explicitly into more traditional models. (Though it should also be remarked that this is not always easy and it has not yet been done for the modal choice model with respect to hysteresis.)

9.6.3 The evolution of transport network structure

A different kind of problem in transport analysis is that of the evolution of networks. Here, we are not concerned with the traditional but very difficult problem of the planning of optimal networks (as for example in Steenbrink, 1974) which are usually large mathematical programming problems, but with the evolution of hierarchical structure within such networks.

Consider the very simple situation of a unidirectional channel as shown in Figure 9.27 (cf. Wilson, 1979). Suppose we are talking about roads. In the first part of the figure, (a), there is a large number of narrow roads; in the second, (b), most of these roads are still present, but there is also one much bigger trunk road. As in the case of retail centres and transport modes, the difference between the two cases is decided on the basis of demand on the one hand and the costs and production functions of supply on the other. There are also non-linear relationships between the costs of building a particular road and its capacity.

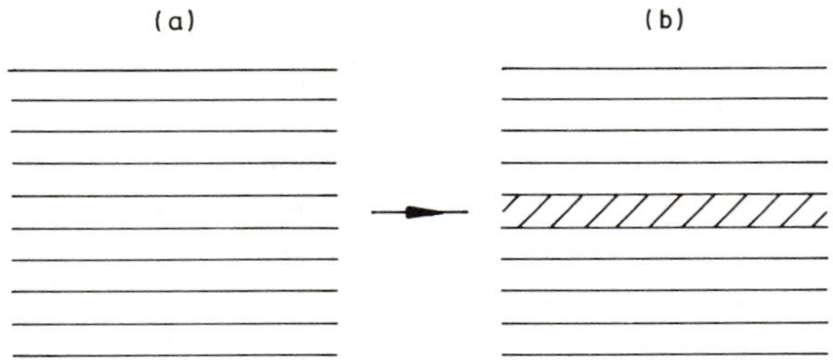

Figure 9.27 Evolution to higher-order facility

Suppose $f(w)$ represents the cost of building a road of width w and that $F(w)$ is its capacity. Suppose also that the demand situation is such that there has to be a minimum density of road across the channel and that capacity has to be met. Let there be n_i roads of type i, each distinguished by width w_i, and then the number of roads of each type can be calculated from an integer mathematical program of the following form:

$$\min_{(n_i)} \sum_i n_i f(w_i) \tag{9.78}$$

such that

$$\sum_i n_i f(w_i) > D \tag{9.79}$$

$$\sum_i n_i > N \tag{9.80}$$

where D is the required minimum capacity and N the minimum number of roads. It can easily be seen that for small D, there will be N of the narrowest roads, but that when D increases smoothly, there will come a point when a trunk facility of the next highest order is 'suddenly' required.

This analysis illustrates two points. Firstly, this kind of jump behaviour arises out of straightforward mathematical programs (and can sometimes be related to what Poston and Stewart (1978) call 'constraint catastrophes'). Secondly, this sort of problem of the evolution of higher-order levels of structure is a very common one, and exhibited in a very simple form here, and the crucial feature of solving it in model form is that it must be possible to specify the form of the production function, given here by the functions f and F, in some suitable form. We should also remark that it is possible in principle to set out this model for the evolution of different orders of shopping centre, and this argument is presented in more detail elsewhere (Wilson, 1978a).

10. *Model comparisons: equivalences, similarities, and differences*

10.1 PHENOMENA, METHODS, MODELS

10.1.1 Introduction

There have been a number of occasions in earlier chapters on which it has been possible to identify formal mathematical equivalences between alternative model formulations for some system of interest. Our purpose in the present chapter is to provide a framework whereby such formal model equivalences may be identified more systematically, and to illustrate the use of this framework with examples of equivalent models for retail sales and shopping-centre structure, spatial economic models, and variants of the Lowry model. Although these later illustrations will be system specific, we argue in more general terms in developing the framework in the earlier sections below.

10.1.2 Phenomena of interest

There are a number of main types of phenomena of common interest to urban and regional modellers: population and economic structure and change, activity location, and spatial interaction (and related network flows). We are concerned with models which represent such phenomena singly or together, in interdependent relationships with each other. These models may be seen to arise from the use of one or more of a number of basic methods (our so-called building-block methods) to portray the phenomena of interest.

10.1.3 Building-block methods

The methods used are selected from a relatively small set; for example, entropy maximizing, utility maximizing, mathematical programming, accounting, graph theory, and dynamical methods. Most of these have already played a prominent role in this book as they are nearly all optimization methods. They can be used singly or in combination, and we recognize that here, as in other parts of the chapter, our categories are not neatly defined. ('Entropy maximizing' could be considered as a subset of 'mathematical programming', for example.) One of the main sources of equivalent, though apparently different, models is the possibility of applying differently ordered combinations of methods to particular phenomena.

10.1.4 A classification of models

When a model is concerned with one phenomenon and is built using one method, then it can be called *simple*. If several methods are needed in combination to model a single phenomenon, then the model can be called *hybrid*. If several phenomena are modelled together, using a combination of methods, then the model is called *composite*. These categories also have a degree of arbitrariness about them because they rely on notions of a 'single phenomenon' or a 'single method'. The notion of a 'single phenomenon' is particularly arbitrary. As a working definition, we associate *one* phenomenon with *one* vector or array of state variables. Broadly speaking, the greater the level of abstraction adopted by the modeller, the wider is the range of phenomena which can be included in one array of state variables—see Barras and Broadbent (1975) for an extreme example.

Examples of the three types of model are given in Figures 10.1 to 10.3.

10.2 SOME CIRCUMSTANCES IN WHICH EQUIVALENCES ARISE

10.2.1 Introduction

In this section we try to characterize, without any hope of being exhaustive, some circumstances in which equivalences of apparently different models arise.

We firstly discuss on an *ad hoc* basis various equivalences that may be identified for our so-called simple models. A more systematic approach is in order for hybrid and composite models, the main models of interest to us below, and three later subsections therefore review three particular sets of circumstances which can typically be expected to lead to interesting results on model equivalences. These are concerned with the nature of large systems of algebraic equations, combinations of mathematical programs, and explorations of dynamical systems theory.

10.2.2 Equivalences in the simpler models

We give an illustrative list of six examples of model equivalences in the case of simple models.

(i) Often, a given model (at the meso scale) can be derived from many different assumptions. A very obvious example is the gravity model of spatial interaction. We list some of the alternative derivations in Figure 10.4. This effect may be called (following Thrift, 1977) 'model equifinality' with respect to assumptions. Even

Model name	Phenomenon represented	Method	Illustrative references
Transportation model	Interzonal spatial interaction	Linear programming	Potts and Oliver (1972)
Doubly constrained gravity model with exponential cost	Interzonal spatial interaction	Entropy maximizing	Wilson (1970)
RAS model	Inter-industry transactions	Non-multiplier account-based	Bacharach (1970)
Matrix inverse input–output model	Total industry outputs	Multiplier account-based	Leontief (1967)
Differential equation model of population growth	Population growth	Dynamical analysis	Rogers (1968)
Population micro simulation model	Population characteristics	Conditional probability methods	Wilson and Pownall (1976)
Shopping model	Shopping flows	Gravity analogy	Lakshmanan and Hansen (1965)
Spatial equilibrium economic model	Inter-industry flows	Linear programming	Takayama and Judge (1972)
Programming residential location model	Residential location	Mathematical programming	Herbert and Stevens (1960)
Economic-base model	Basic service industry ratios	Multiplier account-based	Artle (1961)
Digraph models	Simple cause–effect relationships	Graph-theoretical methods	Roberts (1976)

Figure 10.1. Examples of 'simple' models

Model name	Phenomenon represented	Method	Illustrative references
Dynamic input–output models	Total industry output over time	Accounting coefficients with dynamical analysis	Leontief (1970)
Alternative rectangular input–output models	Commodity × industry flows	Entropy maximizing with accounting coefficients	Macgill (1978a)
Cost minimizing input–output models	Total industry outputs	Mathematical programming with accounting constraints	Intriligator (1971)
Disaggregated trip distribution model	Journey to work by mode and person type	Entropy maximizing with conditional probabilities	Southworth (1977)
Hybrid input–output models	Total industry outputs	Graph-based methods with accounting coefficients	Macgill (1977b)
Dynamic programming models	Various	Dynamical analysis within mathematical programming frameworks	Bellman (1968)
Activity analysis models	Various	Mathematical programming and accounting methods	Takayama and Judge (1972)

Figure 10.2. Examples of hybrid models

Model name	Illustrative references
Lowry model—iterative versions	Lowry (1964) Wilson (1974) Batty (1976) Mackett and Mountcastle (1977) Williams (1979)
—matrix versions	Garin (1966) Macgill (1977c)
—mathematical programming version	Coelho and Williams (1978)
—central place theory version	Wilson (1978a)
Multifacility optimum location model	Leonardi (1978)
Shopping centre size and location	Harris and Wilson (1978)
—iterative version —mathematical progamming version —differential equation version	
Price and industrial location model	Takayama and Judge (1972)
Classic transport package	Bonsall *et al.* (1977)
Linked population–economic model	Schinnar (1976)
Urban dynamics	Forrester (1969)

Figure 10.3. Examples of composite models

more frequently, of course, different assumptions generate 'similar' models, which may be called 'model homeofinality'. The interest here lies in the fact that the degree of similarity can be quite close even though the assumptions are very different.

(ii) There are two well-known distinctive forms of account-based model. The first is exemplified by the population model

$$\mathbf{w}(t+T) = \mathbf{G}\mathbf{w}(t) \tag{10.1}$$

where $\mathbf{w}(t)$ is a vector of population at time t and $\mathbf{G}$ a 'change' operator—which is actually a transposed matrix of rates. The second is illustrated by the well-known input–output model

$$\mathbf{x} = (\mathbf{I}-\mathbf{a})^{-1}\mathbf{y} \tag{10.2}$$

where $\mathbf{x}$ is a vector of total products, $\mathbf{y}$ a vector of final demands, and $\mathbf{a}$ a vector of column-based rates; $\mathbf{I}$ is the unit matrix. Since the model involves the inversion of an often large matrix, the interdependencies of such a model are much more complicated than its population equivalent. This complexity arises partly because one sector, final demand, has been picked out as the driving force of the model (as distinct from the whole initial population, $\mathbf{w}(t)$); and partly because the process modelled involves a sequence of increasingly higher-order effects as spelled out in equation (10.3) below in another context.

We should note that the population model can be cast into the form of equation (10.2) by picking out 'births' to describe it (Stone, 1970). Also, the two kinds of assumptions can be combined (that is, a mixture of row-and-column-based rates are used—Macgill, 1977b) and digraph methods have to be used to

Label	Characteristics	Selected references
Newtonian	By analogy with gravitational force between two masses	
Probablilistic	Based wholly on probabilistic assyumptions as to human behaviour	Harris (1964)
Statistical mechanical (maximum entropy)	Most probable of all possible distributions satisfying origin, destination and interaction cost constraints	Wilson (1967) Wilson (1970)
Information-theoretic (maximum entropy)	Maximally unbiased with respect to origin, destination and cost information	Wilson (1970) Batty and March (1976)
Micro-economic behavioural (constant utility)	Consistant with economic theory of consumer behaviour, with travel as a consumer good	Niedercorn and Bechdolt (1969) Golob and Beckmann (1971)
General share model	Sequential subdivisions from origin totals to destinations, modes, routes	Manheim (1973)
A 'generalized' trip distribution model (macro-behavioural)	Assumptions on aggregate human behaviour to establish sources of variation in data	Cesario (1973)
Spatial discounting	Travellers discount potential interaction opportunities in terms of perceived distances	Smith (1975) Smith (1976)
Random utility (micro-economic behavioural)	Utility and/or cost alternatives regarded as random variables	Williams (1977)
Cost efficiency (macro-behavioural)	Trip patterns with lower total cost are always more likely to occur	Smith (1977)
Entropy-constrained transportation problem	Classic linear programming transportation problem with entropy function as an additional constraint	Coelho and Wilson (1977) Erlander (1977)

Figure 10.4. Alternative derivations of the gravity model of spatial interaction

handle the resulting complexity.

(iii) Equation (10.2) can be expanded to give

$$\mathbf{x} = \mathbf{y} + \mathbf{a}\mathbf{y} + \mathbf{a}^2\mathbf{y} + \mathbf{a}^3\mathbf{y} + \ldots \tag{10.3}$$

This is equivalent to an iterative scheme of the form

$$\begin{aligned} \mathbf{x}^{(0)} &= \mathbf{y} \\ \mathbf{x}^{(n+1)} &= \mathbf{y} + \mathbf{a}\mathbf{x}^{(n)} \end{aligned} \tag{10.4}$$

Some urban models are presented as sets of equations which can only be solved by an iterative scheme of this type. It is then easy to see that such a model can also be cast in the form of equation (10.2). Such an equivalence may be of considerable interest for two reasons: firstly, one form may be computationally more convenient than the other in a particular case; secondly, when a model is cast in the new form, immediate theoretical improvements may suggest themselves. An example of such a model is the Lowry (1964) model (see Chapter 6). In particular, when the model, which is usually presented in iterative form, is presented in the form of equation (10.3) as an input–output model, improvements can be made—for example, to incorporate a full set of inter-industry relations. (Note that the better-known matrix version of the Lowry model, due to Garin, takes a matrix inverse form but is not an input–output model in the usual sense.)

(iv) Certain signed and weighted digraph methods of analysis, although looking completely different, are equivalent to models formulated with difference equations. This is so whenever the signs or weights on the digraph links represent causal relationships between the nodes which they join, specifically when they represent the change in the forward node that will result from a unit change in the backward node (or something equivalent to this) (see Roberts and Brown, 1975). The digraph methods then provide an accessible way of deriving difference equations for the system of interest by methodically picking up the separate interdependencies between each pair of system components.

(v) An important result has recently been derived (Evans, 1976) showing in effect that certain iterative capacity restraint assignment procedures are equivalent to a particular mathematical programming formulation which encompasses both interaction and network variables. This is a result which has important practical consequences (Boyce, 1977; Florian and Nguyen, 1976).

(vi) Our final example here arises in reviewing a number of simple equivalences connected with entropy-maximizing models. Entropy-maximizing models are themselves, of course, mathematical programming models, immediately furnishing equivalences, and making the entire theory of programming available, if required. The RAS matrix scaling method (Bacharach, 1970), and other balancing factor methods used in various accounting contexts, are also all formally equivalent to entropy-maximizing models, given the information used (Macgill, 1975). Macgill (1978b) explains a further equivalence between a conditional probability and entropy-maximizing model in the residential location field, following Senior (1977a).

The previous six examples illustrated equivalences of simple models, and as indicated above the following three subsections in turn review three sets of circumstances in which further equivalences can be expected to arise, this time for hybrid and composite models. Relevant model examples are given in Sections 10.3, 10.4, and 10.5 below.

10.2.3 Large systems of algebraic equations

Most models are finally specified as large systems of algebraic equations which relate the dependent variables of the model—the state variables—to a set of independent variables and parameters. It will often be convenient to introduce large numbers of intermediate variables (and a corresponding number of additional equations), and also to transform some of the variables in various ways. This obviously in itself leads to possibilities of equivalent systems appearing to be very different. It will often be possible to explore the equivalence of algebraic alternatives by trying to reduce them to a common form by manipulation, and geographical insight—such as the identification of basic assumptions in equation form as they affect particular variables—will usually be a considerable help.

The different appearances of equivalent model forms often arise from the use of different solution procedures. Complexity often ensures that there is no simple analytical solution of the equations for the state variables. Possible solution procedures then include: (i) iteration; (ii) matrix inversion; or (iii) noting that the equations arise from a mathematical program. There are some well-known connections between these procedures: iteration usually involves a convergent series, and this may turn out to be a series expansion of an inverse matrix. We will give an example later where an iterative procedure, a matrix inversion formulation, and a mathematical program are all equivalent. If a set of equations can be seen as arising from a mathematical program, then it may be convenient to work with the program directly, rather than the analytical solution. It may also suggest useful variable transformations—for example, the use of dual variables.

10.2.4 Nested and linked mathematical programs

In composite models, the different phenomena will be represented by (linked) submodels, and some of these submodels may arise from optimizing mechanisms and hence mathematical programs. Such programs can either be linked, unidirectionally or both ways—with independent variables in one program being dependent variables in another. A particular form of linking is nesting (or embedding) of one program within another. In these cases, the whole model can sometimes be formulated as a superproblem: that is, as a single problem whose optimality conditions include equations representing each of the submodels. Such superproblems can be formulated in a great variety of ways (and especially because subsets of variables and constraints in one formulation may be replaced by their duals in another), and this obviously leads to different-looking equivalent formulations. In cases where the results of the program are stated analytically, then if different variables are used, the results can look very different indeed, and in such cases the alternative mathematical programming underpinnings may not be recognized immediately.

10.2.5 Equilibrium-point dynamics

Most of the methods mentioned in section 10.1.3 above generate models which represent equilibrium states of systems of interest. Bellman (1968) has indicated in a general way how it is often useful to embed such models within dynamic frameworks. This enables us to learn about the stability of equilibrium points and the behaviour of the system after it has been perturbed. Further, elements of modern dynamical systems theory, such as catastrophe theory and bifurcation theory, emphasize the non-uniqueness of equilibrium states in certain situations.

This argument can be summarized and illustrated as follows. A model whose static equilibrium form is $\mathbf{Ax} = \mathbf{b}$ may be appropriately embedded within a differential equation as

$$\frac{d\mathbf{x}}{dt} = (\mathbf{Ax} - \mathbf{b}).\mathbf{F}(\mathbf{x})$$

(This may also reveal the existence of previously unknown equilibrium states—in this case the solutions of $\mathbf{F}(\mathbf{x}) = 0$.)

This leads to three points which are relevant for the present discussion. Firstly, a dynamic model (stated either in terms of differential equations or difference equations) may have equilibrium properties which coincide with those of a static model, and such an equivalence may not have been noted. Secondly, because dynamical models may be formulated in apparently different, but equivalent, ways, this adds a new layer of complexity to our search for equivalences. Thirdly, we are encouraged to look for alternative equilibrium points and their stability conditions and inter-transition probabilities.

A further related source of dynamic equivalent models is via the Lagrangian expressions of mathematical programming models. In particular, the Lagrangian partial derivatives, the gradients $\partial L/\partial X_i$ (taking X_i as the model state variables), represent the most direct path of return of the model variables to equilibrium ($\partial L/\partial X_i = 0$). It is therefore suitable to use these paths as dynamic trajectories of the variables away from equilibrium, that is, formulating a dynamic model by setting

$$\frac{dX_i}{dt} = -\frac{\partial L}{\partial X_i}.$$

10.2.6 Applications of these analyses

It is useful to anticipate the examples below and discuss in general terms the main useful products of the kind of analyses sketched above.

Firstly, it can be argued that much theoretical insight is gained. Particular (equivalent) formulations of models often have particular strengths and these can be assembled eclectically once it is recognized that the alternatives are equivalent. Also, since methodological habits are often different in different disciplines, it becomes easier to recognize the use of equivalent models in different fields.

Secondly, and perhaps most importantly, a model formulated in one way may be very difficult to modify or extend, while extensions may be made easily, and indeed almost suggest themselves, in another. For example, it may be difficult to add new constraints to a set of simultaneous equations representing an equilibrium point, but it may be easy to do this with a mathematical programming formulation.

Thirdly, one particular representation may be much more convenient computationally than another, and this has important practical consequences.

10.3 EXAMPLE 1: RETAILING FLOWS AND CENTRE STRUCTURES

10.3.1 Basic models

Consider the well-known shopping model of Huff (1964) and Lakshmanan and Hansen (1965):

$$S_{ij} = A_i O_i W_j^{\alpha} \exp[-\beta c_{ij}] \tag{10.5}$$

where

$$A_i = 1/\sum_j W_j \exp[-\beta c_{ij}] \tag{10.6}$$

S_{ij} is the flow of retail sales between residences in i and shops in j; O_i is the spending power in i; W_j is a measure of the attractiveness of shops in j, and will be taken here as measured by size (and hence it is also a measure of shopping-centre supply); c_{ij} is the cost of travel between i and j; α and β are parameters.

This equilibrium model arises from the entropy-maximizing mathematical program (Wilson, 1970):

$$\max_{\{S_{ij}\}} S = -\sum_{ij} S_{ij} \ln S_{ij} \tag{10.7}$$

subject to

$$\sum_j S_{ij} = O_i \tag{10.8}$$

$$\sum_{ij} S_{ij} \ln W_j = \overline{\ln W} \tag{10.9}$$

and

$$\sum_{ij} S_{ij} c_{ij} = C \tag{10.10}$$

for some suitable average value of ln W_j and a total travel cost, C. Various alternative formulations lead to equilibrium equations like (10.5) and (10.6). These are discussed elsewhere (Wilson, 1977; and Chapter 9 above). This, of course, is a 'simple' model.

Suppose now we no longer take the vector $\{W_j\}$ as given. The model now becomes a composite model, since two phenomena are involved characterized by state variable arrays $\{S_{ij}\}$ and $\{W_j\}$ and a mechanism (in the form of additional equations or an optimizing framework, say) has to be added to determine $\{W_j\}$. This leads to some interesting equivalent formulations which illustrate our general argument.

10.3.2 A mechanism to determine $\{W_j\}$

The simplest mechanism is to assume that producers supply shops to balance revenue. In this case, the following equations must be satisfied:

$$\sum_i S_{ij} = kW_j \tag{10.11}$$

for some suitable constant k. The set of equations which determine $\{S_{ij}\}$ and $\{W_j\}$ is then (10.5), (10.6), and (10.11). Alternative procedures exist for solving these equations (as given, for example, in Harris and Wilson, 1978).

10.3.3 A mathematical programming formulation

The next step in the argument is to seek a mathematical programming formulation. It can be shown that the consumers' surplus X associated with this system is

$$X = -\frac{1}{\beta}\sum_{ij} S_{ij} \ln S_{ij} + \sum_{ij} S_{ij}\left(\frac{\alpha}{\beta} \ln W_j - c_{ij}\right) \tag{10.12}$$

Thus, a planning authority could determine $\{W_j\}$ to maximize X in (10.12) subject to equations (10.5) and (10.6) as constraints on the $\{S_{ij}\}$ together with an additional constraint on total provision

$$\sum_j W_j = W \tag{10.13}$$

Three comments can be made immediately about this program: firstly, the solution turns out to be exactly equivalent to that of equations (10.5), (10.6), and (10.11), the so-called 'balancing' procedure. Secondly, because of the non-linearity of both objective function and constraints, no algorithms exist for solution. It is particularly in this latter respect that it pays to seek alternative, but equivalent, programming formulations. Thirdly, the $\{S_{ij}\}$ variables still satisfy equations of the form (10.4) and (10.6) and so consumers are assumed to behave accordingly.

The main clue is in the nature of equations (10.5) and (10.6) and their equivalent programming formulation given by equations (10.7)–(10.10): in effect, the planning authority is seeking to optimize $\{W_j\}$ in equation (10.12) subject to the knowledge that consumers will behave according to the program (10.7)–(10.10). That is, the $\{S_{ij}\}$ program is embedded (or nested) inside the $\{W_j\}$ program. It turns out to be possible to combine the programs in such a way that all the non-linearities are in the objective function and the constraints are linear. The new formulation is

$$\max_{\{S_{ij},\, W_j\}} X = -\frac{1}{\beta}\sum_{ij} S_{ij} \ln S_{ij} + \sum_{ij} S_{ij}\left(\frac{\alpha}{\beta} \ln W_j - c_{ij}\right) \tag{10.14}$$

subject to

$$\sum_j S_{ij} = O_i \tag{10.15}$$

and

$$\sum_j W_j = W \tag{10.16}$$

(where, in effect, we can recognize that the constraints (10.9) and (10.10) of the 'inner problem' have been incorporated via multipliers α and β into the objective function simply so that we can continue to interpret that function as consumers' surplus). This equivalent formulation can now be solved by standard computer programs. It can be shown (Coelho, Williams, and Wilson, 1978) as noted above that $\{S_{ij}\}$ still satisfy equations of the form (10.5) and (10.6) in spite of not being represented explicitly as such.

Two qualifying comments should be made about these equivalences. Firstly, while we noted earlier that alternative formulations of the basic model each generated equations (10.5) and (10.6), this is not true if the model is disaggregated, and so greater caution should be exercised in that case (see Coelho, 1977, for further details). Secondly, the different mechanisms suggested for determining $\{W_j\}$, equations (10.11) and (10.12), stem from assumptions about producers' behaviour and a planning authority acting on behalf of consumers, respectively. It is odd at first sight that such different assumptions should lead to equivalent models. This almost certainly arises from the highly constrained nature of this problem, but though in that sense the equivalence is accidental, it illustrates our general mathematical argument well.

The argument is illustrated further by another interesting equivalent formulation due to Leonardi (1978). He posed the following problem:

$$\max_{\{W_j\}} Z = \sum_i O_i \ln \sum_j W_j \exp[-\beta c_{ij}] \tag{10.17}$$

subject to

$$\sum_j W_j = W \tag{10.18}$$

For obvious reasons he interprets the objective function as a sum of log accessibilities. The solution to this problem is when

$$\frac{\partial Z}{\partial W_j} = \gamma \tag{10.19}$$

where γ is the Lagrangian multiplier associated with (10.18). This derivative can be calculated from (10.17) as

$$\frac{\partial Z}{\partial W_j} = \sum_j \frac{O_i \exp[-\beta c_{ij}]}{\sum_k W_k \exp[-\beta c_{ik}]} = \frac{1}{W_j} \sum_i \frac{O_i W_j \exp[-\beta c_{ij}]}{\sum_k W_k \exp[-\beta c_{ik}]} \tag{10.20}$$

$$= \frac{\sum_i S_{ij}}{W_j} = \rho_j \tag{10.21}$$

where (10.5) and (10.6) are used to obtain (10.21) from (10.20). The optimality condition (10.19) asserts that ρ_j should be equal across space. If we put $\gamma = k$, we see that (10.21) is then equivalent to the balancing equation (10.11) and it turns out that this program is equivalent to the previous formulations. The clue to the equivalence lies in the term $\ln \sum_j W_j \exp[-\beta c_{ij}]$, say v_i, in the objective function. This is the dual variable associated with a constraint of the form (10.8) and $\sum_i O_i v_i$ is, in effect, the dual objective function. So, we have a dual formulation from which the $\{S_{ij}\}$ variables have been eliminated and are only re-introduced in deriving the appropriate optimality conditions (10.21) from (10.20). This illustrates the richness of possible equivalences in apparently very different programming formulations.

10.3.4 $\{W_j\}$ in a dynamic framework

The final step in the argument is to embed the static model into a dynamic framework. This can be done by assuming that $\{W_j\}$ are determined by differential equations of the form

$$\frac{dW_j}{dt} = \varepsilon\left(\sum_i S_{ij} - kW_j\right) \tag{10.22}$$

for a suitable constant ε. This simply says that W_j will grow if revenue exceeds costs, and vice versa. The equilibrium conditions are therefore

$$\frac{dW_j}{dt} = 0 = \sum_i S_{ij} - kW_j \tag{10.23}$$

which are the same as the balancing equation (10.11). This formulation can be used to analyse the *stability* of equilibrium points and this is carried out in Harris and Wilson (1978) and the main results are presented in another context in Chapter 9 above.

10.4 EXAMPLE 2: LOWRY MODELS

10.4.1 The original model

We now turn to families of models based on Lowry's (1964) classic formulation in which employment in the basic sectors of the urban economy stimulates successive locational distributions of resident population and of service employment according to given activity rates and service employment ratios and given land-use and population density limitations. Various developments of this model have not only improved it considerably, but have also gradually revealed a number of interesting properties that had been hidden in the original formulation. We will trace many of these developments in order to pick out cases where different representations turn out to be equivalent; some of these equivalences will be more familiar than others, and some will have already been encountered in Chapters 4 and 6 above.

10.4.2 The straightforwardly 'improved' model

The earliest improvements to the model developed its locational component, introducing explicit spatial-interaction submodels to distribute employees to their zones of residence, and the resulting residential population to service-centre zones. The spatial-interaction terms became important intermediate variables, as their estimation played a key role in determining the aggregate economic (employment) and population totals in each zone. Singly constrained spatial-interaction models were used since in each case the production totals in each zone were known. Alongside these developments the land-use and population density limitations were given more explicit treatment.

The 'Lowry' model could now typically be summarized as follows (clearly a 'composite' model), and we will see below that several rather different looking representations of the model turn out to be equivalent to this form.

The given initial employment in the basic sectors, E_i^B, can be allocated to zones of residence via the following submodel:

$$T_{ij} = \frac{E_i (W_j^H)^\alpha \exp[-\beta^T c_{ij}^T]}{\sum_j (W_j^H)^\alpha \exp[-\beta^T c_{ij}^T]} \tag{10.24}$$

with $E_i = E_i^B$, and $(W_j^H)^\alpha$ an attractiveness term for residential zone j, c_{ij}^T average cost terms and β^T a parameter. From this the total population in each zone, P_j, follows from

$$P_j = g \sum_i T_{ij} \tag{10.25}$$

given some known inverse activity rate g.

This population demands retail (service) goods, thus generating shopping flows

$$S_{ij} = \frac{(W_i^R)^\gamma e_j P_j \exp[-\beta^S c_{ij}^S]}{\sum_i (W_i^R)^\alpha \exp[-\beta^S c_{ij}^S]} \tag{10.26}$$

where S_{ij} is the flow of cash from residents in zone j to retail sectors R in zone i, and e_j is the expenditure per head in j on the output of retail sector k. The zonal distribution of retail employment then follows from

$$E_i^R = c \sum_j S_{ij} + d E_i \tag{10.27}$$

where c and d are home-based and job-based utilisation rates†. These employees create a further demand for residence; thus the sequence through equations (10.24)–(10.27) is repeated until successive rounds produce negligible changes in employment and population totals. Each such round may be subject to land-use and other density limitations as related, for example, in Wilson (1974).

† To simplify the presentation, job based retail utilization will be omitted from the subsequent analysis.

10.4.3 Matrix inverse formulations

A well-known alternative (and equivalent) representation of the model mechanism given in (10.24)–(10.27) above is the matrix representation due to Garin (1966). This may be written as

$$\mathbf{E} = (\mathbf{I} - \mathbf{BA})^{-1}\,\mathbf{E}^B \tag{10.28}$$

where $\mathbf{E}$ and $\mathbf{E}^B$ are vectors of total and basic employment (respectively), $\mathbf{B}$ is a matrix with elements

$$\frac{(W_i^R)^\gamma e_j \exp[-\beta^S c_{ij}^S]}{\sum_j (W_i^R)^\gamma \exp[-\beta^S c_{ij}^S]}$$

and $\mathbf{A}$ a matrix with elements

$$\frac{g(W_j^H)^\alpha \exp[-\beta^T c_{ij}^T]}{\sum_j (W_j^H)^\alpha \exp[-\beta^T c_{ij}^T]}$$

where i are workplace and j residence zones in both $\mathbf{A}$ and $\mathbf{B}$.

The series expansion of (10.28) is

$$\mathbf{E} = (\mathbf{I} + \mathbf{BA} + (\mathbf{BA})^2 + (\mathbf{BA})^3 + \ldots)\mathbf{E}^B \tag{10.29}$$

and it can therefore be shown that the total employment generated after each round of the original formulation is the same as that given by the equivalent number of terms from (10.29). This is a familiar result and is included here for completeness.

A further matrix inverse formulation of the model can be found in Chapter 6 above, following Macgill (1977c); it again produces identical numerical results to the earlier representations but assumes a different form, repeated here for convenience, namely:

$$\mathbf{X} = (\mathbf{I} - \mathbf{D})^{-1}\,\mathbf{Y} \tag{10.30}$$

where the vector $\mathbf{X}$ to be estimated is

$$\mathbf{X} = \begin{bmatrix} \{\mathbf{P}\} \\ \{h\mathbf{E}\} \end{bmatrix} \tag{10.31}$$

with h as the output per employee in the retail sector. The matrix $\mathbf{D}$ is given by

$$\mathbf{D} = \left[\begin{array}{c|c} 0 & \dfrac{(W_j^H)^\alpha \exp[-\beta^T c_{ij}^T]}{\sum_j (W_j^H)^\alpha \exp[-\beta^T c_{ij}^T]} \\ \hline \dfrac{ge_j(W_i^R)^\gamma \exp[-\beta^S c_{ij}^S]}{\sum_i (W_i^R)^\gamma \exp[-\beta^S c_{ij}^S]} & 0 \end{array}\right] \tag{10.32}$$

and the vector **Y** by

$$\mathbf{Y} = \begin{bmatrix} \sum_i \dfrac{E_i^B (W_j)^\alpha \exp[-\beta^T c_{ij}^T]}{\sum_j (W_j^H)^\alpha \exp[-\beta^T c_{ij}^T]} \\ 0 \end{bmatrix} \tag{10.33}$$

It has been noted above that, in contrast to (10.28), (10.30) takes the familiar form of the well-known Leontief economic input–output model, where **Y** would be a 'final demand' vector, **D** a matrix of input–output coefficients, and **X** an endogenous vector of sector outputs. Thus this further equivalent formulation demonstrates that the Lowry model is a special case of a spatially disaggregated input–output model, the case where there are two types of sectors, households and services, and terms **X**, **D**, and **Y** given by (10.31), (10.32), and (10.33). Furthermore, this formulation shows immediately how it would be possible to extend the Lowry model to include full inter-sectoral relations, namely by filling out the input–output coefficient matrix **D**, and possibly also the 'final demand' vector **Y** (see Macgill, 1977c, for more details). The potential for these modifications is not apparent in the earlier versions of the model. Against this advantage, we must also note that Garin's matrix inverse formulation is of considerably smaller dimension than the full input–output version of the model and thus computationally more efficient. This is because it amalgamates, via the product **BA** in (10.28), the individual inter-sectoral interactions, whereas the input–output model represents explicitly each such interaction.

10.4.4 Mathematical programming versions of the Lowry model

In the Lowry model summary in equations (10.24)–(10.27) the singly constrained spatial-interaction submodels that give rise to T_{ij} and S_{ij} in equations (10.24) and (10.26), can be derived, respectively, from the following (simple) mathematical programming formulations:

$$\max_{\{T_{ij}\}} S^1 = \sum_{ij} T_{ij} \ln T_{ij} \tag{10.34}$$

subject to

$$\sum_j T_{ij} = E_i \tag{10.35}$$

$$\sum_{ij} T_{ij}\left(c_{ij}^T - \frac{\alpha}{\beta^T} \ln W_j^H\right) = C^T \tag{10.36}$$

and

$$\max_{\{S_{ij}\}} S^2 = -\sum_{ij} S_{ij} \ln S_{ij} \tag{10.37}$$

subject to

$$\sum_i S_{ij} = e_j P_j \tag{10.38}$$

$$\sum_{ij} S_{ij}(c_{ij}^S - \frac{\gamma}{\beta^S} \ln W_i^R) = C^S \tag{10.39}$$

The resulting estimates for T_{ij} and S_{ij} then determine fresh estimates for P_j and for E_i via equations (10.25) and (10.27). We may now construct a single mathematical programming formulation of the Lowry model by combining (10.25) and (10.27) with (10.34)–(10.36) and (10.24) and (10.26) with (10.37)–(10.39). This produces the following:

$$\max_{\{T_{ij}\}\{S_{ij}\}} S^3 = -\sum_{ij} T_{ij} \ln T_{ij} - \sum_{ij} S_{ij} \ln S_{ij} \tag{10.40}$$

subject to

$$\sum_j T_{ij} = c \sum_j S_{ij} + E_i^B \tag{10.41}$$

$$\sum_i S_{ij} = g e_j \sum_i T_{ij} \tag{10.42}$$

$$\sum_{ij} T_{ij} (c_{ij}^T - \frac{\alpha}{\beta^T} \ln W_j^H) = C^T \tag{10.43}$$

$$\sum_{ij} S_{ij} (c_{ij}^S - \frac{\gamma}{\beta^S} \ln W_i^R) = C^S \tag{10.44}$$

Note that constraints (10.41) and (10.42) interlink the T_{ij} and the S_{ij} more prominently than in the earlier formulations, although they have only been derived directly from equations (10.25), (10.27), (10.35), and (10.37). This implies that residential location is a function both of access to services and of jobs, and it is interesting that this useful result emerges from the technical procedure of formulating the model as a single mathematical program. This result has already arisen in Chapter 4 above in a related context.

A Lagrangian solution to equations (10.40)–(10.44) produces the following model estimates:

$$T_{ij} = A_i (B_j)^{-\theta} (W_j^H)^\alpha \exp[-\beta^T c_{ij}^T] \tag{10.45}$$

$$S_{ij} = (A_i)^{-\phi} B_j (W_i^R)^\gamma \exp[-\beta^S c_{ij}^S] \tag{10.46}$$

with A_i and B_j (transformations of the dual variables associated with constraints (10.41) and (10.42), or alternatively, rather unusual types of 'balancing factors') to be determined to ensure that equations (10.41) and (10.42) are satisfied.

Although this mathematical programming formulation uses the same information and apparently the same assumptions as the earlier formulation, it will not typically give equivalent results (an aspect again raised in Chapter 4 above and to be discussed further below). We can appreciate this by noting that equations

(10.41) and (10.42) will pick up the familiar economic base mechanism in the usual way, but T_{ij} and S_{ij} are now given by (10.45) and (10.46) (effectively doubly constrained estimates) rather than by (10.24) and (10.26) (singly constrained estimates). We can, in fact, enforce an equivalence between this and the earlier formulation if we cancel out the effect of the 'additional' balancing factors ($B_j^{-\theta}$ in (10.45) and $A_i^{-\phi}$ in (10.46)), for example by replacing the W_j^H and W_i^R attractiveness terms in the model given by (10.40)–(10.46) by alternative terms $\bar{W}_j^H$ and $\bar{W}_i^R$ which are such that

$$\bar{W}_j^H B_j^{-\theta} = W_j^H \tag{10.47}$$

$$\bar{W}_i^R A_i^{-\phi} = W_i^R \tag{10.48}$$

This would not be completely unjustified, since the A_i and B_j relate to the shadow prices for workplace and residential zones, respectively, so that the attractiveness terms for these zones now reflect these prices. The desired equivalence is therefore less direct than earlier equivalences.

The above mathematical programming formulation of the Lowry model eliminated stock terms E_i and P_j in deriving constraints (10.41) and (10.42), and therefore produces its results purely in terms of spatial-interaction flow terms T_{ij} and S_{ij}. An alternative approach would be to seek a mathematical programming formulation in which stock as well as flow terms are estimated endogenously. Again we are mainly interested here in such a formulation which can give results which are equivalent to those obtained from earlier 'Lowry' models, but ultimately we would expect that the application of and extensions to this new formulation would be of a slightly different nature. The model to be considered is the following:

$$\max_{\{T_{ij}\}\{S_{ij}\}\{E_i\}\{P_j\}} \quad S^4 = -\sum_{ij} T_{ij} \ln T_{ij} - \sum_{ij} S_{ij} \ln S_{ij} \tag{10.49}$$

subject to

$$\sum_j T_{ij} = E_i \tag{10.50}$$

$$\sum_i S_{ij} = e_j P_j \tag{10.51}$$

$$\sum_{ij} T_{ij}\left(c_{ij}^T - \frac{\xi_j}{\beta^T} \ln e_j P_j \right) = C^T \tag{10.52}$$

$$\sum_{ij} S_{ij}\left(c_{ij}^S - \frac{\zeta_i}{\beta^S} \ln \mathrm{E}_i \right) = C^S \tag{10.53}$$

This simultaneously determines employment and population totals as well as journey to work and shopping flows through the maximization of an entropy function subject to identities (10.50) and (10.51) (reproduced from (10.35) and (10.38)) and modified spatial-interaction cost constraints. The $e_j P_j$ and E_i terms modifying the cost constraints will be interpreted below as attractiveness weights.

An alternative way of introducing these terms (and equivalent for the purpose of model results) would be as weights on the entropy functions, that is, replacing expressions $T_{ij} \ln T_{ij}$ and $S_{ij} \ln S_{ij}$ in (10.49) by

$$T_{ij} \ln \frac{T_{ij}}{e_j P_j} \text{ and } S_{ij} \ln \frac{S_{ij}}{E_i}$$

respectively. The form adopted in constraints (10.52) and (10.53) parallels a procedure adopted earlier in this chapter—trading off zonal attractiveness against spatial-interaction average costs, and the weighted entropy function procedure similarly parallels the use of attractiveness weights in different contexts in Chapter 4 (see also Coelho and Williams, 1978). A third possibility would have been to include current model equivalents of constraints (10.9) in order to allow the model to take account of zonal attractiveness (related in the present case to E_i and P_j terms).

A Lagrangian solution to (10.49)–(10.53) produces estimates of the form

$$T_{ij} = \frac{E_i (e_j P_j)^{\xi_j} \exp[-\beta^T c_{ij}^T]}{\sum_j (e_j P_j)^{\xi_j} \exp[-\beta^T c_{ij}^T]} \tag{10.54}$$

$$S_{ij} = \frac{(E_i)^{\zeta_i} e_j P_j \exp\{-\beta^S c_{ij}^S]}{\sum_i (E_i)^{\zeta_i} \exp[-\beta^S c_{ij}^S]} \tag{10.55}$$

and this model can be shown to be equivalent to the Lowry model summarized in section 10.4.2 if we interpret the attractiveness terms W_i^R and W_j^H in that earlier model in the following way:

$$(W_i^R)^\gamma = (E_i)^{\zeta_i} \tag{10.56}$$

$$(W_j^H)^\alpha = (e_j P_j)^{\xi_j} \tag{10.57}$$

that is, relating them directly to the 'size' of their respective zones, given in terms of total employment opportunities (for service sector zones) and total resident population (for residential zones). Substitutions (10.56) and (10.57) clarify the role of the $e_j P_j$ and E_i terms in (10.52) and (10.53) (as attractiveness weights), and when applied to (10.54) and (10.55) reduce these latter estimates to the form of (10.24) and (10.26). The economic base multiplier mechanism, necessary to complete the Lowry model but well hidden in the model equations (10.49)–(10.53), can be derived from further first-order Lagrangian conditions.

An alternative analysis of Lowry model equivalences based on mathematical programming frameworks has been covered in Chapter 4 above, the main results of which we now summarize. A so-called group entropy model, mathematically equivalent to (10.40)–(10.44) above but slightly different in appearance due to the incorporation of attractiveness weights W_j^H and W_i^R in the entropy functions rather than traded off within the cost constraints, is from a theoretical standpoint closest to the models derived above. However, in Chapter 4 we also encountered models derived from an *alternative* theoretical base, namely rational choice behaviour. One such model, very similar in appearance to the group entropy model, or (10.40)–(10.44) above with the cost constraints absorbed into the objective function, is the group surplus model given by

$$\max_{\{T_{ij}\}\{S_{ij}\}} GS = -\frac{1}{\beta^T}\sum_{ij} T_{ij}\left(\ln\frac{T_{ij}}{W_j^H}-1\right)-\sum_{ij} T_{ij}c_{ij}^T$$
$$-\frac{1}{\beta^S}\sum_{ij} S_{ij}\left(\ln\frac{S_{ij}}{W_i^R}-1\right)-\sum_{ij} S_{ij}c_{ij}^S \qquad (10.58)$$

subject to (10.41) and (10.42). The group surplus function (10.58) is based on the Marshallian–Hotelling definition of consumers' surplus or the random utility concept of expected total surplus. The position of the β^T and β^S terms in this model is critical, and prohibits the direct occurrence of equivalence between this and earlier models (for example the group entropy model, or models given earlier above). It is, however, possible to derive an equivalence between the group surplus model and an iterative (non-mathematical programming) version of the Lowry model known as the probabilistic choice model (again based on rational choice behaviour). The comparison and derived equivalence here between the group surplus and the probabilistic choice model is akin to the comparison and derived equivalence between the mathematical programming ((10.40)–(10.44) and original iterative (10.24)–(10.27)) version of the Lowry model given above. Thus it requires additional weights to be specified within the iterative form of the model in order to nullify the effect of additional shadow prices (dual variables) imputed by the mathematical programming form. For completeness we may add that the iterative probabilistic-choice approach of Chapter 4 and the iterative version of the Lowry model summarized in (10.24)–(10.27) above may also be made equivalent to each other via further judicious specification of attractiveness weights. The manipulation of attractiveness terms to force equivalences between models is of more than just academic significance as it gives some insight into the basic role of these terms.

A further mathematical programming version of the Lowry model based on rational choice behaviour presented in Chapter 4 estimated $\{E_i\}$ and $\{P_i\}$ as well as $\{T_{ij}\}$ and $\{S_{ij}\}$ terms simultaneously, and is therefore comparable to (10.49)–(10.53) above. The constraint set in fact included equations such as (10.25) and (10.28) within a model otherwise similar to (10.40)–(10.44). A further model still (Coelho, 1977) includes the additional feature of endogenously estimating basic employment totals E_i^B which have been assumed to be exogenously specified throughout the analysis above.

Finally in this section we note that although much of the analysis above is paralleled by similar analysis given for shopping models in Section 10.3, it has not been possible to derive a 'Lowry' model which corresponds directly with Leonardi's log-accessibility derivation of the shopping model. This is due, at least in part, to the more highly constrained nature of the Lowry model, though an input–output model developed below, from which a Lowry model may be derived as a special case, makes use of a log-accessibility term within its objective function.

10.4.5 Dynamic models

The final component of the Lowry model analysis is to seek a dynamic equivalent model, that is, a dynamic model whose equilibrium conditions reproduce the Lowry model equations. Burdekin (1977) has already demonstrated one such equivalence, by deriving Lowry model equations as an equilibrium point of a particular type of Forrester-based model. The more complex nature of the Lowry model suggests that the corresponding dynamic possibilities will be richer. A number of available dynamic embeddings of the Lowry model are given below to illustrate our argument. We first choose to embed conditions (10.41) and (10.42) in a dynamic framework. Note that these conditions are implicit in the iterative form of the model in section 10.4.2 (equations (10.24)–(10.27)) as well as being explicit equilibrium constraints within the mathematical programming formulation. They give

$$\frac{dW_i^R}{dt} = \delta_1\left(\sum_j T_{ij} - c\sum_j S_{ij} + E_i^B\right) \tag{10.59}$$

$$\frac{dW_j^H}{dt} = \delta_2\left(\sum_i S_{ij} - ge_j\sum_i T_{ij}\right) \tag{10.60}$$

with

$$T_{ij} = \frac{E_i(W_j^H)^\alpha \exp[-\beta^T c_{ij}^T]}{\sum_j (W_j^H)^\alpha \exp[-\beta^T c_{ij}^T]} \quad \text{say} \tag{10.61}$$

and

$$S_{ij} = \frac{(W_i^R)^\gamma e_j P_j \exp[-\beta^S c_{ij}^S]}{\sum_i (W_i^R)^\gamma \exp[-\beta^S c_{ij}^S]} \quad \text{say} \tag{10.62}$$

Thus, in equation (10.59) it is assumed that retail centre attractiveness (size) will increase until the shopping flows generated $\left(\sum_j S_{ij}\right)$ are in balance with the employment $\left(\sum_j T_{ij}\right)$ in retail centres. Similarly, in equation (10.60) residential zone attractiveness is assumed to increase until the population and hence the spending power of the zone $\left(e_j P_j = \sum_i S_{ij}\right)$ is matched by its residential attractiveness.

The equilibrium conditions are, of course, when

$$\frac{dW_i^R}{dt} = \frac{dW_j^H}{dt} = 0 \tag{10.63}$$

that is, conditions (10.31) and (10.32) from the earlier model are reproduced (the economic base multiplier conditions) and the full Lowry model is completed in this case by equations (10.61) and (10.62), a formulation equivalent to that of section 10.4.2.

Alternative dynamic versions of the Lowry model may be derived from the Lagrangians of the mathematical programming formulations. We have already seen that apparently different mathematical programming formulations can be made equivalent, and it is therefore reasonable to expect that the corresponding dynamic models derived from their Lagrangians can also, at least at equilibrium, be made equivalent.

For the model given by (10.49)–(10.53) we have

$$\frac{\partial L}{\partial E_i} = \lambda_i^1 - \sum_j \frac{\zeta_i}{\beta^S} \frac{S_{ij}}{E_i} \tag{10.64}$$

$$\frac{\partial L}{\partial P_j} = e_j \lambda_j^2 - \sum_i \frac{\xi_j}{\beta^T} \frac{T_{ij}}{P_j} \tag{10.65}$$

(where λ_i^1 and λ_j^2 are the Lagrangian multipliers or dual variables associated with constraints (10.50) and (10.51), respectively), and therefore we may assume

$$\frac{dE_i}{dt} = \left(\sum_j S_{ij} - K_i E_i\right) \tag{10.66}$$

$$\frac{dP_j}{dt} = \left(\sum_i T_{ij} - L_j P_j\right) \tag{10.67}$$

for suitable constants K_i and L_j. It is interesting and reassuring to note that this dynamic model can be made equivalent to that above ((10.59), (10.60)) under exactly the same conditions that rendered their parent mathematical programming models equivalent, namely that conditions (10.50) and (10.51) must hold (thus eliminating E_i and P_j terms from the right-hand side of equations (10.66) and (10.67)), and the attractiveness terms W_i^R and W_j^H can be defined in terms of employment and population terms E_i and P_j.

The corresponding dynamic behaviour of the T_{ij} and S_{ij} terms may be derived from $\partial L/\partial T_{ij}$ and $\partial L/\partial S_{ij}$, respectively, and gives

$$\frac{dT_{ij}}{dt} = -\ln T_{ij} - 1 - \lambda_i^1 - \beta^T c_{ij}^T + \zeta_j \log e_j P_j \tag{10.68}$$

$$\frac{dS_{ij}}{dt} = -\ln S_{ij} - 1 - \lambda_j^2 - \beta^S c_{ij}^S + \zeta_i \log E_i \tag{10.69}$$

Thus the dynamics here are determined by the shadow prices λ_i^1 and λ_j^2 (themselves functions of T_{ij} and S_{ij}) as well as the P_j and E_i terms. The equilibrium conditions, of course, produce estimates for T_{ij} and S_{ij} of the form (10.54) and (10.55).

Any stability analysis for the above systems of equations will be more complex than that required for the corresponding differential equation form for the shopping model which were presented in Chapter 9.

A further differential equation formulation to be presented in section 10.5.4 in the context of input–output models will also be noted to be applicable to Lowry models (since the Lowry model may be identified as a special case of an input–output model: see below). In this case, the interpretation will be in analogy with competing species behaviour from the ecological literature—shopping (service) centres competing for customers and residential areas for population in our case.

10.5 EXAMPLE 3: SPATIALLY DISAGGREGATED ECONOMIC ACCOUNT-BASED MODELS

10.5.1 The Leontief matrix inverse form

We now turn to models underpinned by input–output accounts which tabulate inter-industrial or inter-sectoral flows and corresponding row and column totals (total inputs and outputs of each sector), disaggregated by zone. These models encompass both the shopping models of Section 10.3 and the Lowry models of Section 10.4, and various insights stemming from this aspect will be considered later.

The models to be considered can all be seen as extensions or alternative formulations of a spatially disaggregated version of the classic Leontief form (included in Figure 10.1). They thus involve an input–output multiplier component linking the level of inputs to the level of outputs for each sector, and a spatial-interaction component to represent the relative location of sectors as producers and users of various inputs and outputs.

Taking X_i^m as the total output of sector m in zone i, X_{ij}^{mn} as the output of m in i used by n in j, and Y_i^m as the final demand for the output of sector m in zone i, the following identity must hold:

$$X_i^m = \sum_{jn} X_{ij}^{mn} + Y_i^m \tag{10.70}$$

Thus by defining constant coefficients a_{ij}^{mn} by

$$X_{ij}^{mn} = a_{ij}^{mn} X_j^n \tag{10.71}$$

we get

$$X_i^m = \sum_{jn} a_{ij}^{mn} X_j^n + Y_i^m \tag{10.72}$$

and hence the familiar form

$$\mathbf{X} = (\mathbf{I} - \mathbf{A})^{-1}\mathbf{Y} \tag{10.73}$$

We will be interested below in the case where the coefficients a_{ij}^{mn} consist of technical coefficients specific to the receiving sectors zones, Z_j^{mn} say, amalgamated with an attraction-constrained spatial-interaction model, that is,

$$a_{ij}^{mn} = \frac{Z_j^{mn} f_m(c_{ij})}{\sum_i f_m(c_{ij})} \tag{10.74}$$

Thus (10.72) becomes

$$X_i^m = \sum_{jn} \frac{Z_j^{mn} \exp[-\beta^m c_{ij}^m] X_j^n}{\sum_i \exp[-\beta^m c_{ij}^m]} + Y_i^m \tag{10.75}$$

if $f_m(c_{ij})$ in (10.74) is taken as an exponential cost function.

Under our earlier classification, this would be called a hybrid model, and as before we will present several different-looking formulations that turn out to be equivalent to this model. In addition to the 'equivalent' models we present explicitly, we may note that further spatially disaggregated input–output based models by Leonardi (1978) and Williams (1979) share very strong similarities (but not equivalences) with those given below. We would class the former as a composite model which sequentially solves an input–output identity and then a mathematical programming submodel. The latter is an algorithmic approach, giving a high priority to computational efficiency.

10.5.2 An entropy-maxmizing formulation

We turn straight to a mathematical programming formulation based on the same information and apparently the same assumptions as the model just given, but, in parallel with a similar analysis in Section 10.4, although an equivalence can be forced this model will not typically be equivalent. An entropy-maximizing formulation is used in order to reproduce the exponential cost function of the previous model, and takes the form of

$$\max_{\{X_{ij}^{mn}\}\{X_i^m\}} S = -\sum_{ijmn} X_{ij}^{mn} \ln X_{ij}^{mn} \tag{10.76}$$

subject to

$$X_i^m = \sum_{jn} X_{ij}^{mn} + Y_i^m \tag{10.77}$$

$$\sum_i X_{ij}^{mn} = Z_j^{mn} X_j^n \tag{10.78}$$

$$\sum_{ijn} X_{ij}^{mn} c_{ij}^m = C^m \tag{10.79}$$

Constraint (10.77) is a repetition, for convenience, of the key input–output identity (10.70); constraint (10.78) is based directly on the technical coefficients Z_j^{mn} of the previous model; and constraint (10.79) is the usual spatial-interaction cost constraint.

An analytical solution to this model starts with the Lagrangian expression

$$L = -\sum_{ijmn} X_{ij}^{mn} \ln X_{ij}^{mn} + \sum_{im} \lambda_i^m \Big(X_i^m - \sum_{jn} X_{ij}^{mn} - Y_i^m\Big) + \sum_{jmn} \gamma_j^{mn} \Big(\sum_i X_{ij}^{mn} - Z_j^{mn} X_j^n\Big) + \sum_m \mu^m \Big(\sum_{ijn} X_{ij}^{mn} c_{ij}^m - C^m\Big) \tag{10.80}$$

Thus

$$\frac{\partial L}{\partial X_{ij}^{mn}} = \ln X_{ij}^{mn} - \lambda_i^m + \gamma_j^{mn} + \mu^m c_{ij}^m = 0 \tag{10.81}$$

$$\frac{\partial L}{\partial X_i^m} = \lambda_i^m - \sum_n \gamma_i^{nm} Z_i^{mn} = 0 \tag{10.82}$$

and constraints (10.77), (10.78), and (10.79) must be satisfied to complete the optimality conditions. Putting

$$\exp[-\gamma_j^{mn}] = B_j^{mn} \tag{10.83}$$

equations (10.81) and (10.82) give

$$X_{ij}^{mn} = \prod_{n'} (B_i^{n'm})^{Z_i^{n'm}} B_j^{mn} \exp[-\mu^m c_{ij}^m] \tag{10.84}$$

μ^m may be calculated from (10.79) in the usual way, B_j^{mn} from (10.78), and X_i^m (required in the calculation of B_j^{mn} as well as being of interest in itself) from (10.77). For neatness we may substitute

$$A_i^m = \prod_{n'} (B_i^{n'm}) Z_i^{n'm} \tag{10.85}$$

in (10.84) to give

$$X_{ij}^{mn} = A_i^m B_j^m \exp[-\mu^m c_{ij}^m] \tag{10.86}$$

In order to make this programming model equivalent to that summarized in section 10.5.1, in particular reducing (10.86) to the form

$$X_{ij}^{mn} = \frac{Z_j^{mn} \exp[-\mu^m c_{ij}^m] X_j^n}{\sum_i \exp[-\mu^m c_{ij}^m]} \tag{10.87}$$

we need to take two further steps. Firstly, arrange that B_j^{mn} in (10.86) is given by

$$B_j^{mn} = \frac{Z_j^{mn} X_j^n}{\sum_j \exp[-\mu^m c_{ij}^m]} \tag{10.88}$$

and secondly, neutralize the shadow price A_i^m on the input–output identity (10.77), which has no counterpart in the original model. For this second step it is not appropriate simply to set $A_i^m = 1$ since this would in turn change the value of B_j^{mn} (A_i^m and B_j^{mn} are linked through (10.82) and cannot therefore be treated independently). We can, however, introduce an additional weighting term $(W_i^m)^\alpha$ taking a similar role as the attractiveness terms W_j^H and W_i^R in Section 10.4—see equations (10.47) and (10.48)—such that

$$(W_i^m)^\alpha A_i^m = 1 \tag{10.89}$$

Thus (10.87) would become

$$X_{ij}^{mn} = (W_i^m)^\alpha A_i^m B_j^{mn} \exp[-\mu^m c_{ij}^m] \tag{10.90}$$

and the programming model producing these estimates would be equivalent to the model of section 10.5.1, given conditions (10.88) and (10.89).

10.5.3 Two alternative mathematical programming models

A further equivalent programming model may be derived, building on the analysis of the previous subsection. The shadow price on $\sum_i X_{ij}^{mn}$, the amount of m used by n in zone j (the dual variable from constraint (10.78)) was found to be

$$\ln \frac{Z_j^{mn} X_j^n}{\sum_i \exp[-\mu^m c_{ij}^m]},$$

and therefore

$$\sum_{jmn} \ln \frac{Z_j^{mn} X_j^n}{\sum_i \exp[-\mu^m c_{ij}^m]} \sum_i X_{ij}^{mn}$$

is the total system cost of flows X_{ij}^{mn}. We may therefore formulate a mathematical programming model to minimize these costs and transport costs subject to the input–output identity (10.77). By including an entropy term as a spatial-interaction dispersion term in the objective function (changing the objective function into the group surplus of the producers), we arrive at a further model which can be made formally equivalent to that of section 10.5.1, namely

$$\min_{\{X_{ij}^{mn}\}\{X_i^m\}} S = \sum_{ijmn} X_{ij}^{mn} \ln X_{ij}^{mn} + \sum_{jmn} \ln \frac{Z_j^{mn} X_j^n}{\sum_i \exp[-\mu^m c_{ij}^m]} \sum_i X_{ij}^{mn}$$

$$+ \sum_m \mu^m \sum_{ijn} c_{ij}^m X_{ij}^{mn} \tag{10.91}$$

subject to

$$X_i^m = \sum_{jn} X_{ij}^{mn} + Y_i^m \tag{10.92}$$

It would be possible to construct further different-looking but again equivalent mathematical programming models by further manipulation of the original constraints and their corresponding dual variables, but the above model is considered sufficient demonstration of this in the present chapter.

We construct our second alternative model by using a further set of guidelines offered in section 10.2.3, that is, embedding the spatial-interaction submodels for X_{ij}^{mn} as explicit constraints in a mathematical programming problem (which we have referred to earlier as a superproblem formulation). A formulation that can give an equivalence with earlier input–output models of this section is the following:

$$\max_{\{X_{ij}^{mn}\}\{W_i^m\}} T = -\sum_{ijmn} X_{ij}^{mn} \ln X_{ij}^{mn} + \sum_i \eta_i \sum_m \mu^m \sum_{jn} X_{ij}^{mn}\left(\frac{\alpha}{\mu^m} \ln W_i^m - c_{ij}^m\right) \tag{10.93}$$

subject to

$$X_{ij}^{mn} = \frac{Z_j^{mn}(W_j^n)^\alpha \exp[-\mu^m c_{ij}^m]}{\sum_i \exp[-\mu^m c_{ij}^m]} \tag{10.94}$$

$$X_i^m = \sum_{jn} X_{ij}^{mn} + Y_i^m \tag{10.95}$$

The equivalence arises from the first-order optimality conditions, since we can establish from $\partial t/\partial W_i^m = 0$ that $W_i^m \propto X_i^m$ (given suitable specifications for the parameters) and hence constraints (10.94) and (10.95) reduce to the input–output model summarized in section 10.5.1. It is not obvious how to derive an equivalent log-accessibility version of this latter formulation (corresponding to that given in the shopping-model section); the present input–output models are more difficult to handle in this respect due to their more highly constrained nature and the need to avoid exactly determined constraint equations.

10.5.4 Dynamic models

The simplest dynamic equivalent model is formed in the same way as those of Sections 10.3 and 10.4, that is, by embedding the equilibrium conditions (equations (10.70)) directly into differential equations, namely

$$\frac{dX_i^m}{dt} = \tau(X_i^m - \sum_{jn} X_{ij}^{mn} - Y_i^m) \tag{10.96}$$

with X_{ij}^{mn} given, as before, by (10.87). This has the familiar interpretation that sector outputs will increase until they balance the demands made on them. A corresponding treatment of a similar type of input–output model is given by Leonardi (1978).

We may also derive dynamic input–output models from the Lagrangians of the mathematical programming formulations. For example, model (10.91)–(10.92) gives the following conditions for X_i^m:

$$\frac{\partial L}{\partial X_i^m} = \sum_{jn} \frac{X_{ji}^{nm}}{X_i^m} - \bar{\lambda}_i^m \tag{10.97}$$

(taking $\bar{\lambda}_i^n$ as the Lagrangian multipliers of constraint (10.92)) and hence the dynamic trajectories

$$\frac{dX_i^m}{dt} = \bar{\lambda}_i^m X_i^m - \sum_{jn} X_{ji}^{nm} \tag{10.98}$$

A type of duality between these trajectories for X_i^m and those given by (10.96) may be noted. If we consider a conventional square tabulation of terms X_{ij}^{mn} with row sums $\sum_{jn} X_{ij}^{mn}$ denoting sector outputs by zone, and column sums $\sum_{im} X_{ij}^{mn}$ denoting sector inputs by zone, it may be seen that the dynamic formulation given by (10.98) centres on the column sums (inputs) of the system whereas (10.96) earlier centred on the row sums (outputs).

The corresponding dynamic formulation for X_{ij}^{mn} will follow from

$$\frac{\partial L}{\partial X_{ij}^{mn}} = -\frac{\mathrm{d}X_{ij}^{mn}}{\mathrm{d}t}$$

and as in the case of (10.81) above this in turn will interlock terms X_{ij}^{mn} with what were formerly the dual variables of the mathematical programming formulation. At equilibrium of course,

$$\frac{\mathrm{d}X_{ij}^{mn}}{\mathrm{d}t} = \frac{\mathrm{d}X_i^m}{\mathrm{d}t} = 0$$

automatically reproduces the full Lagrangian solution of the parent mathematical programming formulation.

In the case of the mathematical programming model given by (10.76)–(10.79), the corresponding dynamic model in X_i^m turns out to be

$$\frac{\mathrm{d}X_i^m}{\mathrm{d}t} = \sum_n \gamma_i^{nm} Z_i^{nm} - \lambda_i^m \tag{10.99}$$

We can clarify this model by substituting for Z_i^{nm} from (10.78), giving

$$\frac{\mathrm{d}X_i^m}{\mathrm{d}t} = \lambda_i^m - \sum_n \gamma_i^{nm} \frac{\sum_j X_{ji}^{nm}}{X_i^m} \tag{10.100}$$

$$= (\lambda_i^m X_i^m - \sum_{jn} \gamma_i^{nm} X_{ji}^{nm}) X_i^m \tag{10.101}$$

At equilibrium, an equivalence between (10.101) and (10.98) can be forced, as before (in section 10.5.3). Furthermore, the dynamic trajectories of (10.98) and (10.101) away from equilibrium can also be made to coincide.

A further alternative dynamic embedding of the input–output equations is to present them in the form

$$\frac{\mathrm{d}X_i^m}{\mathrm{d}t} = X_i^m(X_i^m - F(X_1^1, X_1^2, \ldots, X_1^M)) \qquad m = 1, \ldots, M; i = 1, \ldots, I \tag{10.102}$$

with $F(X_1^1, X_1^2, \ldots, X_1^M)$ defined by the r.h.s. of (10.72), and by (10.74). As before, $\frac{\mathrm{d}X_i^m}{\mathrm{d}t} = 0$ reproduces the earlier model equations. Particular forms of equation (10.102) as dynamic representations of a finite number of interacting 'populations' have been studied in depth in the ecological literature (Rescigno, 1968; May, 1971; Strobeck, 1973; Leung, 1976; May and Leonardi, 1975). Smale (1976) has apparently shown that systems of equations that fit the general form of (10.102) can exhibit arbitrary dynamic behaviour (in a generic sense); that is, there are no 'typical' solution paths.

Rather more general than (10.102) is the form

$$\frac{dX_i^m}{dt} = G(X_i^m)(X_i^m - F) \tag{10.103}$$

which appears in May (1971) and could also be used for a dynamic embedding of the type of input–output equations discussed earlier in this section.

Finally, we note that since the Lowry model may be formulated as a special case of an input–output model, equations (10.102) and (10.103) may immediately be used as dynamic representations of the Lowry model.

10.6 CONCLUDING REMARKS

The three examples we have chosen to illustrate our main argument were increasingly general as the chapter proceeded. Thus the shopping model (Section 10.3) is contained within the Lowry model (Section 10.4) and this in turn is a special case of an input–output model (Section 10.5). Although in broad terms we have carried out the same type of equivalence analysis for each class of model (considering in turn matrix/algebraic versions, mathematical programming versions, and dynamic versions), we recognize that the results for each class do not correspond exactly. Thus results from the later sections may be applied to earlier sections, but not vice versa. This reinforces the point that the discussion in Section 10.2 of circumstances in which equivalences may arise offered *guidelines* rather than *rules*, and shows the necessity of this chapter to argue in terms of examples rather than formal theorems.

Bibliography

Adby, P. R., and Dempster, M. A. H. (1974) *Introduction to Optimisation Methods.* Chapman and Hall.

Albright, R. L., Lerman, S. R., and Manski, C. F. (1977) *Report on the Development of an Estimation Program for the Multinomial Probit Model.* Prepared for the Federal Highway Administration, Washington D.C.

Allen, P. M., Deneubourg, J. L., Sanglier, M., Boon, F., and de Palma, A. (1978) *The Dynamics of Urban Evolution*, Volume 1: *Inter-urban Evolution*; Volume 2: *Intra-urban Evolution.* Final Report to the U.S. Department of Transportation, Washington D.C.

Alonso, W. (1964) *Location and Land Use.* Harvard University Press, Cambridge, Mass.

Amson, J. C. (1975) Catastrophe theory: a contribution to the study of urban problems? *Environment and Planning B*, **2,** 177–221.

Anas, A. (1979) The impact of transit investment on housing values. A simulation experiment. *Environment and Planning A*, **11,** 239–55.

Armstrong, A. G. (1975) Technology assumptions in the construction of U.K. input–output tables. In R.I.G. Allen and W.F. Gossling (Eds.), *Estimating and Projecting Input–output Coefficients.* Input–output Publishing Company, London.

Artle, R. (1961) On some methods and problems in the study of metropolitan economics. *Papers of the Regional Science Association*, **8,** 71–87.

Augustinovics, M. (1970) Methods of international and intertemporal comparison of structure. In A. P. Carter and A. Brody (Eds.), *Contributions to Input–output Analysis.* North Holland Publishing Company.

Bacharach, M. (1970) *Biproportional Matrices and Input–output Change.* Cambridge University Press.

Balinski, M. L., and Baumol, W. (1968) The dual non-linear program and its economic interpretation. *Review of Economic Studies*, **35,** 237–56.

Banister, C. E. (1977) A method for incorporating maximum constraints into the Garin-Lowry model. *Environment and Planning A*, **9,** 787–93.

Barras, R., and Broadbent, T. A. (1975) An activity-commodity formalism for socio-economic systems. *Research Paper 18*, Centre for Environmental Studies, London.

Batty, M. (1970) Some problems in calibrating the Lowry Model. *Environment and Planning*, **2,** 95–114.

Batty, M. (1971) Design and construction of a sub-regional land-use model. *Socio-economic Planning Sciences*, **5,** 97–124.

Batty, M. (1972a) Recent developments in land use modelling: a review of British research. *Urban Studies*, **9,** 151–77.

Batty, M. (1972b) Dynamic simulation of an urban system. In A. G. Wilson (Ed.), *Patterns and Processes in Urban and Regional Systems*, Pion, London.

Batty, M. (1976) *Urban Modelling: Algorithms, Calibrations, Predictions.* Cambridge University Press.

Batty, M. (1977) *Operational Models Incorporating Dynamics in a Static Framework.* Department of Geography, University of Reading.

Batty, M., Bourke, R., Cormode, P., and Anderson-Micholls, M. (1974) The Area 8 Pilot Model: Experiments in urban modelling for sub-regional planning. P.T.R.C. Summer Annual Meeting, University of Warwick.

Batty, M., and March, L. (1976) Dynamic urban models based on information minimising. *Geographical Paper No. 48*, Department of Geography, University of Reading.

Beckmann, M. J. (1968) *Location Theory*. Random House, New York.

Beckmann, M. J. and Golob, T. (1971) On the metaphysical foundation of traffic theory: entropy revisited. In *Proceedings of the Fifth International Symposium on the Theory of Traffic Flow and Transportation*. Elsevier, New York.

Beckmann, M. J., McGuire, C. B., and Winsten, C. B. (1956) *Studies in the Economics of Transportation*. Cowles Commission Monograph, Yale University Press, U.S.A.

Bellman, R. (1968) *Vistas of Modern Mathematics*. University of Kentucky Press, Lexington.

Ben-Akiva, M. (1973) Structure of passenger travel demand models. PhD dissertation, Department of Civil Engineering, M.I.T, Cambridge, Mass.

Ben-Akiva, M. (1974) Structure of passenger travel demand models. *Transportation Research Record*, **526,** 26–42.

Ben-Akiva, M. (1977) Passenger travel demand forecasting: applications of disaggregate models and directions for further research. World Conference on Transport Research, Rotterdam.

Ben-Akiva, M., and Lerman, S. (1979) Disaggregate travel and mobility choice models and measures of accessibility. In D.A. Hensher and P. R. Stopher (Eds.), *Behavioural Travel Modelling*. Croom Helm, London.

Ben-Shahar, H., Mazor, A., and Pines, D. (1969) Town planning and welfare maximisation. *Regional Studies*, **3,** 105–13.

Ben-Shahar, H., Mazor, A., and Pines, D. (1970) Analytical methods in town planning. *Long Range Planning*, 42–9.

Blackburn, A. J. (1970) A non-linear model of demand for travel and an alternative approach to aggregation and estimation in the non-linear model. In R. E. Quandt (Ed.), *The Demand for Travel: Theory and Measurement*. Lexington Books, D.C. Heath, Lexington, Mass., pp. 163–96.

Blase, J. H. (1979) Hysteresis and catastrophe theory: empirical identification in transport modelling. *Environment and Planning A*, **11,** 675–88.

Bonsall, P. W., Champernowne, A. F., Cripps, E. L., Goodman, P. R., Hankin, A., Mackett, R. L. Sanderson, I., Senior, M. L., Southworth, F., Spence, R., Williams, H. C. W. L., and Wilson, A. G. (1973) A transport model for West Yorkshire. *Working Paper 46*, Institute for Transport Studies, University of Leeds.

Bonsall, P. W., Champernowne, A. F., Cripps, E. L., Goodman, P. R., Hankin, A, Mackett, R. L., Sanderson, I., Senior, M. L., Southworth, F., Spence, R., Williams, H. C. W. L., and Wilson, A. G. (1977) Models for urban transport planning. In A. G. Wilson, P. H. Rees and C. M. Leigh (Eds.), *Models of Cities and Regions*. John Wiley, Chichester.

Bonsall, P. W., Champernowne, A. F., Mason, A. C., and Wilson, A. G. (1978) *Transport Modelling: Sensitivity Analysis and Policy Testing*. Pergamon, Oxford.

Bouthelier, F., and Daganzo, C. F. (1979) Aggregation with multinomial probit and estimation of disaggregate models with aggregate data: a new methodological approach. *Transportation Research*, **13B,** 133–46.

Boyce, D. E. (1977) Equilibrium solutions to combined urban residential location, modal choice, and trip assignment models. *Discussion Paper 98*, Regional Science Research Institute, Philadelphia.

Brand, D. (1973) Travel demand forecasting, some foundations and a review. In *Urban Travel Demand Forecasting*, Special Report 143, Highway Research Board, Washington D.C.

Broadbent, T. A. (1973) Activity analysis of spatial allocation models. *Environment and Planning*, **5,** 673–91.

Broadbent, T. A. (1975) Comments. In A. Karlqvist, L. Lundqvist and F. Snickars (Eds.), *Dynamic Allocation of Urban Space*. Saxon House, Farnborough, England, p. 285.

Brotchie, J. F. (1969) A general planning model. *Management Science*, **16,** 265.

Burdekin, R. (1977) A simulation and control of urban development. PhD thesis, Department of Control Engineering, University of Sheffield.

Central Statistical Office (1973) Input–output tables for the U.K., 1968. *Studies in Official Statistics No. 22.* H.M.S.O., London.

Cesario, F. J. (1973) A generalised trip distribution model. *Journal of Regional Science*, **13,** 233–47.

Chadwick, G. F. (1971) *A Systems View of Planning*. Pergamon Press, London.

Champernowne, A. F., Williams, H. C. W. L., and Coelho, J. D. (1976) Some comments on urban travel demand analysis, model calibration and the economic evaluation of transport plans. *Journal of Transport Economics and Policy*, **10** (3), 267–85.

Charles River Associates (1972) A disaggregate behavioural model of urban travel demand. Federal Highway Administration, U.S. Department of Transportation, Washington D.C.

Charnes, A., Haynes, K. E., and Phillips, F. Y. (1976) A generalised distance estimation procedure for intra-urban interaction. *Geographical Analysis*, **8,** 289–94.

Chillingworth, D. R. J. (1975) Elementary catastrophe theory. *Bulletin of the Institute of Mathematics and its Applications*, **11,** 155–9.

Chillingworth, D. R. J. (1976) *Differential Topology with a View to Applications*. Pitman, London.

Christaller, W. (1933) *Die zentralen orte in Suddendeustschland*. trans. (1966) *The Central Places of Southern Germany*. Prentice-Hall, New Jersey.

Clark, C. (1961) The greatest of a finite set of random variables. *Operations Research*, **9,** 145–62.

Cochrane, R. A. (1975) A possible economic basis for the gravity model. *Journal of Transport Economics and Policy*, **9,** 34–49.

Coelho, J. D. (1977) The use of mathematical optimisation methods in model based land use planning: an application to the new town of Santo Andre. PhD thesis, School of Geography, University of Leeds.

Coelho, J. D., and Williams, H. C. W. L. (1978) On the design of land use plans through locational surplus maximisation. *Papers of the Regional Science Association*, **40,** 71–85.

Coelho, J. D., Williams, H. C. W. L., and Wilson, A. G. (1978) Entropy maximising submodels within overall mathematical programming frameworks: a correction. *Geographical Analysis*, **10** (2), 195–201.

Coelho, J. D., and Wilson, A. G. (1976) The optimum location and size of shopping centres. *Regional Studies*, **10,** 413–21.

Coelho, J. D., and Wilson, A. G. (1977) An equivalence theorem to integrate entropy maximising models within overall mathematical programming frameworks. *Geographical Analysis*, **9,** 160–73.

Cohen, J. L. and Marks, D. H. (1975) A review and evaluation of multiobjective programming techniques. *Water Resources Research*, **2,** 208–20.

Coventry Transportation Study (1973) *Report on Phase Two*: technical report parts I and II. Coventry City Council.

Crecine, J. P. (1968) *A Dynamic Model of Urban Structures*. Rand Corporation, Santa Monica.

Cripps, E. L., Macgill, S. M., and Wilson, A. G. (1974) Energy and materials flows in the urban space economy. *Transportation Research*, **8,** 293–305.

Daganzo, C. F., Bouthelier, F., and Sheffi, Y. (1977) Multinomial probit and qualitative choice: a computationally efficient algorithm. *Transportation Science*, **11,** 338–58.

Daly, A. J., and Zachary, S. (1978) Improved multiple choice models. In D. A. Hensher and M. Q. Dalvi (Eds.), *Determinants of Travel Choice*. Saxon House, Farnborough, Hants.

Dickey, J. W., and Hopkins, J. W. (1972) Campus building arrangement using TOPAZ. *Transportation Research*, **6,** 59–68.

Dickey, J. W., and Najafi, F. T. (1973) Regional land use schemes generated by TOPAZ. *Regional Studies*, **7,** 373–86.

Dinkel, J., Kochenberger, G. A., and Wong, S. N. (1977) Entropy maximising and geometric programming. *Environment and Planning A*, **9,** 419–28.

Domencich, T., and McFadden, D. (1975) *Urban Travel Demand—A Behavioural Analysis*. North Holland, Amsterdam.

Erlander, S. (1977) Entropy in linear programming—an approach to planning. Linkoping Institute of Technology, Report Lith–Mat–R–77–3. Linkoping, Sweden.

Evans, A. W. (1971) The calibration of trip distribution models with exponential or similar cost functions. *Transportation Research*, **5,** 15–38.

Evans, S. (1973a) The use of optimisation in transportation planning. PhD thesis, University College London.

Evans, S. (1973b) A relationship between the gravity model for trip distribution and the transportation problem in linear programming. *Transportation Research*, **7,** 39–61.

Evans, S. (1976) Derivation and analysis of some models for combining trip distribution and assignment. *Transportation Research*, **10,** 37–57.

Farnsworth, D., Houghton, A. G., Pilgrim, B., and Carter, F. (1969) Systems design project: Macclesfield and district design procedure: progress report. Cheshire County Planning Department, Chester.

Faure, P., and Huard, P. (1965) Resolution des programmes mathematiques a fonction non-lineaire par le methode du gradient reduit. *Revue Française de Recherche Operationnelle*, **9,** 167–205.

Fiacco, A. V., and McCormick, G. P. (1968) *Non-linear Programming: Sequential Unconstrained Minimisation Techniques*. Wiley, New York.

Fletcher, R., and Reeves, C. M. (1964) Function minimisation by conjugate gradients. *Computer Journal*, **7,** 149.

Florian, M., and Nguyen, S. (1976) Recent experience with equilibrium methods for the study of congested urban areas. In M. Florian (Ed.), *Traffic Equilibrium Models*. Springer-Verlag, Heidelberg.

Forrester, J. W. (1969) *Urban Dynamics*. M.I.T. Press, Cambridge, Mass.

Frank, M. and Wolfe, P. (1956) An algorithm for quadratic programming. *Naval Research Logistics Quarterly*, **3,** 269–71.

Friedlander, D. (1961) A technique for estimating a contingency table given marginal totals and some supplementary data. *Journal of the Royal Statistical Society*, Series A, **124,** 412–20.

Garin, R. A. (1966) A matrix formulation of the Lowry Model for intra-metropolitan activity location. *Journal of the American Institute of Planners*, **32,** 361–64.

Ghosh, A. (1958) Input–output approach to an allocative issue. *Economica*, **25** (97), 58–64.

Giarratani, F. (1976) Application of an inter-industry supply model to energy issues. *Environment and Planning A*, **8,** 447–54.

Gigantes, T. (1970) The representation of technology in input–output systems. In A. Brody and A. P. Carter (Eds.), *Contributions to Input–output Analysis*. North Holland Publishing Company.

Gillespie, R. P. (1960) *Partial Differentiation*. Oliver and Boyd, London.

Goldner, W. (1971) The Lowry model heritage. *Journal of the American Institute of Planners*, **37,** 100–10.

Golob, T. F., and Beckmann, M. J. (1971) A utility model for travel forecasting. *Transportation Science*, **5** (1), 79–90.

Goodwin, P. B. (1977) Habit and hysteresis in model choice. *Urban Studies*, **14,** 95–8.

Gustaffson, J., Harsman, B., and Snickars, F. (1977) Housing models and consumer preferences: applications for the Stockholm region. *Papers of the Regional Science Association*, **38,** 125–247.

Hadley, G. (1964) *Non-linear and Dynamic Programming*. Addison-Wesley Publishing Co., Reading, Massachusetts.

Hansen, W. G. (1959) How accessibility shapes land use. *Journal of the American Institute of Planners*, **25,** 76–7.

Harris, A. J., and Tanner, J. C. (1974) Transport demand models based on personal characteristics. Transport and Road Research Laboratory: Supplementary Report 65 UC.

Harris, B. (1961) Some problems in the theory of intra-urban location. *Operations Research*, **9,** 695–721.

Harris, B. (1962) Linear programming and the projection of land uses. Paper No. 20, Penn-Jersey Transportation Study, Philadelphia.

Harris, B. (1964) A note on probability of interaction at a distance. *Journal of Regional Science*, **5,** 31–5.

Harris, B. (1965) A model of locational equilibrium for the retail trade. Institute for Urban Studies, University of Pennsylvania.

Harris, B., and Wilson, A. G. (1978) Equilibrium values and dynamics of attractiveness terms in production-constrained spatial-interaction models. *Environment and Planning A*, **10,** 371–88.

Hausman, J. A., and Wise, D. A. (1978) A conditional probit model for qualitative choice: discrete decisions recognising interdependence and heterogeneous preferences. *Econometrica*, **46** (2), 403–26.

Havers, G., Van Vliet, D. and others (1974) GLTS models: the state of the art. Greater London Transportation Study, Note number 71, Greater London Council, London.

Hawkins, A. F. (1978) Some anomalies due to misspecification in transport models. P.T.R.C. Proceedings: summer annual meeting, University of Warwick.

Herbert, D. J., and Stevens, B. H. (1960) A model for the distribution of residential activity in urban areas. *Journal of Regional Science*, **2,** 21–36.

Hewings, G. J. D. (1971) Regional input–output models in the U.K. Some problems and prospects for the use of non-survey techniques. *Regional Studies*, **5,** 11–22.

Hirsch, M. W., and Smale, S. (1974) *Differential Equations, Dynamical Systems and Linear Algebra*. Academic Press, New York.

Hitchcock, F. L. (1941) Distribution of a product from several sources to numerous localities. *Journal of Mathematical Physics*, **20,** 224–30.

Hotelling, H. (1938) The general welfare in relation to taxation of railway and utility rates. *Econometrica*, **6,** 242–69.

Huff, D. L. (1964) Defining and estimating a trading area. *Journal of Marketing*, **28,** 4–38.

Hyman, G. M. (1969) The calibration of trip distribution models. *Environment and Planning*, **1,** 105–12.

Intriligator, M. D. (1971) *Mathematical Optimisation and Economic Theory*. Prentice-Hall, Englewood Cliffs, New Jersey.

Jenkins, P. M., and Robson, A. (1974) An application of linear programming methodology for regional strategy making. *Regional Studies*, **8,** 267–79.

Koenig, J. G. (1975) A theory of urban accessibility: a new working tool for the urban planner. In *Urban Traffic Models*, P.T.R.C. Summer Meeting, University of Warwick, Coventry.

Koopmans, T. C., and Beckmann, M. J. (1957) Assignment problems and the location of economic activities. *Econometrica*, **25,** 53–76.

Koppleman, F. (1976) Guidelines for aggregate travel predictions using disaggregate choice models. *Transportation Research Record*, **610,** 19–24.

Kraft, G. (1974) Alternative travel behaviour structures: comments. *Transportation Research Record*, **526,** 52–5.

Kuhn, H. W. and Tucker, A. E. (1951) Non-linear programming. Proceedings, 2nd Berkeley Symposium on Mathematical Statistics and Probability, University of California Press, Berkeley, 481–92.

Kullback, S. (1959) *Information Theory and Statistics*. John Wiley, New York.

Lakshmanan, T. R. and Hansen, W. G. (1965) A retail market potential model. *Journal of the American Institute of Planners*, **31,** 134–43.

Langdon, M. (1976) Modal split models for more than two modes. In *Urban Traffic Models*, P.T.R.C. Summer Annual Meeting, University of Warwick.

Lawson, G. P. and Mullen, P. (1976) The use of disaggregate modelling techniques in the Telford transportation study. P.T.R.C. Proceedings, Summer meeting, University of Warwick.

Lee, C. (1973) *Models in Planning: an Introduction to the Use of Quantitative Models in Planning*. Pergamon Press, London and New York.

Leonardi, G. (1978) Optimal facility location by accessibility maximisation. *Environment and Planning A*, **10,** 1287–305.

Leontief, W. (1967) *Input–output Analysis*. Oxford University Press.

Leontief, W. (1970) The dynamic inverse. In A. P. Carter and A. Brody (Eds.), *Contributions to Input–output Analysis*. North Holland Publishing Company, Amsterdam.

Lerman, S. R. (1975) A disaggregate behavioural model of urban mobility decisions. PhD thesis, Department of Civil Engineering, M.I.T., Massachusetts.

Leung, A. (1976) Limiting behaviour for several interacting populations. *Mathematical Biosciences*, **29,** 85–98.

Los, M. (1979) Discrete choice modelling and disequilibrium in land use and transportation planning. Publication 137, Centre de Recherche sur les Transports, University of Montreal.

Lowry, I. S. (1964) *A Model of Metropolis*. R17-4035-RC, The Rand Corporation, Santa Monica.

Luce, R. D. (1959) *Individual Choice Behaviour*. John Wiley, New York.

Lundqvist, L. (1975) Transportation analysis and activity location in land use planning—with applications to the Stockholm region. Chapter 10 in A. Karlqvist, L. Lundqvist and F. Snickars (Eds.), *Dynamic Allocation of Urban Space*. Saxon House, Farnborough, England.

McCormick, G. P. (1970a) The variable reduction method for non-linear programming. *Management Science*, **17,** 147–60.

McCormick, G. P. (1970b) A second order method for the linear constrained non-linear programming problem. In J. Rosen, O. Mangasarian, and K. Ritter (Eds.), *Non-linear Programming*. Academic Press, New York.

McFadden, D. (1973) Conditional logit analysis of qualitative choice behaviour. In P. Zarembka (Ed.), *Frontiers in Econometrics*, Academic Press, New York.

McFadden, D. (1976) The mathematical theory of demand models. In P. R. Stopher and A. Meyburg (Eds.), *Behavioural Travel Demand Models*. Lexington Books, Lexington, Massachusetts.

McFadden, D. (1978) Modelling the choice of residential location. In A. Karlqvist, L. Lundqvist, F. Snickars, and J. W. Weibull (Eds.), *Spatial Interaction Theory and Planning Models*. North Holland, Amsterdam.

McFadden, D., and Reid, F. (1975) Aggregate travel demand forecasting from disaggregate demand models. *Transportation Research Board Record*, **534,** 24–37.

Macgill, S. M. (1975) Balancing factor methods in urban and regional analysis. *Working Paper 124*, School of Geography, University of Leeds.

Macgill, S. M. (1977a) The use of models in the study of energy and commodity flows in cities. PhD thesis, School of Geography, University of Leeds.

Macgill, S. M. (1977b) Simple hybrid input–output models: a graphical approach. *Environment and Planning A*, **9,** 1033–42.
Macgill, S. M. (1977c) The Lowry model as an input–output model and its extension to incorporate full inter-sectoral relations. *Regional Studies*, **11,** 337–54.
Macgill, S. M. (1978a) Alternative rectangular input–output models. *Working Paper 182*, School of Geography, University of Leeds.
Macgill, S. M. (1978b) Multiple classification in socio-economic systems and urban modelling. *Working Paper 212*, School of Geography, University of Leeds.
Macgill, S. M. (1979) Convergence and related properties of a modified biproportional matrix problem. *Environment and Planning A*, **11,** 499–506.
Mackett, R. L. (1977) A dynamic activity allocation transportation model. In P. Bonsall, M. Q. Dalvi and P. J. Hills (Eds.) *Urban Transportation Planning: Current Themes and Future Prospects*. Abacus Press, Tunbridge Wells.
Mackett, R. L., and Mountcastle, G. D. (1977) Developments of the Lowry model. In A. G. Wilson, P. H. Rees and C. M. Leigh (Eds.), *Models of Cities and Regions*. John Wiley, Chichester.
Manheim, M. L. (1973) Practical implications of some fundamental properties of travel demand models. *Highway Research Record*, **422,** 21–38.
Manski, C. F., and Lerman, S. R. (1978) On the use of simulated frequencies to approximate choice probabilities. Cambridge Systematics Inc.
Mathur, P. N. (1972) Multi-regional analysis in a dynamic input–output framework. In A. Brody and A. P. Carter (Eds.), *Input–output Techniques*. North Holland Publishing Company, Amsterdam.
May, R. M. (1971) Stability in multi-species community models. *Mathematical Biosciences*, **12,** 59–79.
May, R. M., and Leonardi, W. J. (1975) Non-linear aspects of competition between three species. *S.I.A.M. Journal of Applied Mathematics*, **29,** 243–53.
Mayberry, J. P. (1970) Structural requirements for abstract-mode models of passenger transportation. In R. E. Quandt (Ed.), *The Demand for Travel: Theory and Measurement*. Heath Lexington Books, Lexington, Mass., pp. 103–25.
Maynard-Smith, J. (1974) *Models in Biology*. Cambridge University Press.
Mazor, A., and Pines, D. (1973) The use of mathematical models in town planning. Papers and Proceedings of the International Congress I.F.H.P., Copenhagen, Denmark.
Morrison, W. I., and Smith, P. (1977) Input–output methods in urban and regional planning: a practical guide. *Progress in Planning*, **7**(2), Pergamon Press.
Moses, L. N. (1955) Stability of interregional trading patterns and input–output analysis. *American Economic Review*, **45,** 803–32.
Murchland, J. D. (1966) Some remarks on the gravity model of trip distribution and an equivalent maximising procedure. Mimeo, LSE-TNT-38, London School of Economics, London.
Murchland, J. D. (1969) Road traffic distribution in equilibrium. *Mathematical methods in the economic sciences*. Mathematisches Forehungsinstitut Oberwolfach, West Germany.
Neuberger, H. L. I. (1971) User benefit in the evaluation of transport and land use plans. *Journal of Transport Economics and Policy*, **5,** 52–75.
Nguyen, S. (1974) A unified approach to equilibrium methods for traffic assignment. In M. A. Florian (Ed.), *Traffic Equilibrium Methods*. Springer-Verlag, New York, pp. 148–82.
Nicolis, G., and Prigogine, I. (1977) *Self-organisation in Non-equilibrium Systems*. John Wiley, New York.
Niedercorn, J. H., and Bechdolt, B. V. (1969) An economic derivation of the gravity law of spatial interaction. *Journal of Regional Science*, **9,** 273–82.
Nijkamp, P. (1972) *Planning of Industrial Complexes by means of Geometric Programming*. Rotterdam University Press, Rotterdam.

Nijkamp, P. (1977) *Theory and Application of Environmental Economics*. North Holland, Amsterdam.
Nijkamp, P. (1979) Gravity and entropy models: the state of the art. In G. R. M. Jansen, P. H. L. Bory, J. P. J. M. Van Est, and F. LeClercq (Eds.), *New Developments in Modelling Travel Demand and Urban Systems*. Saxon House, Farnborough, Hants, pp. 281–319.
Nijkamp, P., and Paelinck, J. H. (1974) A dual interpretation and generalisation of entropy maximising models in regional science. *Papers of the Regional Science Association*, **33,** 13–32.
Ortuzar, J. D. (1978) Mixed mode demand forecasting techniques: an assessment of current practice. P.T.R.C. Proceedings, Summar Annual Meeting, University of Warwick.
Parry-Lewis, J. (1969) Misused techniques in planning: 1—linear programming. Centre for Urban and Regional Research, University of Manchester.
Pilgrim, B., and Carter, R. (1970) Systems design project: a computer program to evaluate urban land use plans—general description. Local Government Operational Research Unit, Reading.
Poston, T.; and Stewart, I. N. (1978) *Catastrophe Theory and its Applications*. Pitman, London.
Potts, R. B., and Oliver, R. M. (1972) *Flows in Transportation Networks*. Academic Press, New York.
Powell, M. J. D. (1964) An efficient method for finding the minimum of a function of several variables without calculating derivatives. *Computer Journal,* **7,** 155–62.
Putman, S. H. (1974) Preliminary results from an integrated transportation and land use models package. *Transportation*, **3,** 193–224.
Quandt, R. E. (1968) Estimation of modal splits. *Transportation Research*, **2,** 41–50.
Quarmby, D. A., and Neuberger, H. L. I. (1969) Transport aspects of land use plan evaluation: a note on methodology. Department of the Environment, M.A.U. Note 140.
Quigley, J. M. (1976) Housing demand in the short run: an analysis of polytomous choice. In S. Winter (Ed.), *Explorations in Economic Research*, **3**(1).
Rassam, P. R., Ellis, R. H., and Bennett, J. C. (1971) The *n*-dimensional logit model: development and application. *Highway Research Record*, **369,** 135–47.
Rescigno, A. (1968) The struggle for life II: three competitors. *Bulletin of Mathematical Biophysics*, **30,** 291–8.
Restle, F. (1961) *Psychology of Judgement and Choice*. John Wiley, Chichester.
Richards, M. G. (1975) An application of disaggregate modelling techniques to transportation planning studies. In *Urban Traffic Models*, P.T.R.C. Summer Meeting, University of Warwick, 246–62.
Ripper, M., and Varaya, P. (1974) An optimising model of urban development. *Environment and Planning A*, **6,** 149–68.
Roberts, F. S. (1976) *Discrete Mathematical Models*. Prentice-Hall, Englewood Cliffs, New Jersey.
Roberts, F. S., and Brown, T. A. (1975) Signed digraphs and the energy crisis. *American Mathematical Monthly*, **82,** 577–94.
Robertson, D. I., and Kennedy, J. V. (1979) The choice of route, mode, origin and destination by calculation and simulation. Department of the Environment, Department of Transport, TRRL Report 877, Crowthorne.
Rockafellar, R. T. (1967) Convex programming and systems of elementary monotonic relations. *Journal of Mathematical Analysis Applications*, **19**(3), 543–64.
Rockafellar, R. T. (1969) *Convex Analysis*. Princeton University Press, USA.
Rogers, A. (1968) *Mdtrix Analysis of Interregional Population Growth and Movement*. University of California Press, Berkeley.
Romanoff, E. (1974) The economic base model: a very special case of input–output analysis. *Journal of Regional Science*, **14,** 121–9.

Rosen, J. B. (1960) The gradient projection method for non-linear programming. *J. Soc. In. Appl. Math.*, **8**(1), 181–217.

Rosenbluth, G. (1968) Input–output analysis: a critique. *Statistiche Heft*, **9**(4), 260.

Ruiter, E. (1973) Analytic structure. In *Urban travel demand forecasting*. Special Report 143, Highway Research Board, Washington D.C.

Samuelson, P. A. (1952) Spatial price equilibrium and linear programming. *American Economic Review*, **42,** 283–303.

Schinnar, A. P. (1976) A multidimensional accounting model for demographic and economic planning interactions. *Environment and Planning A*, **8,** 455–75.

Schinnar, A. P. (1978) A method for computing Leontief multipliers from rectangular input–output accounts. *Environment and Planning A*, **10,** 137–44.

Schlager, K. J. (1965) A land use plan design model. *Journal of the American Institute of Planners*, **31,** 103–11.

Scott, A. J. (1969) Combinatorial programming and the planning of urban and regional systems. *Environment and Planning*, **1,** 125–42.

Scott, A. J. (1971) *Combinatorial Programming, Spatial Analysis and Planning*. Methuen, London.

Senior, M. L. (1977a) Residential location. In A. G. Wilson, P. H. Rees, and C. M. Leigh (Eds.), *Models of Cities and Regions*. John Wiley, Chichester.

Senior, M. L. (1977b) Problems in the integration of land use and transportation models. In P. W. Bonsall, M. Q. Dalvi, and P. J. Hills (Eds.), *Urban Transportation Planning: Current Themes and Future Prospects*. Abacus Press, Tunbridge Wells.

Senior, M. L. (1978) A spatial interaction approach to modelling the housing market: the development and empirical testing of a disaggregated residential location model. PhD thesis, School of Geography, University of Leeds.

Senior, M. L., and Williams, H. C. W. L. (1977) Model based transport policy assessment 1: the use of alternative forecasting models. *Traffic Engineering and Control*, **18,** 402–6.

Senior, M. L., and Wilson, A. G. (1974a) Disaggregated residential location models: some tests and further theoretical developments. In E. L. Cripps (Ed.), *Space-time Concepts in Urban and Regional Science*. Pion, London.

Senior, M. L., and Wilson, A. G. (1974b) Exploration and syntheses of linear programming and spatial interaction models of residential location. *Geographical Analysis*, **7,** 209–38.

Sharpe, R., Brotchie, J. F., and Ahern, P. A. (1975) Evaluation of alternative growth patterns for Melbourne. In A. Karlqvist, L. Lundqvist, F. Snickars, and J. Weibull (Eds.), *Dynamic Allocation of Urban Space*. Saxon House.

Sheppard, E. S. (1974) A conceptual framework for dynamic location-allocation analysis. *Environment and Planning A*, **6,** 547–64.

Smale, S. (1976) Department of Mathematics, University of California, Berkeley: preprint, quoted in R. M. May and W. J. Leonardi (1975), *op cit*.

Smith, T. E. (1975) An axiomatic theory of spatial discounting behaviour. *Regional Science Association Papers*, **35,** 31–44.

Smith, T. E. (1976) A spatial discounting theory of interaction preferences. *Environment and Planning A*, **8,** 879–915.

Smith, T. E. (1977) A cost efficiency principle of spatial interaction. Paper presented at the 17th European Congress, Regional Science Association, Krakow.

Snickars, F., and Weibull, J. (1977) A minimum information principle—theory and practice. *Regional Science and Urban Economics*, **7,** 137–68.

Southworth, F. (1977) Development of trip distribution and modal choice models. PhD thesis, School of Geography, University of Leeds.

Speer, B. D. (1977) Application of new travel demand forecasting techniques to transportation planning: a study of individual choice models. U.S. Department of Transport, F.H.W.A., Office of Highway Planning.

Steenbrink, P. A. (1974) *Optimisation of Transport Networks*. John Wiley, Chichester.

Stone, R. (1970) *Mathematical Models of the Economy*. Chapman and Hall, London.
Stone, R., Bates, J., and Bacharach, M. (1963) Input–output relationships, 1954–1966. University of Cambridge, Department of Applied Economics. *A Programme for Growth*, **3,** Chapman and Hall, London.
Stopher, P. R. (1975) Comfort and convenience in travel demand models: some recent advances. In *Urban Traffic Models*, P.T.R.C. Summer Meeting, University of Warwick, 263–72.
Strobeck, C. (1973) *N*-species competition. *Ecology*, **54,** 650–4.
Strotz, R. H. (1957) The empirical implications of a utility tree. *Econometrica*, **25,** 269–80.
Takayama, T., and Judge, G. (1972) *Spatial and Temporal Price and Allocation Models*. North Holland Publishing Company, Amsterdam.
Thom, R. (1975) *Structural Stability and Morphogenesis*. Benjamin, Reading, Mass.
Thrift, N. J. (1977) Private Communication.
Thurstone, L. (1927) A law of comparative judgement. *Psychological Review*, **34,** 273–86.
Turner, D. G. (1974) The design of urban models for strategic land use transportation studies. Presented at the session on Urban and Regional Models, P.T.R.C. Summer Annual Meeting.
Tyne and Wear Plan (1973) *Urban Strategy*. Report prepared for the Tyne and Wear Plan. A land use/transportation study for Tyneside and Wearside.
United Nations (1968) A system of national accounts. *Studies and Methods Series F*, No. 2. Rev. 3, U.N. Publications 69, XVII, 3, United Nations, New York.
Van Vliet, D. (1976) The choice of assignment techniques for large networks. In M. A. Florian (Ed.), *Traffic Equilibrium Methods*. Springer-Verlag, New York, pp. 396–425.
Victor, P. A. (1972) *Pollution, Economy and Environment*. George Allen and Unwin.
Wardrop, J. G. (1952) Some theoretical aspects of road traffic research. *Proceedings of the Institute of Civil Engineering*, **39,** 433–64.
White, R. W. (1977) Dynamic central place theory: results of a simulation approach. *Geographical Analysis*, **9,** 226–43.
Wigan, M. R. (1977) Demand–supply equilibrium analysis: theory and implementations; Equilibrium assignment. Chapters 17 and 18 in M. R. Wigan (Ed.), *New Techniques for Transport Systems Analysis*. Special Report, No. 10. Australian Road Research Board and Bureau of Transport Economics.
Williams, H. C. W. L. (1976) Travel demand models, duality relations and user benefit measures. *Journal of Regional Science*, **16**(2), 147–66.
Williams, H. C. W. L. (1977) On the formation of travel demand models and economic evaluation measures of user benefit. *Environment and Planning A*, **9,** 285–344.
Williams, H. C. W. L., and Ortuzar, J. D. (1979) Behavioural travel theories, model specification and the response error problem. Presented at the P.T.R.C. Annual Summer Meeting, University of Warwick.
Williams, H. C. W. L., and Senior, M. L. (1977) Model based transport policy assessment 2: Removing fundamental inconsistencies from the models. *Traffic Engineering and Control*, **18,** 464–9.
Williams, H. C. W. L., and Senior, M. L. (1978) Accessibility, spatial interaction and the spatial benefit analysis of land use-transportation plans. In A. Karlqvist, L. Lundqvist, F. Snickars, and J. W. Weibull (Eds.), *Spatial Interaction Theory and Planning Models*. North Holland, Amsterdam.
Williams, I. N. (1979) An approach to solving spatial allocation models with constraints. *Environment and Planning A*, **11,** 3–22.
Wilson, A. G. (1967) A statistical theory of spatial distribution models. *Transportation Research*, **1,** 253–69.
Wilson, A. G. (1969) Entropy maximising models in the theory of trip distribution, mode split and route split. *Journal of Transport Economics and Policy*, **3,** 108–26.
Wilson, A. G. (1970) *Entropy in Urban and Regional Modelling*. Pion, London.

Wilson, A. G. (1971a) A family of spatial interaction models and associated developments. *Environment and Planning*, **3,** 1–32.
Wilson, A. G. (1971b) Generalising the Lowry model. In A. G. Wilson (Ed.), *Urban and Regional Planning*, Pion, London.
Wilson, A. G. (1974) *Urban and Regional Models in Geography and Planning*. John Wiley, Chichester.
Wilson, A. G. (1976a) Retailers' profits and consumers' welfare in a spatial interaction shopping model. In I. Masser (Ed.), *Theory and Practice in Regional Science*. Pion, London, 42–59.
Wilson, A. G. (1976b) Catastrophe theory and urban modelling: an application to modal choice. *Environment and Planning A*, **8,** 351–6.
Wilson, A. G. (1977) Recent developments in urban and regional modelling: towards an articulation of systems theoretical foundations. *Proceedings*, Volume 1. Giornate di Lavoro, A.I.R.O., Parma, Italy.
Wilson, A. G. (1978a) Spatial interaction and settlement structure: towards an explicit central place theory. In A. Kalqvist, L. Lundqvist, F. Snickars, and J. W. Weibull (Eds.), *Spatial Interaction Theory and Planning Models*. North Holland, Amsterdam, pp. 137–56.
Wilson, A. G. (1978b) Towards models of the evolution and genesis of urban structure. In R. L. Martin, N. J. Thrift, and R. J. Bennett (Eds.), *Towards the Dynamic Analysis of Spatial Systems*. Pion, London, pp. 79–90.
Wilson, A. G. (1979) Equilibrium and transport system dynamics. In D. Hensher and P. R. Stopher (Eds.), *Behavioural Travel Modelling*. Croom Helm, London.
Wilson, A. G. (1981) *Catastrophe Theory and Bifurcation: Applications to Urban and Regional Systems*. Croom Helm, London.
Wilson, A. G., and Clarke, M. (1978) Some illustrations of catastrophe theory applied to urban retailing structures. In M. Breheny (Ed.), *Developments in Urban and Regional Analysis*. Pion, London.
Wilson, A. G., and Kirkby, M. J. (1980) *Mathematics for Geographers and Planners* (2nd edition). Oxford University Press.
Wilson, A. G., and Kirwan, R. M. (1969) Measures of benefit in the evaluation of urban transport improvements, *Working Paper 43*. Centre for Environmental Studies, London.
Wilson, A. G., and Pownall, C. A. (1976) A new representation of the urban system for modelling and for the study of micro-level interdependence. *Area*, **5,** 246–54.
Wilson, A. G., Rees, P. H., and Leigh, C. M. (Eds.) (1977) *Models of Cities and Regions*. John Wiley, Chichester and New York.
Wilson, A. G., and Senior, M. L. (1974) Some relationships between entropy maximising models, mathematical programming models and their duals. *Journal of Regional Science*, **14,** 207–15.
Wolfe, P. (1961) A duality theorem for non-linear programming. *Quarterly of Applied Mathematics*, **19,** 239–44.
Wolfe, P. (1965) *The Reduced Gradient Method*. The Rand Corporation, Santa Monica.
Young, W. (1972) Planning—a linear programming model. *G.L.C. Intelligence Unit Quarterly Bulletin*, No. 19, 5–15.
Zangwill, W. I. (1967) The convex simplex method. *Management Science A*, **14**(3), 221–38.
Zeeman, E. C. (1977) *Catastrophe Theory*. Addison-Wesley, Reading, Mass.
Zoutendijk, G. (1960) *Methods of Feasible Directions*. Elsevier, New York.

Author index

Ahern, P. A., 178
Albright, R. L., 74
Allen, P., 227 *et seq.*
Alonso, W., 25
Amson, J. C., 196
Anas, A., 82
Armstrong, A. G., 111, 114, 115, 116
Artle, R., 143, 236
Augustinovics, M., 141

Bacharach, M., 114, 122, 125, 236, 240
Balinski, M. L., 20
Banister, C. E., 155
Barras, R., 235
Batty, M., 85, 88, 89, 92, 110, 155, 238, 239
Baumol, W., 20
Bechdolt, B. V., 239
Beckmann, M. J., 33, 34, 48, 57, 178, 188, 239
Bellman, R., 214, 237, 240
Ben-Akiva, M., 46, 60, 61, 62, 63, 70, 81, 84
Ben-Shahar, H., 177, 178, 191, 193
Blackburn, A. J., 48
Blase, J. H., 232
Bonsall, P. W., 4, 61, 238
Bouthelier, F., 74, 76
Boyce, D. E., 240
Brand, D., 84
Broadbent, T. A., 134, 178, 235
Brotchie, J. F., 158, 169, 178
Brown, T. A., 240
Burdekin, R., 254

Central Statistical Office, 114
Cesario, F. J., 239
Chadwick, G. F., 179
Champernowne, A. F., 4, 41, 82, 164
Charles River Associates, 46, 48, 49, 61, 63
Charnes, A., 31
Chillingworth, D. J., 196
Christaller, W., 173
Clarke, C., 74
Clarke, M., 214, 223
Cochrane, R. A., 46, 48, 49, 50, 82
Coelho, J. D., 107, 108, 159, 164, 169, 214, 238, 239, 245, 253
Cohen, J. L., 193
Coventry Transportation Study, 61, 191
Crecine, J. P., 92
Cripps, E. L., 117

Daganzo, C. F., 74, 76
Daly, A. J., 78
Dempster, M. A. H., 41, 43
Dickey, J. W., 159, 169
Dinkel, J., 31
Domencich, T., 46, 48, 67, 69

Erlander, S., 239
Evans, A., 109
Evans, S. P., 2, 28, 29, 30, 33, 34, 35, 36, 159, 165, 191, 240

Farnsworth, D., 178
Faure, P., 43, 101, 192
Fiacco, A. V., 42
Fletcher, R., 41
Florian, M., 240
Forrester, J. W., 4, 238, 254
Frank, M., 43
Friedlander, D., 122

Garin, R. A., 85, 148, 238, 240, 248, 249
Ghosh, A., 141
Giarratani, F., 141
Gigantes, T., 111, 113, 114, 116, 131
Gillespie, R. P., 15
Goldner, W., 85, 88
Golob, T. F., 48, 57, 239
Goodwin, P. B., 232
Gustaffson, J., 82

Hadley, G., 43, 160, 162

Hansen, W. G., 80, 169, 189, 212, 236, 243
Hansman, J. A., 74
Harris, B., 48, 82, 177, 238, 239, 241, 244, 246
Havers, G., 61, 77
Hawkins, A. F., 80
Herbert, D. J., 25, 82, 236
Hewings, G. J. D., 125
Hirsch, M. W., 209, 220
Hitchcock, F. L., 23
Hopkins, J. W., 159
Hotelling, H., 54, 86
Huard, P., 43, 101, 192
Huff, D. L., 169, 212, 243
Hyman, G. M., 109

Intriligator, M. D., 237

Jenkins, P. M., 178
Judge, G., 112, 236, 237, 238

Kennedy, J. V., 74
Kirwan, R. M., 55
Koenig, J. G., 80
Koopmanns, T. C., 178
Koppleman, F., 76
Kraft, G., 63
Kullback, S., 28

Lakshmanan, T. R., 169, 212, 236, 243
Langdon, M., 74
Lawson, G. P., 78
Lee, C., 92
Leigh, C. M., 1
Leonardi, G., 214, 238, 245, 253, 257, 260, 262
Leontief, W., 236, 237, 249, 256
Lerman, S. R., 74, 81, 82
Leung, A., 262
Los, M., 82, 83
Lowry, I. S., 1, 85, 238, 240, 246
Luce, R. D., 58
Lundqvist, L., 178

Macgill, S. M., 117, 129, 132, 237, 238, 240, 248, 249
McCormick, G. P., 42, 101, 192
McFadden, D., 46, 48, 50, 67, 69, 72, 73, 76, 82, 84
McGuire, C. B., 33
Mackett, R. L., 92, 238
Manheim, M. L., 61, 63, 239
Mansky, C. F., 74
March, L., 156, 239
Marks, D. H., 193
Mason, A., 4
Mathur, P. N., 133
May, R. M., 262
Mayberry, J. P., 58
Maynard Smith, J., 208 *et seq.*
Mazor, A., 158, 175
Morrison, W. I., 125
Moses, L. N., 133
Mountcastle, G. D., 238
Mullen, P., 78
Murchland, J. D., 27, 33, 35

Najafi, F. T., 159, 169
Neuberger, H. L. I., 55, 57, 85, 180
Nguyen, S., 35, 240
Nicolis, G., 227
Niedercorn, J. H., 239
Nijkamp, P., 31, 32, 193

Oliver, R. M., 236
Ortuzar, J. D., 74, 84

Paelinck, J. H., 31
Parry-Lewis, J., 178
Pines, D., 158, 175
Poston, T., 196, 203, 233
Potts, R. B., 236
Powell, M. J. D., 39
Pownall, C., 236
Prigogine, I., 227
Putman, S. H., 85

Quandt, R. E., 48
Quarmby, D. A., 180
Quigley, J. M., 82

Rassam, P. R., 78
Rees, P. H., 1
Reeves, C. M., 41
Reid, F., 76
Rescigno, A., 262
Restle, F., 58
Richards, M. G., 78
Ripper, M., 178
Roberts, F. S., 236, 240
Robertson, D. I., 74
Robson, A., 178
Rockafellar, R. T., 19, 30, 192
Rogers, A., 236
Romanoff, E., 134, 142
Rosen, J. B., 43, 101, 164, 192
Rosenbluth, G., 111

Ruiter, E., 61

Samuelson, P. A., 188
Schinnar, A. P., 111, 115, 116, 238
Schlager, K. J., 177, 178, 191, 193
Scott, A. J., 174
Senior, M. L., 2, 26, 61, 80, 81, 82, 83, 84, 159, 164, 171, 240
Sharpe, R., 178
Sheffi, Y., 74
Sheppard, E. S., 174
Smale, S., 209, 220, 262
Smith, P., 125
Smith, T. E., 239
Snickars, F., 28
Southworth, F., 237
Speer, B. D., 78
Steenbrink, P. A., 232
Stevens, B. H., 25, 82, 236
Stewart, I. N., 196, 203, 233
Stone, R., 111, 238
Stopher, P. R., 78
Strobeck, C., 262
Strotz, R. H., 65

Takayama, T., 112, 236, 237, 238
Tanner, J. C., 48
Thom R., 196 *et seq.*
Thrift, N. J., 235
Thurstone, L., 46
Traffic Research Corporation, 60
Turner, D. J., 191
Tyne and Wear Plan, 190, 191

United Nations, 113, 114

Van Vliet, D., 33
Varaya, P., 178
Victor, P. A., 111

Wardrop, J. G., 33
Weibull, J., 28
White, R. W., 224
Wigan, M. R., 35, 36
Williams, H. C. W. L., 26, 46, 48, 51, 61, 66, 68, 70, 72, 74, 78, 80, 81, 82, 83, 84, 95, 107, 108, 159, 164, 180, 191, 238, 239, 245
Williams, I., 134, 155, 156, 157, 238, 257
Wilson, A. G., 1, 2, 4, 27, 30, 55, 56, 83, 85, 87, 89, 92, 117, 129, 135, 136, 155, 159, 164, 169 *et seq.*, 178, 214, 222 *et seq.*, 232, 233, 236, 238, 239, 244, 245, 246, 247
Winsten, C. B., 33
Wise, D. A., 74
Wolfe, P., 20, 43, 101, 105, 192

Young, W., 178

Zachary, S., 78
Zangwill, W. I., 44, 101, 192
Zeeman, E. C., 196, 200, 210
Zoutendijk, G., 43

Subject index

Absorption matrix, 113, 114, 117, 120, 121
Absorption rate coefficient, 124
Absorption rate equations, 124
Accessibility, 80, 85, 86, 93, 97, 180, 185, 187, 189 *et seq.*
  maximization of, 214
Account-based model, 238
Accounting, 234
  constraints, 36, 101, 181
Activity analysis, 112
Activity location, 85 *et seq.*, 93 *et seq.*, 99 *et seq.*, 108, 164, 177 *et seq.*, 234
Aggregation problem, 3
Aggregation process, 75, 76, 101
Allocation models
  Herbert–Stevens models, 25, 83
  transportation problem, 23 *et seq.*, 26
Allocation process, 23 *et seq.*, 25, 26, 83, 106, 178
Applied catastrophe theory, 199
Assignment, 77
  congested assignment problem, 26, 32 *et seq.*
Attraction constrained, 152, 256
Attractiveness
  factor, composite, 226
  residential, 226
  terms, 251, 252, 253, 258
  weights, 164, 251, 252
Attractors, 205
Attribute correlation, 46, 57 *et seq.*, 63
Attributes, observed and unobserved, 47

Backcloth, retail, 22
Balancing equation, 245, 246
Balancing factors, 31, 89, 130, 132, 135, 250, 251
Balancing procedure, 244
Behavioural approaches, 5
Behavioural basis of models, 3
Behavioural theory (of choice)
  of probabilistic choice models, 47, 85
  of the Lowry model, 88 *et seq.*

Benefit functions, 75 *et seq.*, 79, 180
Benefits, locational, 171
Bifurcation
  and differential equations, 204 *et seq.*
  set, 200
  theory, 241
Branch-and-bound methods, 174
Brussels school, 227
Building block methods, 234
Butterfly catastrophe, 203

Calibration, 109 *et seq.*
Canonical forms of functions, 198
Catastrophe set, 200
Catastrophe theory, 196 *et seq.*, 241
  generalized, 219
Catastrophes, seven elementary, 203
Central place theory, 173
Centre size, hypotheses for, 213 *et seq.*
Chaotic behaviour, 207
Choice
  contexts, 46, 47
  substitutes, 46
Choice model structure, 46, 57 *et seq.*, 69 *et seq.*
  simultaneous and sequential for travel demand models, 59 *et seq.*
Co-dimension, 198
Combinatorial problem of design, 4
Combined distribution-assignment model, 35 *et seq.*
Commodity technology (input–output) model, 115
Comparative statics, 195
  extensions of, 211
Competing species, 256
Competition for resources, 220
Competitive equilibrium, 25, 85
Composite-attractiveness factor, 226
Composite cost and utility functions, 61 *et seq.*, 65, 68, 71, 78 *et seq.*, 93, 95, 104, 105

Composite model, 235, 238, 240, 243, 247, 257
Computational methods (for mathematical programs)
feasible direction algorithms, 43 *et seq.*
Frank–Wolfe method, 43
general approaches for unconstrained problems, 38 *et seq.*
gradient methods for unconstrained problems, 39
gradient projection methods, 45
penalty function methods, 41
reduced gradient method, 44
search methods for unconstrained problems, 39
Conflict set, 199
Congested assignment problem (for road traffic flow), 26, 32 *et seq.*
Constraint catastrophes, 203, 233
Constraints, 5
Consumers' behaviour, theories of, 3
Consumers' surplus, 164 *et seq.*, 174, 214, 244, 253
Consumers' surplus (and economic benefit), 26, 48, 65 *et seq.*, 68, 79, 108
generalized consumer surplus, 54, 56, 86, 100
group surplus, 46, 55 *et seq.*, 97 *et seq.*, 180 *et seq.*
Consumers' welfare maximization, and shopping, 169 *et seq.*
Continuity conditions, for network flow problems, 33
Control manifold, 200
Control plane, 200
Control variables, 200, 212
Co-rank, 198
Critical parameter values, 197, 199
Criticality, 4, 220
Cusp catastrophe, 198, 199 *et seq.*, 203, 231
Cusp point, 219

Degenerate singularity, 197
Delay conventions, 202
Demand-driven model, 140, 148
Demand models, 75 *et seq.*
probabilistic choice models, 46 *et seq.*
random utility models, 47 *et seq.*
spatial interaction models, 32, 77, 85 *et seq.*, 89 *et seq.*, 93 *et seq.*, 109, 129, 152, 178 *et seq.*, 180 *et seq.*, 189, 247, 249, 253
Density, 226
Deterrence functions (in spatial interaction models)
exponential form, 27 *et seq.*
general form, 27
Difference equations, 242
Differential equations, 242, 255
and bifurcation, 204 *et seq.*
strongly coupled, 225
types of solution, 205 *et seq.*
Diffusion phenomena, 211
Disaggregated model, 245
Disaggregation, 75, 76, 101, 108 *et seq.*
in shopping models, 172 *et seq.*
Discrete choice, 46 *et seq.*
Dispersion, 2, 5, 46, 47, 82, 108, 158, 164
of preferences, 181
parameter (for logit models), 50, 80
term, 259
Distribution models, 32, 77, 85 *et seq.*, 89 *et seq.*, 93 *et seq.*, 109, 178 *et seq.*, 180, 189
Divergence, 202
Doubly-constrained (estimates), 251
Dual, 121, 125, 246
of embedding shopping model, 171
programme, and embedding, 165
variables, 121, 154, 241, 246, 250, 253, 255, 259, 260
Duality, 19 *et seq.*, 181, 193, 261
and the group surplus model, 105 *et seq.*
for geometric programme, 31
for linear programme, 22
for plan design problems, 183 *et seq.*
for spatial interaction models, 26, 30 *et seq.*, 98
primal–dual relation for non-linear programmes, 20, 21
Dynamic behaviour, 255
Dynamic embedding, 261, 262
Dynamic model framework, 246, 254
Dynamic trajectory, 242, 260, 261
Dynamical methods, 234
Dynamical models, 242, 254 *et seq.*, 255, 260
framework, 246, 254
Dynamical structure of models, 3
Dynamical systems theory, 5, 10, 194 *et seq.*, 241
Dynamics and shopping models, 172 *et seq.*

Eclecticism, 4
Economic account-based analysis, 111 *et seq.*

Economic base, 181
  mechanism, 177
  multiplier, 252, 254
  theory, 85, 92
Economic change, 234
Economic evaluation measures, 26, 47, 48, 65 *et seq.*, 68, 75 *et seq.*, 79, 83, 86, 98, 108, 180 *et seq.*
  for land use, 85 *et seq.*, 97, 106
Economic structure, 234
Elasticity parameters, 59, 69, 178
Elementary catastrophes, 203
Elliptic-umbilic catastrophe, 203
Embedding, 5
  dynamic, 195, 214
  of shopping model, 170 *et seq.*
  relations, 38
  theorems, 158 *et seq.*
Employment allocation, 165
Entropy, 111
  function, 120, 122, 123, 124, 125, 128, 129, 130, 132, 251, 252, 259
  group, model, 163
Entropy maximizing, 27, 28, 29, 30, 46, 55 *et seq.*, 83, 107 *et seq.*, 111, 128, 152, 153, 154, 234, 240, 243, 257
  principles, 2
Equilibrium, 242, 255, 261
  analysis, 194
  conditions, 246, 254, 260
  constraints, 254
  equations, 243
  'far from', 228
  states, 242
Equilibrium points, 246
  dynamical analyses, 214 *et seq.*
  dynamics, 241
  stability of, 205 *et seq.*
Evolution of transport network structure, 232
Expected total surplus, 253
Exponential cost function, 257
Extinction of species, 220

Fast dynamic, 211
Feedback, 211
Fields, interacting, 224
Firm, theory of, 3
Fletcher–Reeves method, 41, 106
Fluctuations, 211
  order from, 227 *et seq.*
Fold catastrophe, 198, 203, 219
  plot of, 197
Folds, 221
Forecasting, conditional, 3
Forrester model, 254
Functional forms
  concave, 9 *et seq.*, 18
  convex, 9 *et seq.*, 18

Garin's representation, 148, 149, 249
General extreme value (GEV) models, 72 *et seq.*
Generalized cost, 27, 78, 230
Geographical theory, 2
Geometric programs, 31
Gradient systems, 195
Gravity model, 235, 239
Group entropy maximizing model (GEM), 107 *et seq.*
Group entropy model, 163, 252, 253
Group surplus, 46, 55 *et seq.*, 97, 99, 180 *et seq.*, 259
Group surplus maximizing, 46, 55 *et seq.*, 99
  GS model, 99 *et seq.*
  relationship to Lowry model, 104
  relationship to PCM, 104
Group surplus model, 46, 55 *et seq.*, 86, 99 *et seq.*, 253
  function, 253

Habit, and modal choice, 231
Herbert–Stevens model, 25, 83
  for competitive spatial markets, 25, 26
Hierarchical structure in networks, 232
Hierarchy, of shopping centres, 173
Hopf bifurcation, 208
Housing
  cost of, 226
  development, 225
  market, 25, 83
  renewal, 185
Hybrid model, 235, 237, 240, 257
Hyperbolic-umbilic catastrophe, 203
Hysteresis, 202
  and modal choice, 231

Independence form irrelevant alternatives, 58, 68
Industrial location, 81, 84, 85, 86 *et seq.*, 190 *et seq.*
Industry technology (input–output) model, 114
Inertia, in residential development, 227
Information theory, 27, 28, 29, 30, 46, 55 *et seq.*, 57, 83, 107 *et seq.*

Input–output (accounting) framework, 136, 137, 138, 142, 151
Input–output accounts, 113, 116, 136, 256 *et seq.*
Input–output analysis, 111 *et seq.*, 142
Input–output coefficients, 148, 149, 153, 156, 249
Input–output identity, 257
Input–output inverse multipliers, 139, 151
Input–output models, 114, 117, 119, 134 *et seq.*, 140, 148, 149, 238, 240, 249, 253, 256 *et seq.*
Input–output multiplier, 111, 119, 131, 140, 148, 256
Input–output relationships, 118, 119
Input–output tables, 114
Interacting fields, 224
Interaction benefits, 75 *et seq.*, 79, 180 *et seq.*
Inter-activity relations, 134, 149
Interdependence, 3
Inverse multiplier, 147, 148
Iterative linear programming, 158

Journey-to-work model, 136, 143, 251
Jumps, 202
  retail, 223

Kuhn–Tucker conditions, 9, 10, 16 *et seq.*, 21, 22, 27, 32, 160
  for the group surplus model, 102 *et seq.*
Kuhn–Tucker theorem, 18, 21, 30

Lagrangian, 130, 132, 154, 155, 242, 245, 250, 252, 255, 257, 260
  conditions, 9, 10, 22
  multipliers, 130, 132, 154, 155, 242, 245, 250, 252, 255, 257, 260
  theory, 15
Land-use density
  constraints, 155, 247
  limitations, 246
Land-use evaluation, 85 *et seq.*, 79
Land-use models, 81, 84, 85 *et seq.*, 181 *et seq.*
  design, 179 *et seq.*
  TOPAZ, 178
  TRANSLOC, 178
Least squares, 122, 123, 124, 125
Leontief inverse, 112, 256
Leontief-type model, 140, 141, 249
Level of resolution, 3
Limit cycles, 206
Linear programming, 21, 35 *et seq.*, 44, 178, 187, 191, 193
  and constraint catastrophes, 203
  basic feasible solutions, 23
  iterative, 158
  transportation problem, 23 *et seq.*, 26
Local conditions for Thom's theorem, 198
Location, activity, 164
Locational benefits, 171
Location–allocation
  models, 174
  problem, 210
Locational surplus, 26, 48, 65 *et seq.*, 68, 79, 83, 86, 98, 108, 180 *et seq.*
  maximization, 180 *et seq.*
Log accessibility, 245, 253, 260
Logistic growth, 210, 214
Logit model
  binary logit, 54
  cross-correlated logit model, 69, 70 *et seq.*
  logit family, 74, 83
  multinomial logit model, 49, 54, 57, 58 *et seq.*, 69, 75, 94
  nested logit model, 66 *et seq.*, 77 *et seq.*
Lotka–Volterra equations, 208 *et seq.*
Lowry model, 1, 81, 84, 85, 86 *et seq.*, 134 *et seq.*, 212, 227, 234, 240, 246 *et seq.*, 256

Make matrix, 113, 114, 117, 120, 121, 122
Manifold, of equilibrium states, 196
Market clearing, 25, 83, 181 *et seq.*
Marshallian–Hotelling conditions, 253
Mathematical program, 241, 243
Mathematical programming, 111, 112, 116, 126, 127, 128, 132, 133, 134, 152, 154, 155, 156, 234, 235, 240, 241, 249, 250, 251, 253, 254, 255, 259, 261
Matrix inverse, 113, 116, 241, 248, 249, 256
  (multiplier) models, 122, 124, 128, 256
Maximum entropy, 120, 122, 124, 125, 128, 129, 152
Maximum probability, 27, 28, 29, 30, 46, 55 *et seq.*, 83, 107 *et seq.*
Maxwell conventions, 202
Memory variables, 226
Meso scale of analysis, 3
Micro-approach (to model formation), 47
Mixed technology (input–output) model, 115
Modal choice and hysteresis, 231

Modal choice dynamics, 229 *et seq.*
Model comparisons, 234 *et seq.*
Model equifinality, 235
Model equivalances, 154, 234 *et seq.*
Model formation, from nonlinear programs, 26 *et seq.*
Model homeofinality, 238
Model similarities, 154, 234 *et seq.*
Modelling, Zeeman's six steps for, 210
Morse singularities, 198
Multicommodity network flow problem, 26, 32 *et seq.*
Multinomial probit model, 73 *et seq.*
Multiobjective programming, 192, 193
Multiple
  equilibria, 194, 199
  goals, 192
Multiplier
  effect, 125
  mechanism, 133
  model, 154
  reactions, 139, 151
  rounds, 153

Network
  equilibrium, 26, 32 *et seq.*
  evolution, 229 *et seq.*
  flows, 234
  optimal, 232
Newton–Raphson approach, 40, 106, 110
Noise, 211
Normal factor, 202

Objective function, 111, 119, 120, 121, 122, 123, 124, 131, 155, 158, 244, 245, 246, 253, 259
Optimality conditions, 241, 245, 246, 258, 260
  for linear programs, 21 *et seq.*
  for non-linear programs, 9, 10, 16 *et seq.*, 21, 22, 27, 32
  for unconstrained problems, 10 *et seq.*
Optimal networks, 232
Optimization
  and dynamical systems theory, 194 *et seq.*
  in locational and transport analysis, 1 *et seq.*
Optimum
  global, 9
  local, 9
  location and size of shopping centres, 169 *et seq.*
Order from fluctuations, 227 *et seq.*

Parabolic-umbilic catastrophes, 203
Parameter change, consequences of, 196
Partial share models, 47, 63 *et seq.*
Perfect delay convention, 202
Periodic solutions, 206
Plan design, 85, 86, 180 *et seq.*
  activity redistribution, 181, 185 *et seq.*
  computational methods for, 192
  housing reallocation, 181, 183 *et seq.*
  long-run equilibrium, 181 *et seq.*
  mathematical programs for, 181 *et seq.*
  optimal design of land-use plans, 177 *et seq.*
  problems, 181 *et seq.*
  service activity reallocation, 181, 185
Planning constraints, 89, 101, 181
Planning of retail facilities, 172 *et seq.*
Planning process, land use, 177
Population density
  constraints, 247
  limitations, 246
Population structure
  change, 234
  model, 238
Prey–predator systems, 208 *et seq.*
Prior information, 28
Prior distributions, 29
Prior values, 119, 120, 122, 123, 129, 130, 156
Probabilistic choice models, 46 *et seq.*, 253
  location models, 46, 81, 93 *et seq.*
  mode choice models, 46, 77 *et seq.*
  route choice models, 46
Probabilistic choice model (PCM) of activity location, 93 *et seq.*
  its relation to the Lowry model, 96, 97
Producers' behaviour, 245
Producers' surplus, 172
Production constraint, 130
Production functions, 233

Quadratic forms, 14
Quadratic programming, 158

Random utility, 253
Random utility models, 46 *et seq.*, 47 *et seq.*, 108, 181
  generation of (RUM), 47
  independent random utility models, 46, 49 *et seq.*
  numerical integration of choice models, 74
  random coefficient models, 48

Random utility theory, 46 *et seq.*, 47 *et seq.*, 108, 181
RAS
form, 122
method, 240
technique, 125
Rate coefficients, 118, 119, 131
Rate matrix, 238
Rational choice behaviour, 253
Rectangular input–output
accounts, 120
analysis, 111 *et seq.*
approaches, 128
framework, 120
models, 111 *et seq.*, 121 *et seq.*, 134, 154, 156
systems, 111 *et seq.*
Red bus – blue bus problem, 58
Reductionist nature of social science, 3
Repellers, 205
Residential allocation, 167
Residential attractiveness, 226
Residential location, 224 *et seq.*, 240, 250
Residential location model, 46 *et seq.*, 81 *et seq.*, 84, 85, 86 *et seq.*, 93 *et seq.*
Residential zone attractiveness, 254
Resilience, 4
Retail
facilities, planning, 172 *et seq.*
flows, 243 *et seq.*
Retail centres, 212 *et seq.*
attractiveness, 254
structure, 243 *et seq.*
Robustness, 4

Saddle points, 18, 206
Scale, 3
Sensitivity analysis, 4
Separatrix, 206, 220, 229
Series expansion, 248
Service location, 81, 84, 85, 86 *et seq.*, 185
Shadow prices, 253, 259
Shopping centres, 212 *et seq.*, 256
hierarchy, 173
optimum location and size of, 169
Shopping model, 61, 63, 93 *et seq.*, 135, 139, 243, 253, 255, 256, 260
disaggregation and dynamics, 172 *et seq.*
flows, 247, 251, 254
Lowry model, 18, 84, 85, 86 *et seq.*
probabilistic choice models, 46 *et seq.*
probabilistic choice models of activity location, 93 *et seq.*
Signed diagraph, 240
Simple model, 235, 236, 240, 243
Simulation, 74, 76, 179
Simultaneous equations, 242
Singularities, 197, 221
Singly constrained estimates, 251
Singly constrained model, 247, 249
Size benefits of shops, 170
Slow dynamic, 211
Social problems, 2
Spatial equilibrium, 187
Spatial impedance factor, 27
Spatial interaction, 130, 135, 234, 235, 239, 247, 251, 259
Spatial interaction costs, 252
constraint, 130, 131, 132, 251, 257
function, 153
Spatial interaction models, 32, 77, 85 *et seq.*, 89 *et seq.*, 93, *et seq.*, 109, 129, 152, 178 *et seq.*, 180 *et seq.*, 189, 247, 249, 256
and dispersion parameter, 188
family of, 160
fully constrained gravity model, 26
generated from random utility theory, 46
mathematical program for, 20, 98 *et seq.*
numerical solutions for, 41
relationship with transportation problem, 29
Spatial system, 2
Spatially disaggregated models, 129, 131, 132, 148, 149, 153, 249, 256, 257
Special product model, 63
Species, extinction of, 220
Splitting factor, 202
Stability, 4, 246
analysis, 255
of equilibrium, 194
of equilibrium points, 205 *et seq.*
structural, 196
State description, 2
State variables, 194, 212
Static model, 246
Structural singularities, 197, 211
Structural stability, 196
Superproblem, 5, 26, 35, 241, 259
Superprogram, 26, 35, 38
Supply-driven, 141
Swallowtail catastrophe, 203, 219
System components, categories of, 2
Systemic effects, 3
Systems theory, 3

Taylor expansion of a family of functions, 198
Technical coefficients, 112, 113, 114, 256, 257
TOPAZ, 158 *et seq.*
Trajectory sketching in phase space, 205
Transformations, and Thom's theorem, 199
Transport
  analysis, and dynamics, 229 *et seq.*
  network structure, 232
  planning models, 75 *et seq.*, 79 *et seq.*
Transportation problem, 23 *et seq.*, 26
  as a limiting case of fully constrained spatial interaction model, 29
  dual program for, 24 *et seq.*
Travel, disutility of, 170

Unbiased estimates, 122, 123
Unconstrained optimization, 10 *et seq.*, 38 *et seq.*, 106
Unfolding of singularity, 197
Uniform plain assumptions, 2
Universal unfolding, 198
Urban dynamics, Forrrester's, 4
Urban economics, 2
Urban modelling, schools of, 1
User-benefit measures, 26, 47, 48, 65 *et seq.*, 68, 79, 108
Utility distributions, 46, 64
  and economic benefit measurement, 52 *et seq.*, 65 *et seq.*, 68
  and their properties, 50 *et seq.*
  correlation between, 46, 47, 57 *et seq.*, 63, 64
  density functions for, 48
  general extreme value distributions, 73
  normal distribution, 52, 73
  the variance–covariance matrix, 57 *et seq.*, 70 *et seq.*
  translationary invariant distribution, 54
  Weibull distribution, 49, 55, 66 *et seq.*, 94
Utility functions, 46, 64, 79 *et seq.*, 93 *et seq.*
Utility maximizing, 47 *et seq.*, 68, 234
Utility trees, 65

Wardrop's principle in network flow problems, 33 *et seq.*, 36
  equivalent optimization problem for, 34
Weighted digraph, 240
Welfare, 75 *et seq.*, 79, 179, 187
Wolfe's dual formulation, 165
Workplace choice model, 81 *et seq.*

Zeeman's six steps for modelling, 210